D0236965

Electrophysiology of Isolated Mammalian CNS Preparations

Electrophysiology of Isolated Mammalian CNS Preparations

Edited by

G. A. KERKUT and H. V. WHEAL

*School of Biochemical
and Physiological Sciences,
University of Southampton, U.K.*

1981

ACADEMIC PRESS

A Subsidiary of Harcourt Brace Jovanovich, Publishers

London   New York   Toronto   Sydney   San Francisco

ALEXANDER V. Nowicky
KENT 1984

ACADEMIC PRESS INC. (LONDON) LTD.
24/28 Oval Road
London NW1

United States edition published by
ACADEMIC PRESS INC.
111 Fifth Avenue
New York, New York 10003

Copyright © 1981 by
ACADEMIC PRESS INC. (LONDON) LTD.

All Rights Reserved
No part of this book may be reproduced in any form by photostat, microfilm, or
any other means, without written permission from the publishers

British Library Cataloguing in Publication Data

Electrophysiology of isolated mammalian CNS
    preparations.
    1. Brain chemistry
    I. Kerkut, G.A.   II. Wheal, H. V.
    599.01'88   QP376   80-42081

    ISBN 0-12-404680-0

Printed in Great Britain at the Alden Press, Oxford

Contributors

P. Andersen, *Neurophysiological Institute, University of Oslo, Karl Johans Gt 47, Oslo 1, Norway.*

S. Y. Assaf, *MRC Neurochemical Pharmacology Unit, Department of Pharmacology, Medical School, Hills Road, Cambridge CB2 2QD, UK*

J. Bagust, *Department of Neurophysiology, School of Biochemical and Physiological Sciences, Southampton University, Southampton SO9 3TU, Hampshire, UK*

D. A. Brown, *Department of Pharmacology, The School of Pharmacy, 29/39 Brunswick Square, London WC1N 1AX, UK*

V. Crunelli, *Institute Mario Negri, Via Eriterea 62, 20157 Milano, Italy*

J.-M. Godfraind, *Laboratorie de Neurophysiologie, Faculté de Médicine, Université Catholique de Louvain, B-1200 Bruxelles, Tour Claude Bernard UCL 5449, Avenue Hippocrate 54, Belgium*

J. V. Halliwell, *Department of Pharmacology, The School of Pharmacy, 29/39 Brunswick Square, London WC1N 1AX, UK*

J. S. Kelly, *Department of Pharmacology, St. Georges Hospital Medical School, Cranmer Terrace, London SW17 0RE, UK*

G. A. Kerkut, *Department of Neurophysiology, School of Biochemical and Physiological Sciences, Southampton University, Southampton SO9 3TU, Hampshire, UK*

T. A. Langmoen, *Neurophysiological Institute, University of Oslo, Karl Johans Gt 47, Oslo 1, Norway*

Kevin Lee, *Max Planck Institute for Psychiatry, Experimental Neuropathology, Kraepelinstrasse 2, 8 Munich 40, West Germany*

Gary Lynch, *Department of Psychobiology, University of California, Irvine, California 92717, USA*

CONTRIBUTORS

J. J. Miller, *Department of Physiology, Faculty of Medicine, University of British Columbia, Vancouver, British Columbia V6T 1W5, Canada*

Michael Oliver, *Department of Psychobiology, University of California, Irvine, California 92717, USA*

C. D. Richards, *Department of Physiology, Royal Free Hospital School of Medicine, Pond Street, London NW3 2QG, UK*

S. Sawada, *Department of Physiology, Faculty of Medicine, Kanazawa University, Kanazawa 920, Japan*

C. N. Scholfield, *Department of Physiology, Medical Biology Centre, The Queens University of Belfast, 97 Lisburn Road, Belfast BT9 7BL, Northern Ireland, UK*

Frank Schottler, *Department of Psychobiology, University of California, Irvine, California 92717, USA*

Philip A. Schwartzkroin, *Department of Neurological Surgery, School of Medicine, University of Washington, Seattle, Washington 98195, USA*

A. I. Shapovalov, *Laboratory of Physiology of the Nerve Cell, Sechenov Institute of the Academy of Sciences of USSR, Thorez Pr 44, Leningrad 194223, USSR*

B. L. Shiriaev, *Laboratory of Physiology of the Nerve Cell, Sechenov Institute of the Academy of Sciences of USSR, Thorez Pr 44, Leningrad 194223, USSR*

Z. A. Tamarova, *Laboratory of Physiology of the Nerve Cell, Sechenov Institute of the Academy of Sciences of USSR, Thorez Pr 44, Leningrad 194223, USSR*

H. V. Wheal, *Department of Neurophysiology, School of Biochemical and Physiological Sciences, Southampton University, Southampton SO9 3TU, Hampshire, UK*

C. Yamamoto, *Department of Physiology, Faculty of Medicine, Kanazawa University, Kanazawa 920, Japan*

Preface

This book describes in full detail the techniques of preparing thin slices of mammalian brain that will remain alive after isolation for more than 48 hours. The results obtained from such studies are also fully described.

The advantages of these slice preparations are that they allow study of brain activity during the onset and recovery from anaesthetics, allow the investigator to change rapidly the chemical components bathing the brain (a procedure that is very difficult to carry out in the whole brain in the animal due to the blood/brain barrier) and allow application of drugs at known concentrations, together with the study of drug antagonists and agonists. The preparations also provide model systems for the study of long-term facilitation (memory) and the study of epileptic-like discharges induced by the addition of penicillin. The effects of ageing have also been studied by comparing the responses of slices from young and old animals.

Isolated mammalian hemi-spinal cords have been kept alive for several days and these allow study of the organization of the reflex activity of the cord in a manner not possible in the whole animal.

Several different regions of the brain have so far been studied, these include the olfactory cortex, the hippocampus, the lateral geniculate nucleus, the striatal nuclei and the septum. These simple slice preparations will be of value to most neurophysiological research laboratories. The slices can also be used in the pharmaceutical industry for screening the action of drugs on the CNS. They can also be used in teaching classes where the preparation can be set up so that students can stimulate pathways and record the evoked potentials, and study the manner in which these vary with drug or anaesthetic application.

The slice technique ushers in a new era in the electrophysiological investigation of the brain. Careful use of slices through specific anatomical sites will allow the investigator to convert the dense three-dimensional mass of 10^{12} neurons into relatively simple interconnected functional working slices. The exploded diagram of the brain can become a functional reality.

Southampton   GAK
January 1981   HVW

Contents

CONTENTS

1

Introduction

G. A. KERKUT

Department of Neurophysiology, School of Biochemical and Physiological Sciences, Southampton University, Southampton SO9 3TU, Hampshire, U.K.

I. Overview

Isolated slices of pieces of the mammalian central nervous system can be kept alive in artificial salines for more than 48 hours. These preparations give good electrical responses to stimulation and allow study of synaptic potentials. The surrounding medium can be controlled so that ionic concentrations can be altered and the ions involved in the action potential, dendritic potentials, EPSP (excitatory postsynaptic potentials) and IPSP (inhibitory postsynaptic potentials), determined.

Possible synaptic transmitters such as acetylcholine (ACh), noradrenaline, 5-HT (5-hydroxytryptamine; serotonin), dopamine, GABA (γ-aminobutyric acid), glycine, glutamate or aspartate can be added and selective antagonists, at known concentrations, can be applied in the saline. The drugs and antagonists can be washed away and the preparation allowed to recover.

Anaesthetics can be applied at known concentrations and measurement made of the time taken for the synaptic potentials to diminish. The preparation can also be allowed to recover from these anaesthetics. These types of experiment are often difficult to carry out on the central nervous system *in situ* in the whole animal.

The preparations also provide model systems for the study of long-term facilitation (memory) and for the study of epileptic-like discharges induced by the addition of penicillin. The effect of ageing can be studied by comparing the responses of slices from young and old animals.

This chapter will briefly consider some of the work carried out on isolated invertebrate CNS preparations, on the isolated amphibian CNS and on the development of the isolated mammalian CNS preparations. It will conclude by a brief summary of the remaining chapters of the book so as to provide the reader with a general overview.

The preparations used have been slices of the olfactory cortex, hippocampus, lateral geniculate nucleus, striatal nuclei, septum, together with the isolated hemi-spinal cord.

II. Invertebrate CNS Preparations

Since Arvanitaki in 1941 (Arvanitaki and Cardot, 1941) showed that the neurons in the abdominal ganglia of *Aplysia* could remain alive in sea water for many hours and give excellent action potentials, many other molluscan preparations have been used, such as the slug *Arion* (Hughes and Kerkut, 1955), the snail *Helix* (Kerkut and Walker, 1961; Kerkut *et al.*, 1975), *Onchidium* (Hagiwara and Kusano, 1961), *Otala* (Gainer, 1972) and *Tritonia* (Willows, 1961, Dorsett, 1961). The work on *Aplysia* has been extended and developed by Tauc (1969) and his colleagues and by Kandel (1976) and his colleagues.

Other useful invertebrate CNS preparations are those of the leech (Nicholls and Purves, 1972), *Limulus* (Walker and James, 1980), crustacean ganglia (Maynard, 1966; Otsuka *et al.*, 1967; Selverston *et al.*, 1977), insect ganglia (Kerkut *et al.*, 1969).

Much of the literature has been summarized in the books edited by Wiersma (1967), Salanki (1968, 1972, 1976), Usherwood and Newth (1975), Hoyle (1978) and Osborne (1978).

The invertebrates have the advantage that they are cold-blooded and so the respiration of their neurons at 5° or 15°C will be less than that of the mammals at 37°C. Invertebrate brains are relatively small and their nerve cell bodies are usually placed peripherally on the surface of the ganglia. Diffusion of oxygen from the medium to the nerve cells takes place easily.

Invertebrate preparations have provided information about the ionic basis of the action potentials; some of the first evidence for calcium spikes was obtained from the invertebrates. Similarly, they are good material for the study of EPSP, IPSP, heterosynaptic facilitation and the manner in which simple circuits can interact (Selverston *et al.*, 1977). Possibly they have been more use in studying basic nerve-membrane phenomena such as the electrogenic sodium pump (Kerkut and York, 1971; Thomas, 1972) or the isolated perfused neuronal membrane (Kostyuk and Krishtal, 1977), even so, they can show properties that link in with the behaviour of the whole animal (Willows, 1967; Hoyle, 1978). For example, the isolated slug CNS showed a sensitivity to changes in the osmotic pressure of the bathing solution. The neurons were sensitive to a dilution of the surrounding solution from 0.7 Locke to 0.6 Locke solution, the change in solution taking place over 40 minutes (Kerkut and Taylor, 1956). This sensitivity to osmotic concentration is similar to that postulated for the osmotic receptors in the mammalian hypothalamus.

The invertebrate preparations are limited in that there is relatively little information about the anatomical relationships between nerve cells. The vertebrates, and especially the mammals, are much more favourably placed since there is a substantial amount of information on mammalian neuroanatomy and this is essential information in understanding how a CNS works.

III. Amphibian CNS Preparations

The cold-blooded amphibia allowed greater scope than the mammals in the development of isolated CNS preparations. Kato (1934) and Barron and Matthews (1938) carried out recordings from the isolated frog spinal cord preparation. This type of preparation was extended by Katz and Miledi (1963) for microelectrode recordings from the spinal motoneurons. Curtis *et al.* (1960) developed the isolated hemisected spinal cord of the toad for the study of synaptic activity. This preparation has been fully described by Tebecis and Phillis (1969) and the literature on the subject completely summarized by Kudo (1978).

The isolated hemisected spinal cord of the toad can show good electrical activity for up to 8 days when kept at 4°C. This preparation has allowed microelectrode studies of neuronal activity, EPSP, IPSP and AP (action potential), together with external recordings of dorsal and ventral root activity. The cord can be perfused and changes in ionic concentration, drugs and blocking agents established. The perfusate can also be analysed, after the cord has been selectively stimulated, for the presence of increased concentrations of putative transmitters after stimulation.

Schmidt (1976) developed the isolated frog brain-stem preparation. He used male leopard frogs and cut the brain posterior to the olfactory bulbs and made another cut posterior to the brachial nerve. This part of the brain was then either removed and studied as it was or else hemisected. The preoptic nuclear region was stimulated and recordings taken from the larangeal nerve. Stimulation of the preoptic nucleus led to a 2 second burst of activity in the pretrigeminal nucleus and in the larangeal nerve. This activity in the larangeal nerve was correlated with the vocal calling activity in the intact animal.

IV. Mammalian CNS slices

Biochemical studies have been carried out on liver, kidney and brain slice metabolism since the 1920s. The most detailed study on brain slice metabolism was carried out by McIlwain and his group, and they developed the technique of slicing so that the preparations remained viable for many hours (McIlwain, 1961; McIlwain and Rodnight, 1966; McIlwain and Batchelard, 1976). They found that the slices should not be thicker than 0.35 mm and should weigh between 3 and 100 mg. The respiration of the cerebral slices was 60 μmol O_2 g^{-1} h^{-1} which was half that of the brain in vivo. However, if the slices were electrically stimulated while they were in the manometer flasks, with pulses of 4 ms duration at the rate of $10s^{-1}$, the respiration rate increased to 60 μmol O_2 g^{-1} h^{-1}, a rate close to that of the normal brain in vivo (McIlwain and Joanny, 1963). Providing that the glucose concentration was above 2mM, glucose did not limit the rate of respiration, and most saline media have a glucose concentration of 10mM.

Electrical recordings have been made from such brain slices (Hillman et al., 1963) and good recordings of population synaptic activity were obtained from slices of the guinea pig olfactory cortex on stimulation of the lateral olfactory tract (Yamamoto and McIlwain, 1966). This preparation was further developed by Richards and McIlwain (1967) who were able to evoke potentials in the neocortex slices on stimulation of the pial surface. Richards and his colleagues continued with their studies on the olfactory cortex slices and other slices (see Chapter 4). Yamamoto (1972) developed slices of the hippocampus that showed good electrical activity and also seizure-like after discharges that could be followed with intracellular microelectrodes. Hippocampal slices were also studied by Schwartzkroin (1975) and by Schwartzkroin and Anderson (1975) and excellent intracellular recordings were made of the activity of the hippocampal CA1 cells.

These studies have led to the development of many other preparations from the mammalian CNS, such as a hemi-spinal cord, the interpeduncular nucleus, the lateral geniculate nucleus, as well as the hippocampus and the olfactory cortex. These preparations are all described in the chapters of this book.

V. Synoptic View of Chapters in this Book

Chapter 2. Schwartzkroin provides a survey of the advantages and disadvantages of the mammalian slice preparation. Among the advantages, are the technical simplicity of the preparation, control of the conditions of the slice, absence of blood and respiration pulsations and movements, ability to see the tracts and cells in the preparation, direct access to the extracellular space, relatively simple preparation with limited inputs, and the ability to select required neuronal system for study.

Among the disadvantages, are that the slice is separated from its normal sensory input and has no motor output, its excitability is affected by the ionic concentration of the medium, there is always damage done in the slicing and there will be cut axons that could act as a spike-initiation site, the slice will be metabolically running down during the process of observation, and there will be dangers of anoxia both in the preparation of the slices from the whole brain and also in the *in vitro* preparation. Schwartzkroin then discusses the features that are important in the preparation of the slice and what can go wrong in the system, together with the precautions and remedies that the experimentalist can take. He indicates the importance that the slice preparation has played in the study of the membrane properties of CNS neurons (calcium-initiated dendritic spikes), the models of epilepsy with spikes induced by penicillin or by pretreatment of sites of the *in vivo* brain with aluminium, and the post-tetanic potentiation systems in the hippocampus that provide a good model for memory systems.

A series of photographs provide details of how to remove the brain from the guinea-pig, the dissection of the hippocampus from the brain and the preparation of slices from the hippocampus.

Chapter 3. Langmoen and Andersen give full details of the dissection of the hippocampus from the brain, the preparation of slices and the treatment of the slices in the experimental bath. They are particularly concerned with the criteria that enable the researcher to determine whether the slice is truely viable or not, and given details of how to tell if the nerve cells have been damaged during the preparation. They also describe how it is possible to make cuts in the slice so as to limit the number of afferent paths that are available. This makes the interpretation of stimulation experiments more simple.

EPSP and IPSP were found in the CA1 cells and the effects of application of GABA, glutamate, norepinephrine and 5-HT on these cells are described. Penicillin elicits epileptiform bursts of activity from the CA1 and CA3 regions, and the slices provide a good model for the study of epileptiform activity. They also studied the differences in properties of the slices from the brains of young rats and from old rats. Ageing appears to reduce the number of effective afferent fibres.

The slice preparation is thus useful in studying the electrophysiology of inputs and outputs onto the CA cells, investigations into the nature of the synaptic transmitters at specific synapses, the study of long-lasting facilitation, which could be a model for memory systems, penicillin-induced epilepsy, and ageing in the central nervous system.

Chapter 4. Richards describes how working in McIlwain's laboratory in 1967 he developed a preparation of slices of neocortex that allowed study of the electrical potentials following direct stimulation of the pial surface.

Richards then discusses the technical problems of slicing and the choice of thickness of slice, balancing the extent of damage in the thin slices against the limited oxygen diffusion into the thicker slice. It would appear that about 0.4mm thickness is a suitable compromise. Full details are given for the preparation of slices by hand using a razor blade and a thickness guide, and also using a Vibratome. Two types of incubation bath are described, (a) with the tissue partially immersed in the artificial saline, but with the upper surface exposed to the humidified atmosphere, and (b) with the slice totally immersed and superfused on both surfaces by an oxygenated saline.

The properties of two different slice preparations, the olfactory cortex and the hippocampus, are described. The method of analysis of field potentials from the olfactory cortex slice after treatment with halothane is described. The field potentials give information about the average behaviour of a population of neurons, whereas microelectrode studies of single cells indicate what that specific cell is doing. It is necessary to use both techniques in combination to understand the functioning of the cells in the tissue slice.

Chapter 5. Scholfield describes the use of 600μm slices from the guinea pig olfactory cortex. These slices were maintained at 25°C. Good micro-electrode penetration was made into 12μm neurons which could be activated by stimulation of the lateral olfactory tract.

Drugs such as GABA, glutamate, bicuculline or ouabain could be applied and the ionic composition of the Ringer solution altered by changing the calcium, magnesium, sodium, potassium or chloride con-centrations. The preparation showed good mechanical stability and viability.

Chapter 6. Assaf, Crunelli and Kelly show that the isolated hippocam-pal slice has such good mechanical stability in their apparatus that it is possible to keep a microelectrode inside a granule cell (whose size is only 8–10μm) and get good electrical recordings for several hours. These cells respond to both orthograde and retrograde stimulation and also to iontophoresis of putative transmitters. The preparation showed EPSPs and also small amplitude spikes of 4–15mV. It is possible that these small spikes are regenerative action potentials propagated by dendritic hot spots. Such dendritic spikes might enhance the integrative capabilities of the neuron and they could also amplify and enhance the integration of spatially distinct inputs.

Iontophoresis of 5-HT or GABA onto the granule cells produced an increase in membrane conductance and also a depolarization. Preliminary studies indicate that the IPSP may be a depolarizing and not a hyperpolarizing potential. Although it is usual to consider that IPSPs should hyperpolarize the membrane, this is not necessarily so. Research and understanding of insect inhibitory neuromuscular potentials was delayed several years because of failure to appreciate that the inhibitory potentials could often be depolarizing and not just hyperpolarizing. As long as the current flow clamps the membrane potential at a specific value away from that of the action potential threshold, the action potential will be inhibited.

Chapter 7. Lee, Oliver, Schottler and Lynch present a study of the correlation between the electron microscopic (EM) structure and the electrical activity of hippocampal slices. This has two advantages. Firstly, it provides background information about the extent to which the slice maintains its normal histological and cytological structure after isolation and experimentation. Secondly, it allows a study to be made of any structural changes that take place at the synapses after specific stimulation has taken place.

The hippocampus can exhibit very long-lasting synaptic facilitation following a brief burst of high-frequency stimulation. This facilitation can last for several weeks after the brief burst of stimulation. The EM study shows that brief bursts of high-frequency stimulation can produce an increase in the number of synapses located on the shafts of dendrites and also a change in the shape of the dendritic spines. These histological changes agree with similar changes found in the hippocampus of the brain *in vivo* after similar stimulation. This would suggest that the brief burst of stimulation increased the number of synaptic inputs and made the synapses more efficient in transmission, thus contributing to the long-lasting facilitation.

Chapter 8. Wheal describes the use of a circular bath and a rectangular bath for the study of the electrical activity of hippocampal slices from the rat. He compares intracellular recordings from rat CA1 cells with those obtained by other authors from CA1 cells from the guinea pig and rabbit. There is reasonably good agreement between the results for all three animals. The CA1 cells could be stimulated through four different afferent paths and the population spikes are described.

In addition, Wheal describes a silicon chip extracellular microelectrode array of nine electrodes in an area of 2×2 mm (SCEMA 9). Each electrode surface was 10μm and the device incorporated nine buffer amplifiers each leading to an off-chip amplifier and then to a $9:1$ multiplexer which allowed all nine channels to be displayed on one beam of an oscilloscope. The hippocampal slice was placed over the 9-electrode array and the afferent path stimulated. The evoked potentials were picked up by the nine electrodes and displayed showing the simultaneous different potential profiles at the nine different anatomical sites in the slice. The results obtained with this 9-electrode array agreed well with those obtained by external glass microelectrodes. The device should be useful in testing the action of drugs at specific sites in the hippocampal slice.

Chapter 9. Yamamoto and Sawada give full practical details of how to prepare slices from the lateral geniculate nucleus of the guinea pig. They use this isolated preparation to study the alteration in size of the extracellular synaptic potentials following repeated stimulation of the optic tract. This alteration in EPSP amplitude following stimulation of the optic tract is interpreted in terms of an alteration of the proportion of transmitter released from the available transmitter pool. The rate of recovery of the amplitude of the EPSP indicates the rate of replenishment of the transmitter pool.

Chapter 10. Godfraind and Kelly describe the use of slices of the lateral geniculate nucleus (LGN) from rats and cats. The morphology of the

LGN is well known and the principal cells receive a monosynaptic excitatory input from the optic tract. There are also inhibitory interneurons in the LGN.

Approximately 250μm-thick slices were cut, by means of a Vibratome, from the dorsal LGN and placed in the incubation bath. Stimulating electrodes were placed on the optic nerve and recording microelectrodes inserted into the LGN cells. The resting potential was $55 \pm 2mV$ (rat) and $44 \pm 2mV$ (cat). The action potentials were $42 \pm 1.5mV$ (rat) and $38 \pm 4mV$ (cat). The membrane resistance was between 20 and 40MΩ. When the optic tract was stimulated, EPSP could be seen in the cells but IPSP were not seen very often. It is suggested that slicing may have interrupted the inhibitory interneuron circuits since IPSP are very commonly found in the *in situ* preparation. When IPSP were found, they were reversed if the internal chloride concentration of the impaled cell was increased. Long-lasting inhibition was also found, and this suggests that there are two different methods of inhibiting the principal cells, i.e. two different types of inhibitory interneurons. Model systems are presented that suggest anatomical relationship between the two types of inhibitory cells and the principal cell.

Chapter 11. Brown and Halliwell describe the preparation of slices of the rat diencephalon that include tracts from the habenular ganglion running to the interpeduncular nucleus (IPN). This necessitated the slice being made at an angle 40° to the horizontal. The authors tell how they fitted the dorsal surface of the isolated rat brain into a moulded silicone cup so that the afferent tracts from the Fasciculi Retroflexi of Meynert (FRM) running to the IPN are aligned with a pair of guide bars. The brain is then transected just caudal to the IPN and the rostral surface is cut into 500μm slices. The FRM generally crosses the IPN in parallel bundles of fibres though histological studies show sites where the afferent tracts were looped into figures of eight. The effects of stimulating the FRM on the IPN are described.

This account shows how regular slices can be made from interesting regions of the brain provided that the necessary moulds, gigs and tools are available to hold the brain steady at the required angle.

Chapter 12. Miller describes some of the properties of slices from the rat striatal and limbic forebrain. The corpus striatum was cut into slices 350–400μm thick, the slices being cut nearly perpendicular to the longitudinal axis of the striatum, keeping the plane of the section parallel to the fibres radiating from the internal capsule. Extracellular recordings were taken from 243 neurons, 22% fired spontaneously and 78% were silent but could be driven by electrical stimulation. Stimulation of the cortex or globus pallidus region led to spike activity and this could be reduced or eliminated by reducing the Ca^{2+} in the medium. Lowering the chloride concentration in the medium reduced the depression of the population spike resulting from a preceding conditioning pulse. Iontophoretic application of ACh excited the cells and GABA inhibited them. There is evidence for the organization of the striatum into distinct functional regions.

The septum was sliced with a Vibratome since chopper-cut slices tended to be too compressed. The slices contained afferents of the median forebrain bundle, fornix fibres, and dorso-lateral septal nuclei including fimbrial fibre afferents. Of the 350 nerve cells studied, 61% were spontaneously active and the remaining 39% could be electrically driven. Three different populations of spike patterns were found. Substance P was tested on 28 cells and increased the firing rate of most, giving dose-related responses.

The responses of the cells in these isolated slices are compared with those of cells in similar location *in vivo*. The effect of damage and of removal of extrinsic circuits are considered and the advantages of striatal and septal slices discussed.

Chapter 13. Bagust and Kerkut describe the properties of an *in vitro* preparation of the hemi-spinal cord of the mouse. Mice (30–40g) were decapitated and their spinal cords rapidly dissected. The cord was split longitudinally along the mid-line and one-half mounted in the experimental bath. The preparation could show good electrical activity for up to 48 hours after isolation. The dorsal tracts or dorsal roots could be stimulated and activity recorded extracellularly in the dorsal horn, mid-cord and ventral horn. Synaptic activity could be blocked by 2mM Mn^{2+} and the effect of Ca^{2+}, Mg^{2+}, Co^{2+} and Ni^{2+} are also described. The preparation can show a response to a change of the bathing fluid within about 2 minutes. The preparation's activity is affected by perfusion with anaesthetics and dose–response curves are given for the action of urethane, sodium pentobarbitone, halothane and procaine. The concentration of anaesthetic needed to produce a reduction of the synaptic response of the cord were similar to those concentrations used to bring about anaesthesia in the inact animal. The effects of anaesthetic could be studied on non-anaesthetized CNS and the recovery from anaesthetic could also be measured.

This preparation could be of value in basic research on the functional organization of the nerve cord, in screening the action of centrally acting drugs, and in teaching students about the electric electrical activity of the non-anaesthetized spinal cord.

Chapter 14. Shapovalov, Shiriaev and Tamarova describe three different preparations. (a) The isolated perfused hemi-spinal cord of the rat and the kitten. (b) The *in-situ* spinal cord and hind limb of the kitten perfused through the arteries. (c) The *in-situ* arterially perfused kitten medulla.

Good intracellular recordings were obtained from the isolated hemi-spinal cord for up to 5 hours. However, the other preparations showed less mechanical stability and it was not often possible to stay inside the neurons for more than 20 minutes to an hour.

The isolated hemi-spinal cord took 20 minutes for a Ca^{2+}-free, Mn^{2+}-supplemented Ringer to have an effect, whereas in the arterially per-fused spinal cord it took 40–60 minutes for the Ca^{2+}-free, Mn^{2+}-supplemented Ringer to have an effect, although the postsynaptic potentials reappeared after 15 minutes perfusion with standard Ringer solution.

The three preparations make a graded series from the isolated CNS to one in which there is a complete limb and limb innervation attached to the spinal cord, or a preparation of part of the brain which can be perfused with specific salines so allowing the conditions around the neurons in the CNS to be altered.

VI. Conclusions

The study of the electrical activity of isolated brain slices is still fairly difficult with the preparations slowly decaying from their initial *in vivo* condition. There are many major and minor skills in the preparation and maintenance of the slices and the authors of the succeeding chapters will provide information as to how to obtain preparations that will show good electrical activity for many hours. Though there is considerable overlap in the methodologies described (and it may appear repetitious), the development of a system of accepted technical procedures does depend on having different laboratories able to obtain similar results from slices and these results should be similar to those obtained from the preparations in the intact animal. The chapters on the hippocampus and other regions do provide such a series of procedural steps, and the reader wishing to obtain similar results in his own laboratory and coming up against technical difficulties, can, by reading the different accounts, have a greater chance of realizing what is going wrong.

These simple slice and spinal preparations should be of value to most neurophysiological research laboratories and can also be of use in teaching classes where a preparation can be set up so that students can stimulate pathways and record the evoked potentials and study the manner in which these vary with drug or anaesthetic application.

It should also be possible to develop and extend the preparations so that the slices are not just simple I shapes but could be other shapes such as L, C, T, U, O, F, or even X and E shape, the branches bringing in important CNS nuclei and tracts. It should also be possible to dissect out complete pathways with attached input and output nuclei. All that is now required is for the investigator to have a reasonable knowledge of neuroanatomy and neurophysiology; to be adventurous in the exploration of slices, dissected pathways and nuclei of the mammalian central nervous system; and to be courageous.

References

Arvanitaki, A. and Cardot, H. (1941). Contribution à la morphologie due système nerveux des Gastropodes. Isolement à l'etat vivant de corps neuroniques. *C. R. Soc. Biol.* **135,** 965–968, 1207–1211, 1211–1216.

Barron, D. H. and Matthews, B. H. C. (1938). Dorsal root potentials. *J. Physiol. (Lond.)* **94,** 26–27P.

Curtis, D. R., Phillis, J. W. and Watkins, J. C. (1960). The chemical excitation of spinal neurones by certain acidic amino acids. *J. Physiol. (Lond.)* **150,** 656–682.

Dorsett, D. A. (1967). Giant neurones and axon paths in the brain of Tritonia. *J. Exp. Biol.* **46,** 137–151.

Gainer, H. (1972). Effects of experimentally induced diapause on the electrophysiology and protein synthesis of identified molluscan neurones. *Brain Research* **39,** 387–402.

Hagiwara, S. and Kusano, K. (1961). Synaptic inhibition in the giant nerve cell of *Onchidium verruculatum. J. Neurophysiol.* **24,** 167–175.

Hillman, H. H., Campbell, W. J. and McIlwain, H. (1963). Membrane potential in isolated and electrically stimulated mammalian cerebral cortex: effects of chlorpromazine, cocaine, phenobarbitone and protamine on the tissues electrical and chemical responses to stimulation. *J. Neurochem.* **10,** 325–339.

Hoyle, G. (Ed.) (1978). "Identified Neurones and Behaviour of Arthropods." Plenum Press, New York.

Hughes, G. M. and Kerkut, G. A. (1955). Electrical activity in a slug ganglion in relation to the concentration of Locke solution. *J. Exp. Biol.* **33,** 282–294.

Kandel, E. R. (1976). "Cellular Basis of Behaviour." W. H. Freeman, San Francisco.

Kato, G. (1934) "The Microphysiology of Nerve." Maruzen, Tokyo.

Katz, B. and Miledi, R. (1963). A study of spontaneous miniature potentials in spinal motoneurones. *J. Physiol. (Lond.)* **168,** 389–422.

Kerkut, G. A. and Taylor, B. J. R. (1956). The sensitivity of the pedal ganglion of the slug to osmotic pressure changes. *J. Exp. Biol.* **33,** 493–501.

Kerkut, G. A. and Walker, R. J. (1961). The effects of drugs on neurones of the snail *Helix aspersa. Comp. Biochem. Physiol.* **3,** 143–160.

Kerkut, G. A. and York B. (1971). "The Electrogenic Sodium Pump." John Wright-Scientechnica, Bristol.

Kerkut, G. A., Pitman, R. M. and Walker, R. J. (1969). Iontophoretic application of ACh and GABA onto insect central neurones. *Comp. Biochem. Physiol.* **31**, 611–633.

Kerkut, G. A., Lambert, J. D. C., Gayton, R. J., Loker, J. E. and Walker, R. J. (1975). Mapping of nerve cells in the suboesophageal ganglia of *Helix aspersa*. *Comp. Biochem. Physiol.* **50A**, 1–25.

Kostyuk, P. G. and Krishtal, O. A. (1977). Separation of sodium and calcium currents in the somatic membrane of molluscan neurones. *J. Physiol.* **270**, 545–568.

Kudo, Y. (1978). The pharmacology of the amphibian spinal cord. *Prog. Neurobiol.* **11**, 1–76.

Maynard, D. M. (1966). Integration in crustacean ganglia. *Symp. Soc. Exp. Biol.* **20**, 111–149.

McIlwain, H. (1961). Techniques in tissue metabolism. 5. Chopping and slicing tissue samples. *Biochem. J.* **27**, 213–218.

McIlwain, H. and Batchelard, H. S. (1976). "Biochemistry and the Central Nervous System", 4th edition. Churchill Livingston, London.

McIlwain, H. and Joanny, P. (1963). Characteristics required in electrical pulses of rectangular time–voltage relationship for metabolic change and ionic movements in mammalian cerebral tissues. *J. Neurochem.* **10**, 313–323.

McIlwain, H. and Rodnight, R. (1966). "Practical Neurochemistry." Churchill, London.

Nicholls, J. G. and Purves, D. (1972). A comparison of chemical and electrical synaptic transmission between single sensory cells and a motoneurone in the CNS of the Leech. *J. Physiol. (Lond.)* **225**, 637–656.

Osborne, N. N. (Ed.) (1978). "Biochemistry of Characterised Neurones." Pergamon Press, Oxford.

Otsuka, M., Kravitz, E. A. and Potter, D. D. (1967). Physiological and chemical architecture of a lobster ganglion with particular reference to GABA and glutamate. *J. Neurophysiol.* **30**, 725–752.

Richards, C. D. and McIlwain, H. (1967). Electrical responses in brain samples. *Nature (Lond.)* **215**, 704–707.

Salanki, J. (Ed.) (1968). "Neurobiology of Invertebrates." Akademiai Kiada, Budapest.

Salanki, J. (Ed.) (1973). "Neurobiology of Invertebrates: Mechanisms of Rhythm Regulation." Akademiai Kiado, Budapest.

Salanki, J. (Ed.) (1976). "Neurobiology of Invertebrates: Gastropod Brain." Akademiai Kiado, Budapest.

Selverston, A. I., Russell, D. F., Miller, J. P. and King, D. G. (1977). The stomatogastric nervous system, structure and function of a small neural network. *Prog. Neurobiol.* **7**, 215–290.

Schmidt, R. S. (1976). Neural correlates of frog calling; isolated brain stem. *J. Comp. Physiol.* **108**, 99–113.

Schwartzkroin, P. A. (1975). Characteristics of CA1 neurones recorded intracellularly in the hippocampal *in vitro* slice preparation. *Brain Res.* **85,** 423–436.

Schwartzkroin, P. A. and Andersen, P. (1975). Glutamic acid sensitivity of dendrites in hippocampal slices *in vitro*. *Adv. Neurol.* **12,** 45–51.

Tauc, L. (1969). Polyphasic synaptic activity. *Prog. Brain Res.* **31,** 247–257.

Tebecis, A. K. and Phillis, J. W. (1969). The pharmacology of the isolated toad spinal cord. *Exp. Physiol. Biochem.* **2,** 361–395.

Thomas, R. C. (1972). Electrogenic sodium pump in nerve and muscle cells. *Physiol. Rev.* **52,** 563–594.

Usherwood, P. N. R. and Newth, D. R. (Eds.) (1975). "Simple Nervous Systems." Edward Arnold, London.

Walker, R. J. and James, V. A. (1980). The central nervous system of *Limulus polyphemus*; physiological and pharmacological studies. A minireview. *Comp. Biochem. Physiol.*, **66**C, 121–124.

Wiersma, C. A. G. (Ed.) (1967). "Invertebrate Nervous Systems." University of Chicago Press, Chicago.

Willows, A. O. D. (1967). Behavioural acts elicited by stimulation of single identifiable brain cells. *Science* **157,** 570–574.

Yamamoto, C. (1972). Intracellular study of seizure-like after discharge elicited in thin hippocampal section *in vitro*. *Exp. Neurol.* **35,** 154–164.

Yamamoto, C. and McIlwain, H. (1966). Electrical activities in thin sections from the mammalian brain maintained in chemically defined media *in vitro*. *J. Neurochem.* **13,** 1333–1343.

2

To Slice or Not to Slice

PHILIP A. SCHWARTZKROIN

Department of Neurological Surgery, School of Medicine, University of Washington, Seattle, Washington 98195, U.S.A.

I. Introduction

Slices of mammalian CNS tissue, maintained *in vitro*, were shown to be physiologically viable by Yamamoto and McIlwain (1966). Interest in, and productive use of, this preparation, however, have increased explosively over the past 5 years. Investigators using the *in vitro* slice have contributed important findings in several areas of the neurosciences. These contributions have been possible because of the unique advantages of the preparation. In this chapter, I will outline some of the advantages I have found in the slice preparation. I will also try to describe some of the limitations and difficulties of the preparation, as well as some of the superstitions that have arisen in response to these difficulties.

II. Advantages of the Slice Preparation

The slice preparation has provided mammalian neurophysiology with many of the technical advantages previously reserved for invertebrate studies *in vitro*. A neurophysiologist accustomed to the usual struggle of preparing and maintaining a cat for intracellular recording must certainly experience some envy for the simplicity and elegance of experiments using *Aplysia* or crayfish. In the slice preparation, the mammalian neurophysiologist realizes the following features of an *in vitro* preparation.

A. Technical simplicity

Preparation of the slice, and clean-up after the experiment, are rapid. Experiments use relatively inexpensive animals and long multi-day experiments are not called for. Although these rather mundane aspects of the slice preparation may appear unimportant, they are features that help make the investigators' laboratory experience considerably more pleasant. Many CNS experiments *in vivo* require hours of animal preparation (anesthesia, surgery, stabilization); the real experiments may not begin until the investigator is tired and ready to go home. With the slice technique, actual time from animal "anesthesia" to placement of the slices in the incubation–recording chamber may be less than 5 minutes. Since preparation for the experiments is merely a matter of mixing-up the desired bathing medium (10 minutes), and clean-up after an experiment consists principally of flushing the chamber and rinsing a few dissection tools (30 minutes), non-experiment time is reduced to a minimum. Further, the need for technical assistance is obviated (a considerable saving). A further saving is realized by the use of small inexpensive animals (rats, mice, guinea pigs), species which are often difficult to maintain *in vivo*. As preparation time is short, the animals relatively inexpensive and the time of the experiment limited by the naturally occurring "run-down" of the tissue, the investigator is better able to accept "short" 8–10 hour experiments. Especially for the experimenter who works alone, productivity usually declines (even in experiments *in vivo*) by the end of a day; being able to clean up when the preparation (or experimenter) goes bad, and to feel no guilt about the waste of time, effort, and/or money, are important aspects of enjoying research activities.

B. Control over the condition of the preparation

There are, to be sure, concerns about the state of the tissue maintained *in vitro* (see below). Compared to the variables that must be controlled *in vivo*, however, *in vitro* control is relatively simple. There is no blood pressure to monitor, no heart rate to stabilize and no expired CO_2 concentration to maintain. Questions about how much experimental

results differ in anesthetized compared with unanesthetized preparations, under different types of anesthesia, or in different states of
alertness in unanesthetized animals are not problematic in the slice. This
feature may, of course, be viewed as a limitation (see below). However,
there is somewhat less variability *in vitro* than in *in vivo*. The preparation
can be maintained at a constant pO_2, pH and temperature; humoral
states are not altered during the course of an experiment. This constancy
is a welcome feature for many investigators.

C. No pulsation

Of all the technical advantages afforded by the *in vitro* slice, perhaps the
one most appreciated by mammalian CNS neurophysiologists is the fact
that slices do not move up and down with heart beat and respiration.
Stabilization of the slice tissue may pose a technical difficulty (see
below); however, once the bugs have been worked out of the incubation
system, the stability is sufficient to allow intracellular penetrations to be
routinely maintained for long periods.

D. Improved visualization of tissue

Since the tissue is sliced into thin "wafers", it is possible to guide visually
both recording and stimulating electrodes to the desired sites, avoiding
the hazards of stereotaxic techniques. In many tissues, it is possible to
distinguish cellular regions from fiber tracts, and even somatic from
dendritic cell regions (Skrede and Westgaard, 1971). Recent advances
in optical systems now make the direct visualization of single cells
possible (Yamamoto and Chujo, 1978b). One may then impale selectively cell somata or dendrites with recording pipettes, and stimulate
discrete synaptic or antidromic pathways (Wong *et al.*, 1979). This
ability visually to control electrode placement also allows accurate
laminar analyses and precise independent localization of iontophoretic
electrodes. In addition, microsurgery can be carried out on slices so that
critical pathways may be tested (Schwartzkroin and Prince, 1978).

E. Direct access to extracellular space

A major advantage of *in vitro* preparations is the fact that bathing
medium may be changed easily. Since the process of slicing breaks down
pial barriers to normal diffusion, it has been assumed that the ionic and
drug concentrations in the bathing medium equilibrate with the extracellular space. Thus, extracellular ion concentrations, essential for cell

spiking activity, for synaptic release of transmitters and for mediation of postsynaptic potentials, can be accurately controlled and easily altered (Richards and Sercombe, 1970). Putative transmitter substances can be tested in the bathing medium as can other drugs (Yamamoto, 1974; Schwartzkroin and Andersen, 1975; Spencer *et al.*, 1976). Drugs can also be discretely applied to specific slice regions (Mesher and Schwartz-kroin, 1980). Perfusate from the slice, perhaps containing transmitters released by stimulation of selected pathways, may also be collected by various techniques (Wieraszko and Lynch, 1979).

F. Relatively simplified preparation

In removing the tissue from the rest of the brain, and in slicing it into thin wafers, the investigator effectively simplifies the preparation. The tissue is no longer subject to neurohumoral effects or to inputs from distant brain regions. Further, the number of cells involved in gener-ation of activity in the slice is much smaller than in intact preparations. Thus, it is feasible to record simultaneously from cells synaptically connected (MacVicar and Dudek, 1978), or even to impale different parts of the same cell (Wong *et al.*, 1979).

G. Neural organization

Although mammalian CNS neurophysiologists may envy the beautiful simplicity of *in vitro* invertebrate technology, they continue to work on the mammalian brain because it is organized to perform such interesting and complex functions. Although mammalian CNS tissue may be stud-ied *in vitro* using cell-culture or organ-culture techniques, the normal complex organization is lost or drastically altered. The *in vitro* slice preparation combines many of the technical advantages of invertebrate simplicity with the normal complex organization of mammalian CNS tissue.

III. Limitations

The advantages of the slice preparation have been sufficiently attractive to lure an ever-increasing number of neuroscientists to adopt this tech-nique. Before committing time and effort to this preparation, investiga-tors should be made aware of the drawbacks and limitations of the slice. I have tried to list below some of those limitations that I find most troubling. In general, these considerations all revolve about the following major flaw: the tissue is not in a *normal* state. Of course, this point is obvious, and may well be the reason that the technique was so reluctantly accepted by many investigators. The tissue is abnormal in the following senses.

(i) It is separated from its normal inputs. Tonic excitatory and/or inhibitory influences are lost, as well as the natural patterning of synaptic events. Long feedback loops are lost, and local circuits oriented out of the plane of the slice may be interrupted. These separations mean that although the local *capabilities* of the tissue may be studied, one can never use the preparation to find out how a particular group of cells behaves *in situ*.

(ii) Activity in the slice tissue has no functional/behavioral consequence. In addition to being cut off from inputs, the targets of outputs of the tissue under study are lost. Again, one never knows what the tissue really does in the intact animal.

(iii) The excitability level of the tissue is determined primarily by the ionic concentrations in the bathing medium. For example, cellular excitability and/or instability can be increased by raising $[K^+]_o$ and lowering $[Ca^{2+}]_o$; excitability can be decreased by the opposite alteration of these ion concentrations. In a sense, these ion changes play a role similar to anesthesia in intact animal preparations. Although one can measure the "correct" ion concentrations of, say, cerebrospinal fluid, the composition of extracellular space fluid is somewhat more difficult to measure. The advantage, then, of biasing neuronal activity to a desired level is offset, in part, by an ignorance of where it normally "should" be.

(iv) Inevitably, some of the tissue will be damaged by the slicing process. Cellular processes will be severed, and cells may undergo degeneration. Neurons that normally project to sites outside the sliced tissue may show abnormalities due to retrograde degenerative processes; cells with extensive dendritic trees may be damaged in the cutting, and subsequent physiological, pharmacological and/or anatomical descriptions of neurons in the slice must be suspect because of the possibility of such damage. In addition, slicing may selectively affect a certain population of neurons if their normal orientations are not parallel to the plane of the cut. In many CNS structures, it will be difficult to maintain intact circuitry because cell input–output orientation is not contained within the lamella of the slice.

(v) The preparation will, sooner or later, "run down". Although it is now possible to keep the tissue alive for several days, the life of the tissue is limited. This factor, perhaps, does not differentiate the slice preparation from *in vivo* preparations; however, since the run-down is a gradual process, it is often difficult to know when the process is far enough advanced to invalidate the data being collected. Live tissue becomes "abnormal" not only during late stages of the preparation, but also during early stages when the tissue is "recovering" from the trauma of the slicing procedure. Recognition of the slices' stable period is therefore very important.

(vi) The tissue is subject to an anoxic period during preparations of the tissue (Lipton and Whittingham, 1979). Resistance to anoxic damage is determined by a number of factors (age of the animal, speed and care of the dissection, composition and temperature of the medium), but the ultimate health of the tissue is discernable only when one starts to record. Again, recognition of abnormal activity in the tissue, even if it is alive, is essential.

(vii) Whether, and how well, the slices "work" are determined in large part by luck, superstition and an ignorant adherence to what has worked for others in the past. Very little has been done to determine optimal bathing medium ion concentrations, proper bath temperature and O_2 pressure. We do not know how critical are the dissection speed, care and temperature at which it is carried out. Because of these unknowns, when experiments do not work well, it is difficult to pinpoint the problem; we are reduced to "fiddling", changing anything that might be suspect.

IV. Superstitions and Unknowns

Despite the large number of unknown factors, there are superstitions that have grown up that do appear to make the preparation "work". These procedures vary somewhat from laboratory to laboratory, so that the comments that follow should be viewed as my own prejudices.

A. Dissection: time, care and temperature

Intuitively, it would seem that the faster one transfers the tissue from animal to slice chamber, the healthier the tissue will be. A quick preparation results in less possibility of anoxic damage to the tissue. Although I am sure there is an upper time limit beyond which the tissue becomes sick and dies, there appears to be a considerable margin within which viable healthy slices are obtained. Care taken during the dissection appears to be at least as important as timing; tissue that has been traumatized, cut into, or otherwise injured in the dissection, or roughly handled during transfer into the chamber, is likely to display abnormal or subnormal activity. In preparing hippocampus tissue from guinea pig, it normally takes me 3–4 minutes to kill the animal, dissect out hippocampus, cut the slices, and transfer them to the chamber. Dissections of animals with thin, brittle skulls are easier than dissections of large animals (large guinea pigs, adult rabbits and cats). Dissection of immature animals, although difficult because of the soft, mushy brains, has the advantage of the tissue being somewhat more resistant to hypoxia.

Investigators have generally found that keeping tissue cold during the dissection improves its viability and makes the dissection easier (the tissue remains firm, and does not become "sticky"). This can be accomplished by immersing the whole brain in 4° C oxygenated bathing medium, either briefly before dissecting the desired structure, or for the entire course of the dissection. It should be noted, too, that a cold solution will maintain more oxygen; O_2 comes out of solution as it is warmed. The tissue should not be allowed to dry out.

One variable which may be very important (my own prejudice) is the time that stagnated blood remains in the tissue. There is some evidence from clinical studies of stroke victims that the contact with stagnant pools of blood, rather than loss of an oxygenated blood supply, results in the most damage. An attempt to drain at least some blood from the brain at the start of a dissection via a quick cut across the throat to sever the jugular veins may improve tissue viability.

B. Composition of the bathing medium

In general, the ionic composition of the bathing medium is best determined by the design of the experiment. Manipulation of potassium will cause changes in cell excitability (Oliver et al., 1978; Prince and Schwartzkroin, 1978), as will changes in calcium and magnesium. Calcium/magnesium levels will also affect size of synaptic potentials, membrane stability and "leakiness", and perhaps calcium-mediated potentials (Wong and Prince, 1978). To my knowledge, there has been no study that has specified the "proper" normative concentrations for in vitro studies. Choice of buffer has also not been studied well, although phosphate/bicarbonate is most often used. Phosphate can be omitted (without causing any obvious changes in cell activity) if an ion such as barium or manganese is to be added to the medium (they will precipitate immediately with phosphate) (Hotson and Prince, 1980). Effects of addition of amino acids, vitamins, or other nutrients to the balanced salt solution are questionable. No obvious improvement in cell health or slice longevity was noted in experiments in which a bathing medium containing amino acids and vitamins was used. However, such additions *may* be important to long-term (over 1 day) maintenance of slices.

C. Temperature of the bathing medium

Since our purpose is to study "normal" mammalian CNS tissue, it makes intuitive sense to try to mimic the brain environment as best we can. For this reason, many investigators maintain their bath temperature near body temperature (37° C). However, a number of laboratories have reported that lowering bath temperature to 30° C, or even 25° C, results in a longer-lasting slice (White *et al.*, 1978; Scholfield, 1978; Alger and Nicoll, 1979). Presumably, the catabolic rate in the tissue is somewhat slowed. No careful study has been carried out to investigate differences in neural activity at various temperatures, although it has been briefly reported that overheated tissue shows unstable bursting activity (Skrede and Westgaard, 1971). Until this matter is systematically studied, it will remain a subject of some concern. In a limited number of studies carried out in my laboratory, bath temperature was lowered to 25° C. Although evoked field potentials (hippocampal CA1) appeared normal at this temperature, intracellularly recorded activity was grossly abnormal; spikes were broad, input resistance low and membranes very fragile.

D. Static bath *vs.* continuous circulation

Both types of chamber arrangements have been used successfully. In general, if the bath volume is large compared to the tissue, a static bath will support the slices. It should be noted, however, that fluid in a static chamber is difficult to heat and oxygenate. For experiments requiring quick changes of bathing medium (e.g., to add drugs or to replace ions), a constant-circulation chamber has obvious advantages. For experiments in which there is an interest in assaying slice effluent, a chamber which allows at least some static period might be desirable. On this point, the experiment may be the best determinant of the bath construction.

E. Equilibration time

Investigators using the slice have generally found that there is a silent period between slicing and stable recording from the tissue. It appears as though the tissue uses this time to recover from the trauma of the procedure. Curiously, during this silent recovery time, not only are there no spontaneous spikes and no synaptic drive, but there is also little evidence of cell membrane potentials. Visually, one may see the tissue swell during this period, with distinctions in tissue lamination becoming blurred. As neural activity returns, the tissue appears to lose the excess fluid. Cell spiking in most cases precedes synaptic activity.

The silent recovery period may vary considerably, depending on such factors as the degree of trauma during preparation, temperature difference between the dissection and bathing media and anesthesia used on the animal. This last factor has not been systematically studied, but seems to make quite a difference. In preparations started by stunning the animal with a blow to the back of the neck, equilibration time may be as little as 10–15 minutes. Ether does not substantially lengthen recovery, but barbiturate anesthesia may extend "recovery" for hours.

F. Submersion *vs.* suspension at a gas–liquid interface

The belief that slices are not viable when submerged below the fluid surface can no longer be maintained; a number of laboratories now use submerged slice preparations (Richards and Tegg, 1977; Oliver *et al.*, 1977; Llinas and Sugimori, 1978). What difference this makes to the tissue is not clear, but there are a number of pros and cons for each type of system. The submerged preparation should provide a more constant environment free from perturbations in fluid flow, drainage, changes in "environmental" O_2 pressure and appearance of O_2 bubbles. Slices are exposed to medium on two sides, so that diffusion from medium into the slices should be faster than when only one surface contacts the fluid. Major disadvantages of submersion include difficulties in holding the tissue down (maintaining stable recordings) and in accurately guiding electrodes to their targets. In addition, in the submerged slice, local application of drugs to the slice surface is impossible.

G. Slice thickness and oxygenation requirements

Clearly, the thicker the tissue, the more difficult it will be to assure a sufficient O_2 supply to cells within the slice. Slices as thick as 700 μm have been used successfully (White *et al.*, 1978) (i.e., long-term activity has been recorded), and slices as thin as 60 μm are being used in experiments in which direct cell visualization is obtained with Nomarski optics (Yamamoto and Chujo, 1978b). Slice thickness must take into account the probability of some degenerative changes on the cut surfaces of the slices (Bak *et al.*, 1979). For experiments in which cell circuitry is of interest, thicker slices are desirable. Studies of cell electrical properties must employ tissue thick enough to contain major processes of the cells to be studied.

Insufficient cell oxygenation will quickly lead to pathological activity, and eventually to neural silence. It is possible that the abnormal bursting activity seen in hypoxic tissue is due to an initial selective effect on inhibitory interneurons. Loss of O_2 pressure above or below the slices also causes a marked and rapid decrease in cell input resistance. This damage appears, for the most part, to be irreversible.

H. Chamber cleanliness

Since glucose is a constituent of the bathing medium, the perfusion and chamber parts in contact with the bathing fluid should be carefully rinsed after each experiment. However, it appears that too thorough a cleaning (i.e. with detergent or acid) is disruptive. New chambers, and chambers completely cleaned or renewed, often require an "ageing" period. Running distilled water or a saline solution through the chamber for a couple of days before starting experiments is often helpful. If experiments are started before the chamber has been properly aged, cellular activity appears less than optimal, and slice longevity is reduced.

V. Technical Difficulties

The slice preparation is, in principle and in fact, very simple. There are, however, a number of technical "bugs" that must be worked out of every system. Some common problems are given below.

A. Maintaining smooth drainage of circulating bathing medium

In chambers in which the slices sit at a fluid–gas interface, smooth constant removal is critical for maintaining stable recordings. Because the surface tension of the fluid is considerable, fluid tends to accumulate beneath the mesh support and is then suddenly released. A few tricks are helpful in dealing with this problem. First, if the mesh is stretched tautly across the chamber, there will be less opportunity for gradual rises and sudden falls of its level to occur. Second, if something is used to continually break fluid surface tension, drainage is smooth; I routinely drape a small piece of tissue paper (Kimwipes) over the edge of the mesh to maintain constant drainage. Suction is often used to carry away circulated fluid. Especially when using a "wick" to encourage drainage, vacuum level should be adjusted so that it is just above the threshold for removing the fluid; high vacuum pressure may cause drafts over the slices, and may even suck fluid out of the bathing chamber.

A number of investigators have attempted to solve the smooth circulation problem by using a push–pull pumping system. In principle, this should work well; however, when small chamber volumes are used, it is important that the output *exactly* balances the input, and this balance is difficult to achieve (White *et al.*, 1978).

These problems with smooth drainage are less important if a submerged slice system is used. In such chambers, slices do not ride on the varying fluid surface, so some degree of discontinuous flow may be acceptable.

B. Elimination of oxygen bubbles

In many chambers, bathing medium is oxygenated at a reservoir site, and pumped (or gravity-fed) into the slice chamber where it is warmed to the desired bathing temperature. During the warming, O_2 gas escapes the fluid, forming bubbles which may rise beneath the slices. These bubbles can cause mechanical disturbance when they form, and they tend to be trapped by the mesh. If large, they can isolate tissue from the bathing medium. Bubbles are less likely to form if the bathing medium is pre-warmed, in the reservoir bottle, to the bath's final temperature. In submerged preparations, the bubbles may present little problem. Small bubbles can sometimes be removed with a pipette or tissue wick without disturbing the surrounding tissue.

C. Rapid replacement of medium

The replacement of control medium with medium containing an experimental drug (or different ion composition) should be carried out smoothly and rapidly. Time for replacement of fluid in the slice chamber will depend on the volume of the chamber, rate of fluid flow and the dead space in the delivery system. In most chamber designs, there appears to be a trade-off between speed and smoothness in a bath change; decreasing chamber volume, increasing fluid flow and decreasing tubing volume will generally result in a more direct transmission of switching artifacts to the slice. A slight jolt at the time of making a switch may dislodge a cell penetration. Flow rates of 1.5 to 2.0 ml min^{-1} are normally employed with chamber volumes 1–3 cc. An instantaneous switching mechanism, with no pause in fluid flow, will greatly facilitate these procedures (Dore and Richards, 1974).

D. Spread of stimulating current

This problem is essentially no different in slice preparation from that in *in vivo* preparations. If it is necessary to restrict stimulus current to a specific pathway, bipolar stimulating electrodes with small (< 0.5 mm) spread between tips should be used. Local stimulation can be easily accomplished in the slice if care is taken to use minimal current strengths. It is useful to use a constant-current stimulation source if it is important that one strictly control the stimulus. Especially when delivering repetitive pulses, electrodes may polarize, reducing the effective current output from a constant-voltage source.

E. Localized application of drugs

In addition to bath application of various substances, local application of small amounts of drugs may be applied to the slice iontophoretically or via pressure ejection from coarse electrodes. Since the tissue is constantly bathed in fluid, diffusion of substances across the slice surface may occur very rapidly. Controls must be taken to assure that excessive diffusion does not occur. One neat means of locally applying drugs is to dissolve the substance in agar, and contact the slice surface with an electrode filled with this agar (Mesher and Schwartzkroin, 1980). Very discrete applications can be made this way, although the effective concentration of the drug in the tissue cannot be well controlled.

In general, we have assumed that drug and ion concentrations in bathing medium will rapidly equilibrate with extracellular space concentrations. This assumption is still in need of rigorous testing. There is some evidence that changes in extracellular $[K^+]$ are slow to reach equilibrium within the slice (a 10 minute delay between bath and center of a 350 μm slice) (Benninger *et al.*, 1980).

F. Maintenance of long-term healthy penetrations

Cellular penetrations for long periods are readily accomplished in the slice. There are a number of factors I have found helpful in intracellular recordings. Penetration and maintenance of healthy neurons are facilitated if pipettes are bevelled and of as low resistance as is consistent with a sharp tip. A stepping microdrive, with micron-step control, is an effective means not only for initial penetration of cells, but also of subtle adjustments to improve penetration. In some cases, injection of hyper-polarizing current into an injured neuron will help it repair; however, such a maintenance current should not be used while actually determining cell properties. A brief squeal from the negative-capacitance adjustment on the amplifier is often useful for impaling neurons, improving "dirty" recordings, or cleaning-out clogged pipettes.

G. Manufacture of recording electrodes

It is often difficult to determine whether or not a slice is healthy, and there is little to be done if it is not. For these reasons, activities for improving recording quality during an experiment center primarily on improving electrode quality. The magic involved in making good electrodes is emphasized in slice experiments, since we know so little about improving the quality of the tissue. Electrode manufacture has always been an important aspect of neurophysiological experiments, with each experimenter developing his own method and superstitions. Continued attention to this aspect of slice studies is essential.

VI. Response to Limitations

The limitations of the slice technique are substantial, and cannot be ignored unthinkingly. Like all experimental preparations, the *in vitro* slice must be used cautiously and appropriately. The following guidelines may be helpful in deciding what experiments to do using the slice.

(i) Choose experiments that are not directly compromised by the *in vitro* and limited nature of the preparation. Obviously, the slice preparation is not appropriate for studies involving behavioral, sensory or motor correlates. However, it may be useful in studying suspected neural bases of various behaviors (e.g., memory or breathing) if one is extremely cautious in interpreting results. As noted above, the slice is also inappropriate for studies that require extensive intact connections between distant regions, since such connections are normally severed during slicing.

(ii) Choose tissue that can be studied to advantage using the slice. Laminated structures, such as hippocampus, cerebellum and neocortex, are obvious candidates given their regular architecture, well-defined input and output pathways, and identifiable cell groupings. In addition, tissue that is almost impossible to study *in vivo*, may prove more approachable in slices. For example, if it is necessary to study tissue in isolation from the rest of the brain, the slice technique would be one approach to consider; similarly, if tissue is situated *in vivo* so that interfering structures make it inaccessible, an *in vitro* technique would be useful.

(iii) Take advantage of whatever has been accomplished *in vivo* to guide the initial experiments *in vitro*. Establish criteria by which to judge the "health" of the preparation and of recordings. Since the tissue may be in a different state from that studied *in vivo*, it is important to determine whether conventional criteria are applicable. Attempt to compare normal properties defined *in vivo* with those that you find *in vitro*. If these characteristics are similar, the slice investigator then has a good foundation on which to build using experimental data. However, if properties differ drastically, the investigator should be aware that the slice studies will always be suspect; an explanation for the differences, and a justification of the slice data will be required. This point has not been brought up to suggest that all *in vivo* work must be replicated *in vitro*. On the contrary, the slice should be used to explore new areas not investigated *in vivo*. A bridge with intact animal experimental data, however, provides a good beginning for many studies; if the connection can be made as experiments progress, the data take on a considerably richer significance.

(iv) Try to make slice studies relevant to a real problem. This suggestion is not, of course, restricted to slice studies, but is applicable to all experimentation. However, there is a tendency in *in vitro* work to do experiments triggered by the coherent elegance of the technique or the preparation. Such a temptation is to be resisted; there is no end to the number of studies that should be carried out to answer real questions. I am not trying to make an argument here for applied, as opposed to basic, research, but rather suggesting that slice research should be problem, not preparation, oriented.

(v) Take advantage of interdisciplinary approaches. The slice technique lends itself to collaborative effort between physiologist, pharmacologist, anatomist and biochemist. The same tissue that is studied with microelectrodes may be stained, or analyzed for transmitter content. Unlike *in vivo* studies, obtaining information about the tissue from a variety of approaches is not only feasible, but relatively easy. Correlates of electrical activities, of specific patterns of growth, or biochemical specializations or abnormalities can and should be explored at every opportunity. For example, using intracellular injections of dyes, physiological–morphological correlates can be studied.

VII. Types of Suitable Experiments

Given the advantages and limitations of the slice preparation, investigators have developed active and useful research programs in a number of areas of study. The following list describes some of the major areas of investigation for which the slice is a particularly appropriate technique and in which work is currently progressing. The list is obviously biased toward electrophysiological techniques; there are, to be sure, numerous studies not mentioned here that can be carried out using the slice preparation.

A. Study of single cell membrane properties

Membrane characteristics of single cells can be studied using intracellular recording techniques. Such studies have been carried out in hippocampus, cerebellum and olfactory cortex. Parameters such as cell time-constant, input resistance, contribution of active dendritic membrane to cell activity and conductances to various ions, can be determined (Johnston and Hablitz, 1979; Hotson and Prince, 1980; Schwartzkroin, 1975, 1977, 1978; Wong and Prince, 1978; Schwartzkroin and Slawsky, 1977; Schwartzkroin and Prince, 1980b; Llinas and Sugimori, 1978; Sugimori and Llinas, 1979; Scholfield, 1978). The latter determinations may be carried out by changing ion concentrations in the bathing medium and/or adding selective channel blockers (e.g., tetrodotoxin for sodium, tetraethylammonium for potassium, and manganese or cobalt for calcium). Voltage clamp experiments may also be feasible for some types of neurons. Action-potential characteristics (such as waveform and after-potentials) and cell spiking properties (including rheobase, accommodation and rhythmical firing determinants) can also be carefully described.

B. Study of simple circuits

Synaptic connections between neurons can be traced in local circuit studies. Excitatory and inhibitory postsynaptic potentials (EPSPs and IPSPs) can be evoked with discrete stimuli (Richards and Sercombe, 1968; Skrede and Westgaard, 1971; Dudek *et al.*, 1976), and their ionic bases exposed. In some connections, miniature PSPs may be recorded (Brown *et al.*, 1979), and quantal analyses carried out. Since the slice tissue is simplified and contains a relatively small number of cells, it is possible to obtain simultaneous recordings from cell pairs in synaptic contact to study the nature of their connections (excitatory, inhibitory, electrical, etc.) (MacVicar and Dudek, 1978).

C. Iontophoresis–transmitter studies

The slice is ideal for iontophoretic experiments, and for other techniques which approach the question of what neurotransmitter is used by a particular system. Iontophoretic electrodes may be precisely positioned under visual control, independently of the recording electrode; they can even be moved while maintaining intracellular penetrations so that regional differences in sensitivity to a putative transmitter can be tested (Schwartzkroin and Andersen, 1975; Spencer et al., 1976; Alger and Nicoll, 1979). Transmitter candidates, and their antagonists, can also be studied in bath perfusions (Schwartzkroin and Prince, 1980a). Ionic bases of transmitter mechanisms can be studied. The slice preparation also allows collection of tissue perfusate, so that substances released during orthodromic stimulation can be analyzed (Wieraszko and Lynch, 1979). Pressure-ejection techniques, to study various substances that are difficult to eject electrophoretically, are also easily adapted to the slice preparation.

D. Studies of physiology–morphology correlates

With the extensive use of intracellular markers, it is now possible to study both the physiology and morphology of the same cell (Schwartz-kroin and Mathers, 1978; Turner and Schwartzkroin, 1980). This procedure is especially attractive in slice preparations since recovery of marked neurons is not difficult. Such correlated studies should provide more detailed information about cell-membrane properties, and about what morphological specializations mean in terms of cell function. The morphological aspects of these studies may employ fluorescence, phase-contrast and electron microscopic techniques, as well as conventional light microscopic analysis.

E. Developmental studies

Developmental studies of the mammalian CNS are extremely difficult *in vivo* given the fragile nature of the young animals. Slice studies can be used to great advantage to study physiological, morphological, pharmacological and biochemical aspects of development. Since immature tissue tends to be somewhat more resistant to hypoxic damage than the adult, the preparation of slice tissue is particularly facilitated with young animals (Schwartzkroin and Altschuler, 1977).

F. Plasticity

This general area is perhaps that of greatest present activity in laboratories using slice preparations (Schwartzkroin and Wester, 1975; Andersen et al., 1977; Lynch et al., 1977; Yamamoto and Chujo, 1978a; Alger and Teyler, 1976; Dunwiddie et al., 1978; Dunwiddie and Lynch, 1978; Teyler and Alger, 1976; Misgeld et al., 1979). In vitro tissue displays many of the long-term potentiation and depression phenomena exhibited by intact animals; parameters influencing physiological changes, as well as mechanisms underlying such changes, can be accurately determined. In addition to stimulus-induced changes initiated in the slice tissue, it is also possible to induce changes in the intact animals before slicing (e.g., lesioning a specific input to a cell population of interest) and to study effects of that manipulation in a subsequent slice study. Morphological correlates of these changes may, of course, be studied in parallel.

G. Study of disease models

Mechanisms underlying abnormal cell behavior may be studied in slice models of the abnormal state, or in slices taken from abnormal brain. Virtually all of the lines of study outlined above can then be employed to determine what changes have occurred in the tissue in parallel with the pathology. Not only "diseases" (e.g., epilepsy) (Schwartzkroin and Prince, 1977; Ogata et al., 1976; Ogata, 1975, 1978; Yamamoto, 1972; Yamamoto and Kawai, 1968; TerKeurs et al., 1973; Voskuyl et al., 1975; Voskuyl and TerKeurs, 1978; Oliver et al., 1977), but also genetic mutations affecting the nervous system may be studied in this way.

H. Studies of hormone and other "modulator" effects

Since the slice tissue is bathed in a fluid medium, with free access to neural elements, effects of hormones or other modulator substances (e.g., opiates) can be tested in vitro. In addition, cells that secrete such substances may be studied (e.g., hypothalamus) to determine what influences hormone output (e.g., positive or negative feedback controls). Investigations of neurons involved in feedback systems (e.g., cells with neuro-humoral receptors) may be carried out in the absence of interfering neural modulation. Such studies are obvious candidates for an interdisciplinary approach.

I. Measurement of extracellular milieu

Ion concentrations in the extracellular environment are important determinants of neuronal and glial activity. With recent advances in the development of ion-sensitive microelectrodes, changes in extracellular ion concentrations of K^+, Ca^{2+} and Cl^- may be monitored while cellular activity is observed (Fritz and Gardner-Medwin, 1976; Benninger *et al.*, 1980). Such studies in the slice eliminate such problems as pial diffusion barriers. In addition, bath concentrations of various ions may be controlled, and active neural mechanisms (i.e., pumps) for controlling ion concentrations then studied.

VIII. Slice Studies

To illustrate the possible applications of the slice preparation, I will present a brief outline of some of the work carried out in my laboratory. All of the studies I will mention have been carried out using intracellular recordings in the hippocampus (of mouse, guinea pig, rabbit, cat, monkey and man). This choice was made for historical as well as experimentally determined reasons. Clearly, the regularity of lamination of the structure, its simple architecture and the possibility of maintaining extensive intact circuitry within thin lamellae have made the hippocampus very popular in slice laboratories. It is also a relatively large structure which is easily dissected and which has been implicated in a number of complex behaviors. Finally, hippocampal neurons display a wide variety of interesting activities which attract study from many approaches.

A. Methods

Experiments were generally carried out as follows. Guinea pigs were stunned with a blow to the back of the neck (with a heavy file) (light ether anesthesia was used for preparing larger animals such as rabbits), and a quick cut was made with a scalpel across the throat to open the jugular vein (to exsanguinate the animal) (Fig. 2.1A). The scalp was opened along the midline and pulled down laterally (Fig. 2.1B). The spinal cord was cut at the level of the obex (Fig. 2.1C); scissors were inserted under the bone at this point, and a slit was clipped in the skull case along the midline (Fig. 2.1D). A pair of rongeurs were then used to peel away the bone and dura (Fig. 2.1E). A small spatula was slipped behind the cerebellum and under the brain and the brain was lifted gently out of the skull case (Fig. 2.1F). Cranial nerves were cut where necessary, and the freed brain was then dropped into a petri dish of cold, oxygenated (4° C) bathing medium (Fig. 2.1G).

To dissect out the hippocampus, the brain was laid on its dorsal surface on a dissection stage soaked in cold bathing medium. The cerebellum and brainstem were pushed up and out of the way (Fig. 2.2A). The brain was cut along the midline and the thalamic regions were pushed medially so that the fimbria could be seen. A small spatula was inserted under the fimbria (into the ventricular opening) and moved rostrally and caudally to cut the commissural fibers at the poles of the hippocampus (Fig. 2.2B). The hippocampus was then flipped laterally and the final connections with the entorhinal cortex were cut so that it was freed entirely from the rest of the brain (Fig. 2.2C). The hippocampus was rinsed quickly in cold bathing medium, laid on the stage of a tissue chopper (Fig. 2.3A) (dorsal side up), and slices were cut transverse to the long axis (Fig. 2.3B) (a Sorvall TC-2 Tissue Chopper was used for most experiments). A fine brush was used to remove each slice from the cutting blade and to put it in a petri dish with cold bathing medium (Fig. 2.3C). Slices were 350–400 μm thick, and taken from the middle third of the hippocampus. Between 6 and 10 slices were transferred to the incubation chamber using a large-mouthed dropper pipette (Fig. 2.3D). The incubation chamber temperature was set at 36.5° C and controlled to $\pm 0.5°$ C via a thermister-feedback unit.

The slices sat on a nylon mesh support at a fluid–gas interface. Bathing medium circulated below the mesh and contacted the lower surface of the slice; warmed, moistened gas (95% O_2/5% CO_2) saturated the atmosphere above the slices. The bathing medium (124 mM NaCl, 5 mM KCl, 1.25 mM $NaH_2PO_4 \cdot H_2O$, 2 mM $MgSO_4 \cdot 7H_2O$, 2 mM $CaCl_2 \cdot 6H_2O$, 26 mM $NaHCO_3$, 10 mM dextrose) entered the chamber via a gravity flow system (1.5 ml min^{-1}), flowed over the mesh net and was removed by suction. The chamber (Fig. 2.4) (West Coast Instrumentation, Moss Beach, CA) was designed with an outer heated-water bath (DC current of 4–5 A through a 1–1.5 ohm wire coil) through which wound the bathing medium input tube. The O_2/CO_2 gas mixture also bubbled through the bath to be warmed and moistened.

Slices were left about 20 minutes to recover from the trauma of the slicing procedure. Electrodes were then prepared and positioned over the slice to be studied (Fig. 2.5). Recording electrodes (containing fibers) were back-filled with 4 M potassium acetate; output from these electrodes was led through a high-input impedance microelectrode amplifier with internal bridge circuit (for current passage). Stimulating

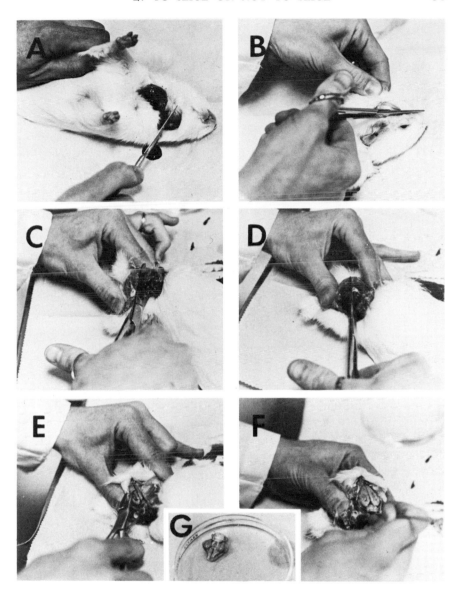

Fig. 2.1. Photographic sequence showing removal of the brain from a guinea pig. A, The animal was first stunned with a blow to the back of the neck. Here, the throat was cut to allow blood to drain from the brain. B, The scalp was cut along the midline. C, The spinal cord was severed from the brain. D, Scissors were inserted at the spinal cord opening and the skull cut along the midline. E, Rongeurs were used to peel away the bone. F, A small spatula was inserted under the brain to pry it out. G, The guinea pig brain (ventral side up) in a petri dish.

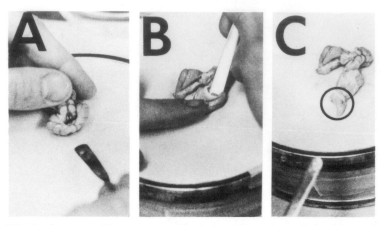

Fig. 2.2. A photographic sequence depicting dissection of the hippocampus from the brain. A, The brain lay ventral side up on the dissection stage. The brainstem and cerebellum were lifted and pushed out of the way. B, After detecting the fimbria, a small spatula was inserted into the ventricular opening, and moved back and forth to sever commissural connections. C, The hippocampus was then flipped out laterally and the last connections with the rest of the brain were cut.

electrodes were generally made from sharpened tungsten wires (insulated to the tip, and cut off at about 10 μm) and used in either monopolar or bipolar configurations. Most studies have concentrated on activities in the CA1 region where stimulation in stratum radiatum evokes orthodromic responses and stimulation of the alveus evokes antidromic responses.

Drugs and ions were tested in two ways. They could be added to the bathing medium so that the entire slice was exposed to the change. This procedure was chosen when an accurate measure of ion or drug concentration was necessary. Alternatively, test substances could be applied focally via a coarse microelectrode. For this procedure, the microelectrode tip was filled and positive pressure applied to the back of the electrode to expel a small droplet onto the slice.

Other procedures used for slice experimentation were similar to those used by experimenters working *in vivo*. I have found a stepping microdrive to be particularly effective for cell impalements in the slice. Overcompensation on the capacitance neutralization adjustment (which results in a shrill ringing) is also helpful in impaling neurons or cleaning out clogged electrodes. The exact procedures undoubtedly include a large superstitious component.

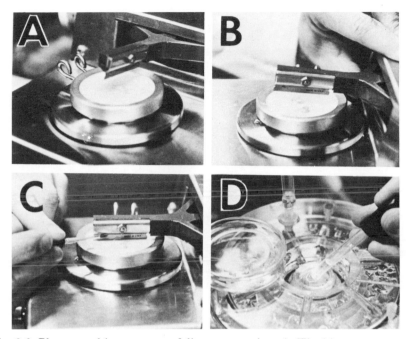

Fig. 2.3. Photographic sequence of slice preparation. A, The hippocampus was stretched across the tissue chopper stage, transverse to the blade. B, Slices were cut (350–400 µm thick) from the middle third of the hippocampus. C, Slices were removed from the chopper blade individually with a fine brush. D, Slices were transferred to the incubation chamber with a large-mouthed pipette.

B. Results

1. *Characteristics of normal hippocampal pyramidal cells*

Studies of CA1 hippocampal pyramidal cells initially focused on a comparison between characteristics recorded *in vitro* (Schwartzkroin, 1975, 1977, 1978), and those previously described *in vivo* (Kandel *et al.*, 1961; Kandel and Spencer, 1961; Spencer and Kandel, 1961a,b). Cell resting potential, spike amplitude and waveform, input resistance, and time constant were comparable; subsequent studies resulted in improvement of our technique and increased cell health as mirrored in these parameters. Also studied were current threshold and accommodation properties and rhythmical repetitive discharge characteristics. EPSPs and IPSPs were recorded in response to orthodromic stimulation, and antidromic spikes were elicited by stimulating axons of the CA1

Fig. 2.4. Top (A) and side (B) views of slice chamber. Bathing medium was led to the chamber via a gravity feed system. It was circulated through a heated-water bath before entering the inner chamber where contact was made with the slices. Electrodes were directed toward the slices through a hole in a clear plastic cover over the inner chamber. A 95% O_2/5% CO_2 gas mixture was bubbled through the outer water bath and formed the atmosphere over the slices in the inner chamber.

neurons. The fast prepotential (dendritic spikes) and depolarizing after-potentials (DAPs) of hippocampal pyramidal cells were suggestive signs of active dendritic participation in regulating cell output. Since dendritic spiking had been shown to be calcium-mediated in cerebellar Purkinje cells, we started to study the ionic bases of activity in hippocampal pyramidal cells. These experiments revealed a calcium-conductance (Fig. 2.6) (Schwartzkroin and Slawsky, 1977) and calcium-dependent potassium conductance (Fig. 2.7) (Hotson *et al.*, 1977) in CA1 neurons. Many of these experiments were carried out with ion channel blockers such as tetrodotoxin (TTX) and tetraethylammonium (TEA). Parallel experiments on CA3 neurons have been carried out by other investigators.

In addition to the physiological characterization of CA1 neurons, correlated physiology–morphology studies have been done. Using intracellular injection of HRP dye into physiologically characterized

A

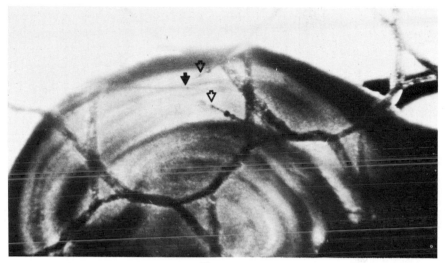

B

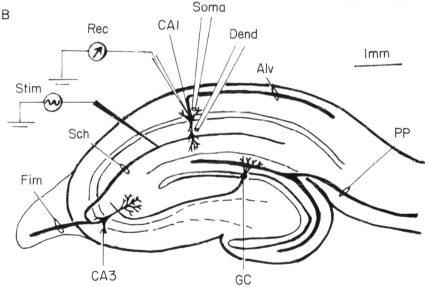

Fig. 2.5. A, Picture of transverse hippocampal slice, as seen in the incubation chamber. Dark arrowhead points to recording electrode entering stratum pyramidale (cell body region) in the CA1 area; open arrowheads point to iontophoresis pipettes positioned in stratum pyramidale (upper) and stratum radiatum (lower). The slice sits on a nylon mesh, the strands of which are also seen in this field. B, Diagrammatic representation of slice with electrodes for stimulation (Stim), recording (Rec) and drug application in somata (Soma) or dendrites (Dend). Other abbreviations: Alv, alveus; Sch, Schaffer collaterals; Fim, fimbria; CA1, CA1 pyramidal cell layer; CA3, CA3 pyramidal cell layer; GC, granule cells in the dentate region; PP, perforant path.

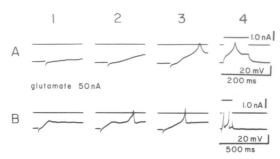

Fig. 2.6. Elicitation of calcium spikes from CA1 pyramidal cells. A and B show similar events at different oscilloscope sweep speeds. Tetrodotoxin (TTX) (5 μM) was added to the bathing medium to block sodium action potentials. Sweeps 1–3 show effects of glutamate iontophoresis (50 nA) in the cell's dendritic region; the glutamate electrode position was carefully adjusted (1–3) so as to produce a more rapid membrane depolarization. The sweeps in trace 4 show effects of intracellular injection of a depolarizing current (current monitor above trace).

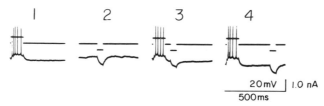

Fig. 2.7. Demonstration of conductance increase associated with the hyperpolarization following repetitive spike firing. In sweep 1, a depolarizing current pulse elicited repetitive spiking followed by a long after-hyperpolarization (AHP). A hyperpolarizing constant-current pulse, injected to measure cell resistance (trace 2), caused a smaller voltage deflection when it fell during the AHP (traces 3 and 4).

neurons, a model of the CA1 pyramidal cell was developed that takes into account morphological as well as physiological cell properties (Turner and Schwartzkroin, 1980).

2. *Development of the CA1 pyramidal cell region*

Longitudinal developmental studies have been carried out on kitten (Schwartzkroin and Altschuler, 1977) and rabbit hippocampus. The CA1 cell area has been studied to determine single-cell membrane properties and synaptic characteristics during development; physiologi-

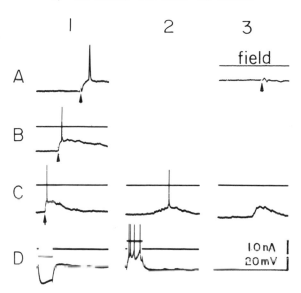

Fig. 2.8. Recordings from a pyramidal cell in a hippocampal slice from a 2-day-old rabbit. A1, B1 and C1 show cell response to stimulation in stratum radiatum (at arrowhead); note the very large EPSP and absence of any hyperpolarizing IPSP. C2 and C3 illustrate spontaneous depolarizing events. In D1 and D2, hyperpolarizing and depolarizing current pulses were injected; cell input resistance was extremely high. Trace A3 shows an extracellular field potential, recorded outside this cell, in response to radiatum stimulation. The time calibration bar (in D3) is 100 ms for A and B, 200 ms for C, and 500 ms for D.

cal findings have been correlated with findings at the light and electron microscopic level of cell and synapse development. Most striking in these studies has been the late development of inhibitory synaptic activity in the young animals. Excitatory input is intact (in cat and rabbit) on the day of birth, although it is undoubtedly not entirely normal (i.e. mature) (Fig. 2.8). Work is underway in which fetal hippocampus is being studied during the time of initial synapse and spiking development (Fig. 2.9). Such fetal studies may be a useful bridge to cell-culture studies *in vitro*.

3. *Local circuit studies*

Experiments have been carried out to identify physiologically and morphologically (intracellular HRP injection) interneurons in hippocampus. A cell population has been described that is obviously different

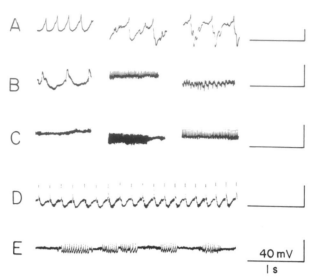

Fig. 2.9. Recordings from hippocampus of a 20–21-day fetal rabbit. A–E show various spontaneous potentials. Cells had resting potentials of 15–25 mV; no activity was seen unless hyperpolarizing DC current was injected (and maintained) into these cells (0.5–1.0 nA).

from the CA1 pyramidal cell (Schwartzkroin and Mathers, 1978), and which may well mediate inhibition in hippocampus. These cells apparently receive direct afferent input (i.e., their function is not strictly "recurrent"), have high spontaneous activity, and respond with burst discharges to orthodromic input (Fig. 2.10). Their processes reach into the CA1 dendritic region (stratum radiatum) but may as well show intrapyramidale endings (suggestive of basket cells).

4. *Transmitter sensitivities of CA1 neurons*

Glutamate and γ-aminobutyric acid (GABA) have been tested electrophoretically on CA1 neurons (Schwartzkroin and Andersen, 1975). Cells were shown to be exquisitely sensitive to glutamate, depolarizing sharply after even small amounts of glutamate were applied. No particular cell region was found to be particularly sensitive to glutamate application. GABA induces a hyperpolarization-mediated inhibition when applied near cell somata and may also cause membrane depolarization. IPSPs have been shown to be antagonized by the GABA blocker bicuculline and by penicillin.

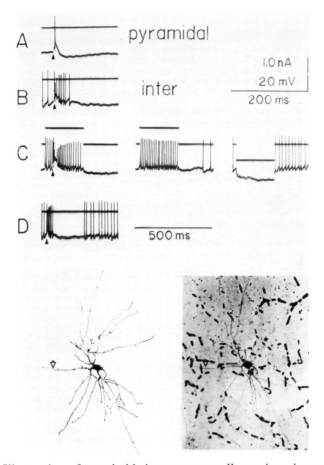

Fig. 2.10. Illustration of a probable interneuron cell type in guinea pig hippo-campus. A, Typical pyramidal cell response to radiatum stimulation. B, Inter-neuron response to the same stimulus, at the same time scale. In C, depolarizing current was injected in the first two sweeps. Superimposition of an ortho-dromic input during the current-induced firing (first trace) revealed little inhibition; however, an inhibitory AHP did follow the current-induced firing. The third trace in C shows effects of a hyperpolarizing current pulse. The trace in D shows, at a slow time scale, the cell response to stimulation; an inhibitory pause clearly followed the stimulus-evoked burst. The figure at the bottom shows a probable interneuron filled intracellularly with horseradish peroxidase (actual photo on right, camera lucida drawing on left). An open arrow indicates a process running parallel to the pyramidale layer.

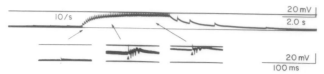

Fig. 2.11. Glial cell recording. Glial response to repetitive stimulation (10/s for this cell) was a slow depolarization resulting from summation of small depolarizing responses to each stimulus pulse. The three traces below show faster sweeps of individual responses at indicated times during the long depolarization. Note, too, the extremely slow return of membrane potential to baseline after the stimulus train.

5. *Glial recordings*

Glial cells may be impaled and studied in the hippocampal slice (Schwartzkroin and Prince, 1979). Glial cell characteristics *in vitro* are similar to those described *in vivo*: low input resistance, no active response and slow depolarizing response to repetitive stimulation (Fig. 2.11), presumably due to the effect of neuronal potassium release. Repetitive

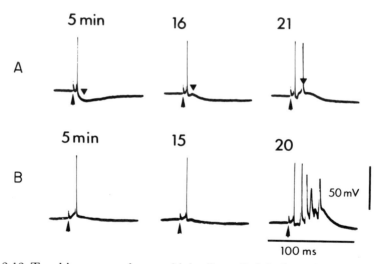

Fig. 2.12. Two hippocampal pyramidal cells studied during addition of penicillin to the bathing medium. Cells are responding to stimulation in stratum radiatum. Numbers above each trace indicate time from addition of penicillin to the reservoir solution. A, The control response consisted of a spike followed by a large IPSP. The IPSP gradually diminished, and a depolarizing potential appeared which eventually triggered additional spikes. B, This cell showed a smaller control IPSP. A small after-depolarization did develop, and was then replaced by a full-scale, all-or-none burst.

stimulation results in a slow depolarization of the glial membrane which reaches a plateau at a maximum of 25 mV above resting potential.

6. *Studies of bursting and epileptiform activity*

The *in vitro* hippocampal slice has been productively employed as an acute epilepsy model (Schwartzkroin and Prince, 1976, 1977, 1978, 1980a; Prince and Schwartzkroin, 1978; Schwartzkroin and Wyler, 1980; Schwartzkroin and Pedley, 1979). Application of the epilepto-genic agent penicillin to the bathing medium results in synchronized cell bursting activity. Our studies have indicated that the paroxysmal depolarizing shift (PDS) underlying bursting "epileptic" neurons is not due to "giant EPSP" input. Penicillin, however, appears to enhance inva-sion of dendritic activity into the soma, and is certainly a blocker of GABA-mediated IPSPs (Fig. 2.12). These data, together with our knowledge of calcium and potassium activity during bursting and repetitive action potential discharge, have suggested that the PDS is generated by a dendritic calcium influx.

In addition to these results from the acute *in vitro* penicillin model, recordings have been obtained from slice studies of tissue from chronic epileptic foci in monkey neocortex (alumina foci) and human hippo-campus. The data from these experiments are still preliminary; bursting neurons have been observed, but their importance or the mechanism underlying such bursts have not been determined.

IX. Conclusions

This chapter has been an attempt to survey the advantages and disad-vantages of the slice preparation, and to give some examples of the types of studies that may be profitably pursued using the preparation. It is certainly my bias that the preparation represents an important tool for neurosciences research, and that its development has already resulted in important contributions. Future development of the technique will assuredly open up new possibilities for study of the mammalian CNS.

Acknowledgements

The author is supported by grants NS 15317, NS 00413 and N504053 from NINCDS, NIH, and by grant BNS 79-15115 from NSF. I would like to express my appreciation to all those who have collaborated with me in the slice studies described in this paper. A special thanks to Per

Andersen (who introduced me to the slice technique) and to David A. Prince. I am grateful, too, for the exceptional technical and secretarial support I have received in Oslo, at Stanford, and at the University of Washington.

References

Alger, B. E. and Nicoll, R. A. (1979). GABA-mediated biphasic inhibitory responses in hippocampus. *Nature (Lond.)* **281,** 315–317.

Alger, B. E. and Teyler, T. J. (1976). Long-term and short-term plasticity in the CA1, CA3 and dentate regions of the rat hippocampal slice. *Brain Res.* **110,** 463–480.

Andersen, P., Sundberg, S. H., Sveen, O. and Wigström, H. (1977). Specific long-lasting potentiation of synaptic transmission in hippocampal slices. *Nature (Lond.)* **266,** 736–737.

Bak, I. J., Misgeld, U. and Weiler, M. H. (1979). Morphological evaluation of neostriatal slices used for electrophysiological experiments. *Neurosci. Abstr.* **5,** 67.

Benninger, C., Kadis, J. and Prince, D. A. (1980). Extracellular calcium and potassium changes in hippocampal slices. *Brain Res.,* **187,** 165–182.

Brown, T. H., Wong, R. K. S. and Prince, D. A. (1979). Spontaneous miniature synaptic potentials in hippocampal neurons. *Brain Res.* **177,** 194–199.

Dore, C. F. and Richards, C. D. (1974). An improved chamber for maintaining mammalian brain tissue slices for electrical recording. *J. Physiol.* **239,** 83–84P.

Dudek, F. E., Deadwyler, S. A., Cotman, C. W. and Lynch, G. (1976). Intracellular responses from granule cell layer in slices of rat hippocampus: perforant path synapse. *J. Neurophysiol.* **39,** 384–393.

Dunwiddie, T. and Lynch, G. (1978). Long-term potentiation and depression of synaptic responses in the rat hippocampus: localization and frequency dependency. *J. Physiol.* **276,** 353–367.

Dunwiddie, T., Madison, D. and Lynch, G. (1978). Synaptic transmission is required for long-term potentiation. *Brain Res.* **150,** 413–417.

Fritz, L. C. and Gardner-Medwin, A. R. (1976). The effect of synaptic activation on the extracellular potassium concentration in the hippocampal dentate area, *in vitro. Brain Res.* **112,** 183–187.

Hotson, J. R. and Prince, D. A. (1980). A calcium activated hyperpolarization follows repetitive firing in hippocampal neurons. *J. Neurophysiol.,* **43,** 409–419.

Hotson, J. R., Schwartzkroin, P. A. and Prince, D. A. (1977). Calcium activated after-hyperpolarization in the hippocampal slice maintained *in vitro. Neurosci. Abstr.* **3,** 218.

Johnston, D. and Hablitz, J. (1979). Voltage-clamp analysis of CA3 neurons in hippocampal slices. *Neurosci. Abstr.* **5,** 292.

Kandel, E. R. and Spencer, W. A. (1961). Electrophysiology of hippocampal neurons. II. After-potentials and repetitive firing. *J. Neurophysiol.* **24**, 243–259.

Kandel, E. R., Spencer, W. A. and Brinley, F. J. Jr. (1961). Electrophysiology of hippocampal neurons: I. Sequential invasion and synaptic organization. *J. Neurophysiol.* **24**, 225–242.

Lipton, P. and Whittingham, T. S. (1979). The effect of hypoxia on evoked potentials in the in vitro hippocampus. *J. Physiol. (Lond.)* **287**, 427–438.

Llinas, R. and Sugimori, M. (1978). Dendritic calcium spiking in mammalian Purkinje cells: in vitro study of its function and development. *Neurosci. Abstr.* **4**, 66.

Lynch, G. S., Dunwiddie, T. and Gribkoff, V. (1977). Heterosynaptic depression: a postsynaptic correlate of long-term potentiation. *Nature (Lond.)* **266**, 737–739.

MacVicar, B. A. and Dudek, F. E. (1978). Local synaptic circuits in rat hippocampal slices: Interactions between pyramidal cells. *Neurosci. Abstr.* **4**, 581.

Mesher, R. A. and Schwartzkroin, P. A. (1980). Can CA3 epileptiform discharge induce bursting in normal CA1 hippocampal neurons? *Brain Res.*, **183**, 472–476.

Misgeld, U., Sarvey, J. M. and Klee, M. R. (1979). Heterosynaptic postactivation potentiation in hippocampal CA3 neurons: Long-term changes of the postsynaptic potential. *Exp. Brain Res.* **37**, 217–229.

Ogata, N. (1975). Ionic mechanisms of the depolarization shift in thin hippocampal slices. *Exp. Neurol.* **46**, 147–155.

Ogata, N. (1978). Possible explanation for interictal–ictal transition: evolution of epileptiform activity in hippocampal slice by chloride depletion. *Experientia* **34**, 1035–1036.

Ogata, N., Hori, N. and Katsuda, N. (1976). The correlation between extracellular potassium concentration and hippocampal epileptic activity *in vitro*. *Brain Res.* **110**, 371–375.

Oliver, A. P., Hoffer, B. J. and Wyatt, R. J. (1977). The hippocampal slice: A system for studying the pharmacology of seizures and for screening anticonvulsant drugs. *Epilepsia* **18**, 543–548.

Oliver, A. P., Hoffer, B. J. and Wyatt, R. J. (1978). Interaction of potassium and calcium in penicillin-induced interictal spike discharge in the hippocampal slice. *Exp. Neurol.* **62**, 510–520.

Prince, D. A. and Schwartzkroin, P. A. (1978). Non-synaptic mechanisms in epileptogenesis. *In* "Abnormal Neuronal Discharge" (N. Chalazonitis and M. Boisson, eds.), pp. 1–12. Raven Press, New York.

Richards, C. D. and Sercombe, R. (1968). Electrical activity in guinea-pig olfactory cortex *in vitro*. *J. Physiol. (Lond.)* **197**, 667–683.

Richards, C. D. and Sercombe, R. (1970). Calcium, magnesium and the electrical activity of guinea-pig olfactory cortex *in vitro*. *J. Physiol. (Lond.)* **211**, 571–584.

Richards, C. D. and Tegg, W. J. B. (1977). A superfusion chamber suitable for maintaining mammalian brain tissue slices for electrical recording. *Br. J. Pharmac.* **59,** 526P.

Scholfield, C. N. (1978). Electrical properties of neurones in the olfactory cortex slice in vitro. *J. Physiol. (Lond.)* **275,** 535–546.

Schwartzkroin, P. A. (1975). Characteristics of CA1 neurons recorded intracellularly in the hippocampal in vitro slice preparation. *Brain Res.* **85,** 423–436.

Schwartzkroin, P. A. (1977). Further characteristics of hippocampal CA1 cells in vitro. *Brain Res.* **128,** 53–68.

Schwartzkroin, P. A. (1978). Secondary range rhythmic spiking in hippocampal neurons. *Brain Res.* **149,** 247–250.

Schwartzkroin, P. A. and Altschuler, R. J. (1977). Development of kitten hippocampal neurons. *Brain Res.* **134,** 429–444.

Schwartzkroin, P. A. and Andersen, P. (1975). Glutamic acid sensitivity of dendrites in hippocampal slices *in vitro. In* "Physiology and Pathology of Dendrites. Advances in Neurology", Vol. 12 (G. W. Kreutzberg, ed.), pp. 45–51. Raven Press, New York.

Schwartzkroin, P. A. and Mathers, L. H. (1978). Physiological and morphological identification of a nonpyramidal hippocampal cell type. *Brain Res.* **157,** 1–10.

Schwartzkroin, P. A. and Pedley, T. A. (1979). Slow depolarizing potentials in "epileptic" neurons. *Epilepsia* **20,** 267–277.

Schwartzkroin, P. A. and Prince, D. A. (1976). Microphysiology of human cerebral cortex studied *in vitro. Brain Res.* **115,** 497–500.

Schwartzkroin, P. A. and Prince, D. A. (1977). Penicillin-induced epileptiform activity in the hippocampal in vitro preparation. *Ann. Neurol.* **1,** 463–469.

Schwartzkroin, P. A. and Prince, D. A. (1978). Cellular and field potential properties of epileptogenic hippocampal slices. *Brain Res.* **147,** 117–130.

Schwartzkroin, P. A. and Prince, D. A. (1979). Recordings from presumed glial cells in the hippocampal slice. *Brain Res.* **161,** 533–538.

Schwartzkroin, P. A. and Prince, D. A. (1980a). Changes in EPSPs and IPSPs leading to epileptogenic activity. *Brain Res.*, **183,** 61–76.

Schwartzkroin, P. A. and Prince, D. A. (1980b). Effects of TEA on hippocampal neurons. *Brain Res.*, **185,** 169–181.

Schwartzkroin, P. A. and Slawsky, M. (1977). Probable calcium spike in hippocampal neurons. *Brain Res.* **135,** 157–161.

Schwartzkroin, P. A. and Wester, K. (1975). Long-lasting facilitation of a synaptic potential following tetanization in the in vitro hippocampal slice. *Brain Res.* **89,** 107–119.

Schwartzkroin, P. A. and Wyler, A. R. (1980). Mechanisms underlying epileptiform burst discharge. *Ann. Neurol.*, **7,** 95–107.

Skrede, K. K. and Westgaard, R. H. (1971). The transverse hippocampal slice: a well-defined cortical structure maintained *in vitro. Brain Res.* **35,** 589–593.

Spencer, H. J., Gribkoff, V. K., Cotman, C. W. and Lynch, G. S. (1976). GDEE antagonism of iontophoretic amino acid excitations in the intact hippocampus and in the hippocampal slice preparation. *Brain Res.* **105,** 471–481.

Spencer, W. A. and Kandel, E. R. (1961a). Electrophysiology of hippocampal neurons. III. Firing level and time constant. *J. Neurophysiol.* **24,** 260–271.

Spencer, W. A. and Kandel, E. R. (1961b). Electrophysiology of hippocampal neurons. IV. Fast prepotentials. *J. Neurophysiol.* **24,** 272–285.

Sugimori, M. and Llinas, R. (1979). A non-inactivating voltage-dependent sodium conductance in mammalian Purkinje cell somata studied in vitro. *Neurosci. Abstr.* **5,** 107.

TerKeurs, W. J., Voskuyl, R. A. and Meinardi, H. (1973). Effects of penicillin on evoked potentials of excised prepiriform cortex of guinea pig. *Epilepsia* **14,** 261–271.

Teyler, T. J. and Alger, B. E. (1976). Monosynaptic habituation in the vertebrate forebrain. The dentate gyrus examined *in vitro. Brain Res.* **115,** 413–425.

Turner, D. A. and Schwartzkroin, P. A. (1980). Steady-state electrotonic analysis of intracellularly stained hippocampal neurons. *J. Neurophysiol.* **44,** 184–199.

Voskuyl, R. A. and TerKeurs, H. E. D. J. (1978). Excitability increase of neurons in olfactory cortex slices of the guinea pig after penicillin administration. *Brain Res.* **156,** 83–96.

Voskuyl, R. A., TerKeurs, H. E. D. J. and Meinardi, H. (1975). Actions and interactions of dipropylacetate and penicillin on evoked potentials of excised prepiriform cortex of guinea pig. *Epilepsia* **16,** 583–592.

White, W. F., Nadler, J. V. and Cotman, C. W. (1978). A perfusion chamber for the study of CNS physiology and pharmacology *in vitro. Brain Research* **152,** 591–596.

Wieraszko, A. and Lynch, G. (1979). Stimulation-dependent release of possible transmitter substances from hippocampal slices studied with localized perfusion. *Brain Res.* **160,** 372–376.

Wong, R. K. S. and Prince, D. A. (1978). Participation of calcium spikes during intrinsic burst firing in hippocampal neurons. *Brain Res.* **159,** 385–390.

Wong, R. K. S., Prince, D. A. and Basbaum, A. I. (1979). Intradendritic recordings from hippocampal neurons. *Proc. Natl. Acad. Sci. U.S.A.* **76,** 986–990.

Yamamoto, C. (1972). Intracellular study of seizure-like afterdischarges elicited in thin hippocampal sections *in vitro. Exp. Neurol.* **35,** 154–164.

Yamamoto, C. (1974). Electrical activity recorded from thin sections of the lateral geniculate body, and the effects of 5-hydroxytryptamine. *Exp. Brain Res.* **19,** 271–281.

Yamamoto, C. and Chujo, T. (1978a) Long-term potentiation in thin hippocampal sections studied by intracellular and extracellular recordings. *Exp. Neurol.* **58,** 242–250.

Yamamoto, C. and Chujo, T. (1978b). Visualization of central neurons and recording of action potentials. *Exp. Brain Res.* **31,** 299–301.

Yamamoto, C. and Kawai, N. (1968). Generation of the seizure discharge in thin sections from the guinea pig brain in chloride-free medium *in vitro*. *Jap. J. Physiol*. **18,** 620–631.

Yamamoto, C. and McIlwain, H. (1966). Electrical activities in thin sections from the mammalian brain maintained in chemically-defined media *in vitro*. *J. Neurochem*. **13,** 1333–1343.

3

The Hippocampal Slice *In Vitro.* A description of the technique and some examples of the opportunities it offers

I. A. LANGMOEN AND P. ANDERSEN

Neurophysiological Institute, University of Oslo, Karl Johans gt 17, Oslo 1, Norway

I. Introduction

The machinery of the mammalian brain does not readily expose itself to its explorers. The experimenter's wish to change the environment of brain cells is often effectively counteracted by homoeostatic mechanisms: the blood/brain barrier prevents the admission of many important drugs and components, and strong homoeostatic regulations for pH, blood pressure and temperature control make changes of these parameters difficult. Further, pulsations and vibrations make it difficult to record extra- or intracellularly from identified neurons for long periods of time. A complicated histology makes it time consuming and often difficult to establish which cells or cell structures have been manipulated.

In preparations without such obstacles, like many invertebrate preparations, several important questions have been resolved. The successful use of these *in vitro* preparations has also greatly influenced scientists working with the mammalian brain. If small parts of the brain could be kept alive *in vitro*, the stability would improve, and many other experimental procedures would be easier. Anaesthesia would no longer interfere with the responses. One would be able to place electrodes more exactly, and also manipulate the environment significantly more than is allowed by intact animal preparations.

In the middle of the 1960s, it was demonstrated that reliable synaptic field potentials could be evoked from brain slices maintained *in vitro* (Yamamoto and McIlwain, 1966). The interest in this technique has increased spectacularly after the first demonstration of high-quality intracellular recordings from brain slices (Yamamoto, 1972; Schwartzkroin, 1975).

It is our purpose to present the experience we have obtained from slice preparations since the earliest reports from our laboratory (Skrede and Westgaard, 1971; Schwartzkroin, 1975; Schwartzkroin and Andersen, 1975). In addition, we will give a short survey of some of the results we have obtained with such preparations.

II. Methods

A. The chamber

The slice chamber which we use most frequently (Fig. 3.1 and 3.2) is based on the early construction of Li and McIlwain (1957) with certain modifications made by Richards and Sercombe (1970) and ourselves. Since the first intracellular studies were made in our laboratory (Schwartzkroin and Andersen, 1975; Schwartzkroin, 1975) it has not undergone major changes.

In principle, it consists of two different parts, an inner bath where the slices are incubated in artificial cerebrospinal fluid (ACSF) and an outer bath containing distilled water. The temperature in the outer bath is kept at a desired value and produces a temperature in the inner bath which is about 2°C lower. The ACSF flows through a polyethylene tube (inner diameter 2–3 mm). The tube is coiled in the outer bath for at least 20–30 cm before it exits in the inner chamber. This heats the fluid to the desired temperature. The gas (95% O_2/5% CO_2) which oxygenates the tissue bubbles through the outer chamber. This is important for two

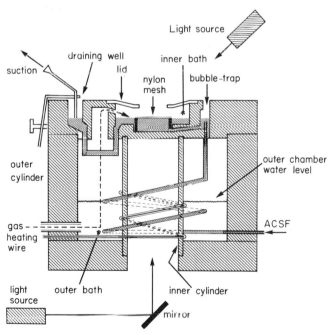

Fig. 3.1. Diagram of the slice chamber. For a further description, see the text.

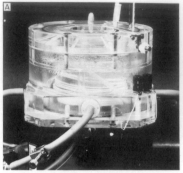

Fig. 3.2. A, Photograph of the slice chamber. B, Photograph of the top of the chamber. Note the inner chamber with the nylon mesh on which the slices rest during the experiment. Under the mesh a part of the ground electrode can be seen. The thermometer is pushed through a hole in the top plate and placed in the water in the outer chamber. Note also the draining well to the left of the thermometer. The needle is connected to a negative-pressure device during the experiment and the level of the needle tip (regulated by the white knob in the lower left) thus determines the level of ACSF in the draining well and the inner chamber.

reasons. First, the gas is humidified, which is necessary to prevent the slices from drying out, and second, the gas is heated.

The outer chamber consists of a space between the walls of a small and a large cylinder. The bottom of the chamber is a circular plate on which these cylinders rest. The bath has a hollow cylinder in the middle, which extends to the bottom of the inner bath, allowing illumination of the slices from below. The distilled water in the outer bath is heated by a heating element consisting of an insulated resistance wire coiled around the inner cylinder. The gas enters the water jacket through a dispersing filter and bubbles through the water to form a gas pocket on top of the fluid in the outer bath. From the top of the outer chamber the gas is led to the inner chamber through eight conveying channels providing an atmosphere of high humidity, O_2 and CO_2 content for the slices. If these channels are too narrow or too few in number, water may condense in them, resulting in drying of the slices.

The top of the chamber consists of a circular plate with a central well representing the inner chamber. A low wide cylinder (height 10 mm, diameter 25 mm) is covered with a nylon net (mask width about 1 mm). After being warmed in the coiled polyethylene tube, the ACSF is delivered to the inner bath where it ascends towards the net with the slices. After flowing through the net, the ACSF is collected in a surrounding trough from which it flows through a U-shaped tube to a draining well where it is removed by suction. The fluid level in the chamber is adjusted by changing the vertical position of the suction needle. The inner chamber is covered by a lid having a cone-shaped depression with a central hole for electrodes. The lid is necessary to maintain the particular climate required for survival of the slices.

Holes in the top of the bath allow insertion of a thermometer and the addition of fresh liquid. The outer chamber should be a quarter filled with distilled water.

The nylon mesh on the top of the cylinder in the inner bath is covered by a piece of lens paper. The fluid level of the inner chamber is adjusted in the following way. The suction needle is adjusted to a level that makes the lens paper float so that it can easily be moved horizontally with a fine painting brush or a spatula. Then the needle tip is lowered until it is barely possible to move the lens paper. The correct fluid level is now obtained by further lowering the fluid level about 1 mm. The slices may then be placed on the lens paper.

The fluid is supplied from a bottle hanging about 1 m above the chamber. The content of the bottle is perfused with 95% O_2/5% CO_2 throughout the experiment. An ordinary intravenous infusion set connects the bottle to the input of the chamber. The set allows the flow to be calibrated as the number of drops per minute. In our inner chamber, measuring 2 ml, the perfusion rate is usually 1–3 ml min^{-1}. A more precise valve is to be preferred when intracellular studies are performed to secure a stable flow.

Recently, another simpler chamber has been constructed by Haas *et al.* (1979). The main advantage of this chamber is that the delay of the fluid exchange is shorter. Its main principle is that the "dead space" of the bath is reduced by placing the nylon mesh directly on the floor of a single compartment. The fluid flows as a thin film from one end to the other where it drips over an edge. The chamber is simple to make.

B. The artificial cerebrospinal fluid (ACSF)

The composition of the ACSF is presented in Table 3.1. The pH should be kept at 7.4. As shown in the table, the ionic concentrations are close to those found in *in vivo* measurements. A comment should be made on the K^+ concentration. A rather high K^+ level has often been used in slice experiments, but a potassium concentration of 3.25 mM, which agrees with physiological values, appears to have a beneficial effect on the slices. Although synaptic responses seem to be smaller in a small number of slices, both extracellular and intracellular potentials are regularly of higher quality.

Table 3.1. The composition (in mM) of the artificial cerebrospinal fluid (ACSF) compared to the composition of CSF from the cat cisterna magna (Ames *et al.*, 1964) and the rabbit lumbar CSF and arterial plasma (Davson, 1967).

	ACSF	Cat cisterna magna CSF	Rabbit lumbar CSF	Rabbit arterial plasma
Na	150 (153)	158	149	148
K	6.25 (3.25)	2.7	2.9	4.3
Ca	2	1.5	1.24	2.8
Mg	2	1.3	0.87	1.01
Cl	133	144	130	106
HCO_3	26	18		25
SO_4	2			
PO_4	1.25			
Glucose	10		5.35	8.3

When fluids are changed, osmolarity differences of ± 5 mosm seem to be unimportant, although preferentially compensations should be made to avoid precipitation when changes include addition of divalent cations.

The temperature of the incubation fluid seems to be important. Bursting activity and faster deterioration seem to be associated with temperatures above about 37°C. We commonly adjust the temperature to 32–34°C.

The inner bath and the bottle and tubing system should be washed with distilled water after each experiment and should be kept filled with distilled water when not in use.

C. Oxygenation/humidification

Gas containing 95% O_2/5% CO_2 flows over the slices after being humidified in the outer chamber. This serves to provide the slices with sufficient O_2 and to prevent them from drying out. The CO_2 in the gas is adjusted to the bicarbonate concentration in the ACSF to buffer the pH to 7.4.

The gas flow seems to us to be a critical parameter for experimental success. If the gas flows too slowly, the slices dry out. If the flow is too fast, the slices also deteriorate, perhaps because of O_2 intoxication. We normally use gas flow rates in the range of 0.1 ml min^{-1}. The right value must be found by observing the slices, the electrodes and the lid covering the inner chamber. Slices with a granulated surface that scatters the light are in danger of drying out. Translucent slices are often already dead. We also have the impression that slices that appear very white are of inferior quality and that this might be due to a too intense oxygenation. A moderate condensation on the electrodes and the lid covering the inner bath will occur when the gas flows with sufficient rate. We have at present not tested gas with lower O_2 content. It is possible that such gas mixtures could increase the range of tolerance between the lower and upper gas flow rates.

D. Preparation of the slices

The equipment needed include a surgical knife, scissor, forceps and bone forceps, two small spatulas, a wide-bored pipette, a fine painting brush, a tissue chopper and two small glass or plastic dishes, one of them with a lid. The spatula should be blunt, not to destroy the tissue, and, if possible, slightly curved. We prefer a spatula made of plastic to one made of steel. They can be made from plastic knives by just removing the sharp edge with fine sandpaper. The wide-bored pipette is used for the transfer of the slices from the temporary incubation in the ACSF in one of the plastic cups to the slice chamber. It is best made from a glass tube with inner diameter of 7–10 mm. The lower end must be carefully flamed to remove the sharp edge. We use a Sorvall tissue chopper to cut slices, but they can be cut manually. Indeed, there is some evidence that manual cutting improves the quality of the slices (Garthwaite *et al.*, 1979). The first of the two cups is used to wash the blood from the brain and should therefore be sufficiently large. The second cup serves as a temporary incubation chamber for the slices until cutting is finished. Just before the operation is started both cups should be filled with oxygenated ACSF, preferably of a temperature similar to that of the chamber. After removing the brain from the skull, the tissue should rest on filter paper moistened with ACSF during all surgical procedures. This includes the cutting where the filter paper is glued to an underlying plastic plate. The amount of ACSF used to moisten this filter paper should, however, not be so large that fluid flows on the surface.

Preferably, there should not be more than 3 minutes between the death of the animal, or the cessation of the blood supply to the brain, and the deposition of the slices in the chamber. This time requirement is, however, not absolute. Good results may also be obtained provided the delay does not exceed 5 minutes.

The animal is anaesthetized with ether in a glass box with a tight lid. The field potentials tend to be better if the animal is deeply anaesthetized. A fatal dose is better than a dose that is too low. When the animal is properly anaesthetized, it is removed from the box and a long sagittal skin incision is made with the surgical knife. The skull is then cut along the midline with the scissors and each half of the dorsal skull removed with bone forceps. The brainstem is divided below the cerebellum. The hemispheres are lifted with a spatula and the cranial nerves and blood vessels are cut to allow the brain to fall gently into the large cup prepared for this purpose. Here the blood is washed away. The brain is then placed on a filter paper moistened with ACSF and divided into its two halves. The hippocampal formation can now be dissected out. This must be done carefully. Equipment with sharp edges should be avoided. The hemisphere is placed with the lateral side down. The brain stem is lifted with the forceps while a spatula gently pushes the occipital cortex downwards until the whole medial and inferior side of the hippocampal formation can be seen. Both ends of the structure are then freed and the hippocampus is rolled upwards and backwards towards the occipital pole by carefully inserting spatulas at both sides under its superior surface. The hippocampal formation now lies with the dentate area down and the alveus up. The fimbria can be cut and the tissue transferred to the chopper with a spatula.

During slicing, the part to be used during the experiment should point upwards. That is, if recordings are to be made from the CA1 region, the tissue should rest on the dentate area and *vice versa*. Before cutting, the angle must be adjusted. If the right hippocampus is used, rotate clockwise, if the left hippocampus is used, rotate anticlockwise. An angle of 15° from the transverse plane is sufficient, but the recurrent inhibitory pathway is preserved better if the angle is 30–45°. The first cut removes 2–3 mm from one end of the structure. By proceeding from this site five to eight slices are cut. The slices are transferred from the chopper in the following way. After the knife has penetrated the tissue, a wet (ACSF), fine painting brush is inserted between the knife and the slice. If the painting brush is rolled between the fingers while this is done, the damage to the slice is less. The slice can then be transferred to the cup containing ACSF on the tip of the painting brush. A lid should be kept on this cup until the arrival of the first slice in order to reduce O_2 evasion. When the last slice has been cut, the slices are transferred to the chamber by the pipette. The thickness of the slices influences their quality to some extent, as the proportion of healthy slices increases with their thickness at least up to 650 μm.

The tissue needs about 1 hour to recover after this procedure. After this, however, recordings usually improve for a further hour. The ideal time to start the experiment is therefore approximately 2 hours after preparation.

From a theoretical view point, one would expect better slices if the metabolic processes were slowed by cooling during operation. We have therefore done some experiments where isolation and cutting of tissue was done at a temperature between 0 and 5°C. The hippocampal formation was then placed in a dish resting in ice. A block of ice with a hole somewhat larger than the tissue was placed around it and the hole was filled with agar. The block of ice was then removed and the dish was filled with ACSF at 4°C. Thereafter, the tissue was sliced manually being fixed to the agar at the bottom of the dish. However, the described procedure did not greatly improve the slices as judged by our physiological records. It is possible, however, that some of the pathological signs observed in the tissue (see below) are diminished. Alternatively, the animal could be killed by perfusing it with cold (4°C) saline or ACSF.

As mentioned above, the slices need some time to recover after sectioning. This is correlated to morphological changes in the tissue (Andersen *et al.*, 1980d). After 5 minutes incubation, electron micrographs show vacuolization of the tissue. There is swelling of the dendrites and many axons and boutons show an appearance which looks like degeneration. There are numerous enlarged cisterns in the cell bodies and synaptic boutons. These changes are gradually reduced, however, during the first hours of incubation. Even with good recordings, however, the tissue is not restored to the condition seen with good perfusion fixation.

E. Stimulation

Several types of electrodes can be used for this purpose. We have most often employed electrodes made from tungsten needles with a shaft diameter of 0.3 mm. They are electrolytically sharpened and insulated with lacquer. The resistance is in the range of 0.3–3 MΩ. Glass-coated platinium–iridium needles also work satisfactorily and glass pipettes filled with Wood's metal prevent shunting of the stimulus current through the fluid covering the slices, which can sometimes be a problem. The electrodes are placed under visual control through a dissection microscope with micromanipulators. In good slices, population spikes of 5 mV or more can be evoked with a stimulation current of less than 50 μA (0.05–0.1 ms duration).

The afferent input to the cell can be restricted by placing cuts perpendicular to the cell body layer, leaving intact only a small "bridge" of afferent fibres. Knives for this purpose can be made from chips of razor blades. A razor blade is broken in small pieces and the chips are glued (with epoxy resin) to a shaft under visual control through a microscope. When cutting, the knife is moved by a micromanipulator. The "bridges" appear clearly immediately after the sectioning, but are often hard to see when some time has passed. They can, however, be visualized by putting some diluted Alcian blue on the knife before the lesion is made. If left too long, the dye tends to reduce the quality of the slice and the procedure should, therefore, be used with caution. Current density analysis has shown that the synaptic input to pyramidal cells is restricted to the dendritic tree corresponding to the "bridge" (Andersen et al., 1980b).

The strength of the afferent input is best controlled by recording the fibre volley, usually called the presynaptic volley (Andersen et al., 1978). This is a short-lasting diphasic potential appearing early in the field potential when an extracellular recording electrode is placed at the site of the afferent input. Its amplitude depends on the number of activated fibres when the electrode is placed appropriately.

Antidromic stimulation is performed by stimulation of the alveus caudal to the recording site. Contamination of an orthodromic component can be a problem at higher stimulation intensities. This can usually be avoided by placing a cut over the afferent fibres running in stratum oriens near the alveus.

Stimulation of single cells can also be performed by glutamate iontophoresis (Dudar, 1974; Schwartzkroin and Andersen, 1975). Care must then be taken to place the glutamate electrode accurately in the dendritic tree since the sensitive areas are well localized.

F. Recording

Extracellular recording of field potentials (field EPSP, pre-volley and population spike) is made with glass-micropipettes filled with 4 M potassium acetate or citrate, or 4 M NaCl. Pipettes for intracellular recording are pulled from fibre-filled glass tubes and backfilled with the solutions. The electrode resistance is measured by passing alternate positive and negative current pulses of equal magnitude with a frequency of 1 KHz through the electrode. Although we have obtained quite stable intracellular recordings with electrodes of 40 MΩ and less, the stability is better with electrodes with higher resistance. We normally use electrodes with resistances between 80 and 150 MΩ. With higher electrode resistance, above 150 MΩ, the noise recorded through the electrode becomes bothersome. In addition, it is difficult to balance the bridge when current pulses are delivered through the electrode.

Extracellular records can be made of the population spike, extracellular EPSP (excitatory postsynaptic potential) and presynaptic volley (Fig. 3.3). The *population spike* is obtained from the pyramidal cell layer in response to orthodromic stimulation of afferent fibres in stratum radiatum or oriens. It is produced by summation of the individual action potentials of many neighbouring neurons (Andersen *et al.*, 1971).

The size of the population spike depends on factors like the synchrony of discharge, the distribution of the activated cell population and the tissue resistance. Other things being equal, however, the amplitude is

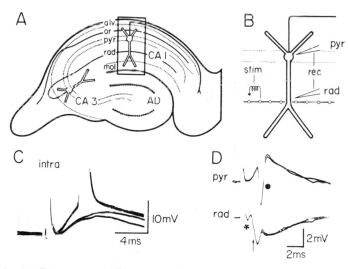

Fig. 3.3. A, Diagram of the transverse hippocampal slice. Nomenclature according to Blackstad (1956). Abbreviations: alv, alveus; or, stratum oriens; pyr, stratum pyramidale; rad, stratum radiatum; mol, stratum moleculare; AD, area dentata; CA1 and CA3, cornu Ammonis, field 1 and 3. B, The boxed-in area in part A showed in higher magnification. The unmyelinated fibres employed for orthodromic stimulation with their *en passage* boutons are symbolized by lines and circles. The positions of stimulation electrodes (stim) and recording (rec) electrodes are shown. C, Typical intracellular record from a CA1 pyramidal cell. Superimposed traces of responses to two subthreshold stimuli giving rise to EPSPs and two suprathreshold stimuli giving rise to a spike followed by a DAP. D, Typical extracellular responses to orthodromic stimulation in stratum radiatum recorded in the positions indicated in part B. Upper traces: record of population spike (●) in stratum pyramidale. Lower traces: record of prevolley (*) and field EPSP (arrow) at the site of the afferent input. The positive deflection in the middle of the field EPSP is due to the current flow to the somata from the dendrites during the synchronous discharge.

linearly related to the number of discharging cells. Directly or antidromically activated population spikes frequently occur with high stimulation strength in slices of inferior quality. The latency should always be observed to avoid this. The *field EPSP* is a negative potential obtained at the site of the afferent input to the dendritic tree. It is due to the activated synaptic current flowing into the dendritic branches near the electrode. When the stimulation strength is sufficient to evoke a population spike, this appears as a sharp positive deflection in the middle of the negative field EPSP. This is due to current associated with the synchronous action potentials flowing out of the dendrites and towards the cell bodies. The presynaptic volley or *prevolley* is a sharp diphasic deflection seen before the field EPSP. It represents the synchronous firing of afferent fibres (Andersen *et al.*, 1978). The recording of this potential is the best way to control the size of the afferent input to a cell or cell population. It may be difficult to find since the activated fibres often are localized within a small area. However, systematic search at the site of the afferent input to the cell or cell population in question will reveal it in slices of high quality.

The input resistance is an important parameter for estimating cell quality. The amplifier should, therefore, have a bridge circuit that allows current to be passed through the intracellular electrode. This also offers a possibility for direct stimulation with depolarizing current pulses. Whenever the bridge is employed, the bridge balance should be checked frequently. Square-wave pulses can not be used to detect fast changes in the input impedance. An alternating current (AC) bridge circuit can then be used to impose low-amplitude sine-wave current across the membrane (Katz, 1942; Hagiwara and Tasaki, 1958; Dingledine and Langmoen, 1980). The frequency of the AC must be adjusted so that the current passes through the membrane resistance and not the membrane capacity. This calls for a compromise in which both the frequency of the event in question and the time constant of the cell must be considered. In hippocampal pyramidal cells where the time constant is in the range of 15 ms, a satisfactory sensitivity can be provided with 10 Hz or less. Frequencies up to 30 Hz can, however, often be used depending on the degree of the conductance change.

Before impalement, high-frequency square-wave pulses (1 KHz) should be passed through the electrode. If the capacity adjustment does not make the pulses appear as square waves when displayed on an oscilloscope, the electrode should be discarded. The same applies to electrodes in which rectification of the high-frequency square-wave pulses appears (i.e. where the direction of the positive deflection is larger than the negative or vice versa).

The electrode is connected to a micromanipulator. When the tip of the electrode approaches a cell, small, usually positive, deflections in the DC recording can be seen due to spontaneous activity and/or orthodromic stimulation. With a good electrode, the impalement occurs suddenly, signalled in the usual way by a steady negative DC shift of 50–65 mV, with action potentials of about 60–80 mV. However, usually impalement is more gradual. Two tricks can then be employed to penetrate the cell membrane. The first is to "whistle" the circuit by a sudden short-lasting increase of the capacity compensation. This produces small movements of the electrode tip and often leads to successful impalement. This procedure usually fails when the electrode has partially penetrated the cell membrane (i.e. when positive spike potentials of 20–50 mV are seen). Alternatively, a negative DC current (0.1–0.3 nA is usually sufficient) may be applied through the electrode. This appears to attract the electrode to the cell membrane and, more important, it stabilizes the cell after impalement. Slowly increasing membrane potential, spike amplitude and input resistance are then often observed in conjunction with reduction in spontaneous activity. The DC current can then be reduced gradually. If the cell does not improve in spite of such a constant negative current injection, small movements of the electrode, both upwards and downwards should be tried. When the cell has become stable, it should be evaluated by the criteria described below.

G. Evaluation of the slices

Evaluation of the slices are made both visually and by judging the extracellular recordings. The visual evaluation requires some experience, but is time-saving. Evidence for damage to the slices often appears within 20 minutes if the preparation has failed or the slice environment is not satisfactory. The first sign of perhaps the most common problem, dried-out slices, is a slightly mottled surface. The preparation can often be saved if the gas flow is increased appropriately. Minor adjustments are often satisfactory, whereas too large a flow is harmful. If no adjustments are made, the thin ACSF film on the slice surface is lost and the preparations appear translucent. Clearly translucent slices are, as a rule, damaged, but not necessarily due to drying. Even if large population spikes can be recorded from some of them for a short while, few cells are penetrated, and the size of the potentials is thus more likely caused by intracellular oedema and an increased extracellular tissue resistance. Another certain sign of damage is an increased whiteness of the slices, which also occurs with too large a gas flow.

When recovery has taken place after the preparation of the slices, these can be evaluated by the response to orthodromic stimulation. A good preparation shows a population spike of 5 mV or more in response to orthodromic stimulation of less than 50 μA (50 μs, 1 Hz). Slices that require higher stimulus intensity can be very useful, but it should be pointed out that the frequency with which good cells are encountered is inversely proportional to the current required for evoking extracellular potentials of acceptable size. If proper evaluation of electrodes and slices made for intracellular purposes are made, much unsuccessful cell-hunting can be avoided.

H. Evaluation of the cells

To distinguish normal from abnormal cells is, of course, impossible as long as no standards are set. In order to evaluate the penetrated cell before the experiment starts one therefore needs some standards to determine whether the cell should be considered normal or injured. This has often been done by rather arbitrarily defined lower limits for spike size and input resistance. In order to make the evaluation somewhat less arbitrary, we have used the following criteria to select the cells to be analysed. (i) Cells that show stable membrane properties and firing behaviour for long intervals (at least 1 hour) are less injured than others. (ii) Cells in which the electrode can be moved several micrometers up and down without measurable changes in any parameter are well impaled and less disturbed by the presence of the electrode than others. The same applies to cells that tolerate mechanical disturbance. (iii) High-resistant electrodes make less injury than low-resistant ones. Cells penetrated with the former category are therefore more "normal." (iv) Cells that are encountered in tissue that has been subject to damage are less "normal" than others. (v) Slices showing a single population spike of 5 mV or more to stimulation currents below 50 μA show significantly less pathology than slices where much higher stimulation intensity is required to evoke even small potentials (I. A. Langmoen and P. Hurlen, unpublished observations). The same applies to slices where multiple population spikes are easily evoked (at 33°C). Cells that are impaled in slices judged as healthy by these criteria are therefore considered closer to "normal" than others.

Even if no definite standards can be set by these criteria, evaluation of cells by these means has given us a fairly strong impression of what can be considered as normal firing behaviour. In addition, we think it has allowed us to set limits for certain parameters, like spike amplitude and input resistance, that are somewhat less arbitrary but, in fact, do not differ much from those employed earlier.

When the described criteria are satisfied: (i) stimulation of afferent fibre systems evokes intracellular EPSPs which with higher stimulation strength elicit one action potential; (ii) spike discharge from baseline levels and multiple spikes are seldom seen; (iii) weak depolarizing square-wave pulses elicit one long latency spike. With increasing current, the latency decreases and the number of spikes increases. When the criterion is not fulfilled, one or more of the following phenomena can be encountered. (a) Some cells do not discharge due to depolarizing currents or elicit only one or two spikes with short latency. (b) The first spike takes off from the baseline level. (c) Depolarizing currents evoke variable long-lasting depolarizing potentials. (d) As expected, the membrane potential, spike amplitude and input resistance were lower when the criteria were not fulfilled.

On this basis, we suggest the following criteria for identification of pyramidal cells.

(i) The cells should be antidromically identified. A typical response is described in Section III A3, p. 67.

(ii) When the cell is orthodromically activated, it should possess an EPSP which elicits a single action potential at higher stimulus intensities.

(iii) Depolarizing pulses should evoke a train of spikes which decreases in latency and increases in number with an increase of current. Cells in which spike train ceases before the end of the pulse should not be accepted when current strength is less than 1.5 nA.

(iv) The spike amplitude should be above 60 mV.

(v) The variability in input resistance is surprisingly large and a lower limit is therefore difficult to set. Values lower than 15 MΩ are, however, seldom seen in otherwise healthy cells.

(vi) The membrane potential should measure 55 mV or more.

I. Important factors for experimental success

Experiments should be made in slices showing good extracellular responses, since such slices show significantly less pathology and cells are much more easily found. Proper evaluation of slices and electrodes warrants successful impalement. One or more of the following factors are usually involved when the preparation does not work.

1. *Gas flow*

Dried-out slices (too low gas flow) can be a serious problem. Too intense gas flow is harmful as well since too high pO_2 is known to be detrimental to brain tissue. Careful adjustment of gas flow and the ACSF level in the inner bath of the chamber as described earlier help to avoid these problems.

2. *Mechanical destruction*

The tissue should be handled as gently as possible. Direct mechanical contact should be avoided. The slices should normally not be touched after they are placed on the net. If required, they may be moved by increasing the fluid level so that they float freely and may be gently pushed around by a brush touching less important parts of the slice. Such re-arrangement is only possible within the first 5 minutes or so. After this time, the tissue fastens to the cellulose fibres of the lens paper.

3. *Speed of preparation*

This factor is not so imperative as the two above, but the slices should preferentially lie in the chamber within 3 minutes after the blood supply to the brain was cut, but results may be obtained in slices with preparation times of up to 5 minutes.

III. Results

A. Characteristics of hippocampal pyramidal cells

1. *General properties*

A sample of 166 CA1 pyramidal cells from guinea pigs was selected on the basis of the criteria stated above. The cells had a mean spike amplitude (measured from spike threshold) of 71 ± 9 mV (s.d.). The input resistance was 31 ± 12 MΩ as measured from the voltage deflection due to a hyperpolarizing current pulse. The membrane potential was 63 ± 6 mV. Similar data have been obtained by Schwartzkroin (1975, 1977). The width of the action potential, measured at the spike base ranged from 1.6 to 1.9 ms. The cells usually showed a depolarizing after-potential (DAP) when stimulated orthodromically, although this could be absent, especially in cells with spikes taking off from near the baseline level. The DAPs declined faster in cells with prominent IPSPs (inhibitory postsynaptic potentials) and could be increased in amplitude by applying hyperpolarizing current. It is uncertain whether this was due to a direct effect on the DAP itself or due to a decreased amplitude of the IPSP which often contaminates the DAP (see Section III B2, p. 72). The cells showed anomalous rectification in that the input resistance increased near the spike threshold, but this was not systematically studied by us (see Hotson *et al.*, 1979).

2. *Excitatory postsynaptic potentials*

Stimulation of afferent fibres in stratum radiatum and stratum oriens, with the exception of the outer one-fifth of the radiatum, produced an EPSP. At higher stimulus current, the EPSP triggered an action potential. In different cells, the time to peak of the EPSP varied from 3 to 10 ms. Only a small fraction of the variation in time to peak of the EPSP may be explained by an asynchronous afferent volley.

An equally wide distribution was observed for the firing level. Spikes usually took off at a membrane potential of 3–10 mV more positive than the resting potential (in 6.25 mM K^+). In a small number of cells, the spike took off from or near the baseline level. Other observations from these cells, however, raised the question whether these cells could be considered as "healthy".

The EPSP amplitude increased with hyperpolarizing current. This was, however, not always the case when the slices were cut nearly transversely to the longitudinal axis of the hippocampus, the IPSP then being smaller or even absent. EPSPs were then surprisingly unaffected by hyperpolarization. An observed increase in the EPSP amplitude with increasing membrane potential is thus probably due to interference with the IPSP (see Section III B2, p. 72).

Normally, orthodromic stimulation triggered only one action potential at 33°C but in slightly deteriorated slices, two to three action potentials were often seen. In more seriously deteriorated slices, long bursts could sometimes be observed.

3. *Antidromic activation and inhibitory postsynaptic potentials*

One of the most frequently used tests for identification of CA1 pyramidal cells is antidromic activation by alveus stimulation. Suprathreshold alveus stimulation elicited a single action potential which took off from the baseline level. This is due to antidromic invasion of the cell and therefore shows considerably shorter latency, a higher degree of regularity and follows high-frequency stimulation more easily than the orthodromically evoked spike. Threshold stimulation characteristically showed an all-or-nothing behaviour. Subthreshold alveus stimulation evoked a negative potential deflection which is associated with a conductance increase and inhibition of spike discharges. It is therefore referred to as an IPSP (see Section III B4, p. 75).

Repetitive firing due to antidromic stimulation was only observed in cases where activation of neighbouring afferent fibres due to spread of stimulus current could not be ruled out.

4. Current injection and spontaneous activity

Depolarizing square-wave current pulses elicited action potentials when strong enough. Low current intensity triggered only one long latency spike. With increasing current intensity, this spike appeared at progressively shorter latencies and was followed by others at regular intervals, the interspike interval being inversely proportional to the current intensity. Strong (1–3 nA), long-lasting (1–5 s) pulses gave a high frequency burst, but with progressive reduction of spike amplitude and increase in spike duration throughout the pulse. This was associated with a strong reduction in input resistance.

Injection of a hyperpolarizing current abolished spontaneous activity and spike discharge due to orthodromic activation. Hyperpolarization decreased and eventually reversed the IPSP and often increased the amplitude of the apparent EPSP and DAP (but see Section III B2, p. 72). Spike discharges associated with the turn-off of a hyperpolarizing pulse was seldom observed. A small steady hyperpolarizing current (0.1–0.3 nA) was used to stabilize cells after impalement, and this was gradually removed as the spike amplitude increased and spontaneous activity faded.

Spontaneous activity was present in only 10% of stable cells. When present, it was invariably of low frequency (1–3 Hz). Spontaneous bursts were not seen (at 33°C) when both the cell itself and the slice could be considered "healthy". This, however, is more often seen in CA3 pyramidal cells.

B. Properties of postsynaptic potentials

1. Relative efficiency of distal versus proximal synapses

Afferent fibres make synaptic contact with all parts of the dendritic tree of cortical nerve cells. Synaptic potentials produced in one part of the dendritic tree will be passively conducted to other parts. The synaptic potentials are thereby attentuated when recorded from the soma, the degree of which is determined by the cable properties of the dendrites. The attenuation is related to the length which a given potential propagates. If the cell has a single spike trigger zone, synapses near this zone would be expected to have a larger influence than more distal ones.

Is it possible that there is a mechanism that compensates for this "disadvantageous" position of distal synapses? The hippocamal slice technique is ideal for studying the efficiency of distal versus proximal synapses (Andersen *et al.*, 1980b). First, the afferent fibres are known to run parallel to each other and perpendicularly to the dendritic axis. Second, it is possible to cut the afferent fibre system and thus restrict the input to a small identifiable part of the cell. This is done by a double cut, sparing only a small portion of afferent fibres. The synapses belonging to this fibre population are situated at the level of the gap between the lesions, the "bridge". The distance from the recording site to the site of afferent input can then be measured. Synapses at two different locations can be compared by having a similar gate at the other side of the recording area.

Experiments in which the afferent input was restricted in this way showed that even small populations of fibres were remarkably effective in eliciting action potentials in pyramidal cells. With tissue bridges as narrow as 35 μm (8% of the length of the apical dendritic tree) 20 out of 22 bridges tested were able to drive the cells with a firing probability of 1.0. Damage to fibres from compression by the knife and local oedema probably reduced the amount of fibres passing the 35 μm gate to about 5% of all fibres passing through the apical dendritic tree, i.e. 3% of the fibres contacting the whole neuron.

Activation of fibres synapsing on the outer third of the basal and the distal fifth of the apical dendritic tree has, however, not been able to trigger CA1 cells. This may be related to the reduced density of asymmetrical synapses and the increased number of the symmetrical type in these regions (see below).

The synaptic efficiency may be estimated by measuring the extracellular population spike, which is an indication of the number of activated cells. However, for two different inputs, a given stimulus current or prevolley size does not necessarily give rise to the same synaptic current. One may get an index of the synaptic current by measuring the amplitude of the field EPSP. Its rising phase is assumed to be linearly related to the injected synaptic current. The population spike, the firing probability and the intracellular EPSP amplitude were plotted against the amplitude of the extracellular EPSP. The proximal and distal synapses had a remarkably similar efficiency. This applied both when the amplitude of the population spikes and the firing probability were used as indices. In Fig. 3.4, two tissue bridges (100 μm wide) are located 70 and 380 μm from the middle of the pyramidal layer. The two inputs were both able to discharge the cell with a probability of 1.0. Although the stimulus current at the discharge threshold and at the saturation point was different for the two inputs, the curves approached each other when the firing probability of the cell was related to the size of the corresponding field EPSP.

For a further comparison, the intracellular EPSPs produced by proximal and distal synapses were compared with regard to their rise time, half-width and the slope/amplitude index. The synchrony of the intracellular EPSP was in part dependent on the synchrony of the afferent volley. Care was therefore taken to make the conduction distance for the two inputs as similar as possible. When this precaution was taken, the time to peak for the proximal and distal EPSP was remarkably similar. This was true irrespective of whether the time to peak or time from 10 to 90% of the EPSP amplitude was used as an index. In contrast, the rise time of the EPSP varied from 3.2 to 10 ms between different cells. The

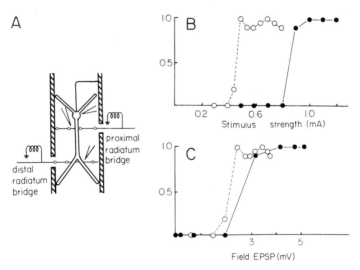

Fig. 3.4. A, Diagram of the experimental set-up. Two tissue bridges, each 100 μm wide, had their nearest borders 70 and 380 μm from the middle of the pyramidal cell layer. The field EPSP for each input was measured at the site of the activated synapses. B, Probability of discharge of a CA1 cell as a function of the stimulus current for the proximal (o) and the distal (●) input. C, Probability of discharge of the cell as a function of the size of the field EPSP for the proximal (o) and the distal (●) bridge.

duration of the EPSP at half-maximum amplitude also was roughly similar for the two inputs, although in some cells it was longer for the distal input. The third shape index (the slope/amplitude index) is the ratio between the rate of rise and the amplitude of the EPSP. This parameter also showed similar values for the two inputs, although there was a considerable scatter of the data.

It thus appears that distal and proximal dendritic synapses are functionally similar. Three possible explanations may be offered. First, the density of synapses could be larger distally, thus giving more intense synaptic current. Second, there could be different trigger points for the action potential. Third, the efficiency of the distal synapses could be explained by an ability to create local responses or dendritic spikes (the so-called prepotentials or d-spikes) which secondarily could induce a full-fledged spike in the soma membrane.

In order to test the first hypothesis, the main types of synapses were defined and their density was calculated at different locations along the dendritic tree. About 95% of the synapses did contact the spines. The number of spine synapses remained nearly constant at about 42 per 100 μm^2 throughout the main part of stratum radiatum and oriens. These synapses were of the asymmetrical type (Type I of Gray, 1959). Towards the end of the dendrites this number decreased and the number of synapses contacting the dendritic branches directly increased to 30% (Fig. 3.5). The synaptic density hypothesis is therefore probably not valid.

The second hypothesis could be rejected on the basis of threshold measurements. If there are different trigger points in the same cell, one would expect the action potential to be elicited at different levels of the soma membrane potential. With one trigger point the firing level would be expected to be constant. The firing level was compared for two inputs with the stimulus strengths adjusted to the threshold for cell discharge. There was no systematic difference suggesting the existence of a single trigger point. Because the firing level for action potentials elicited by depolarizing pulses was the same, the trigger point is probably situated near to the recording site in the cell body.

The third hypothesis appears unlikely, since d-spikes were observed in a small proportion of the cells only. Furthermore, it was relatively uncommon to unmask such responses by blocking the full action potential by hyperpolarizing the soma membrane.

A tentative explanation for the observed phenomenon is that the internal longitudinal resistance experienced by the injected current at the spine heads is similar for the two inputs. Assuming an average diameter of 0.1 μm for the spine neck, 1 μm for the secondary dendrite

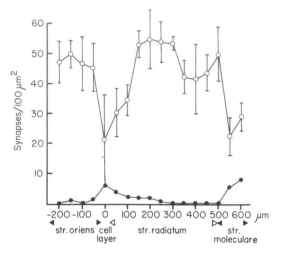

Fig. 3.5. Number of asymmetrical synapses with active sites (vesicle aggregation at the synaptic cleft and postsynaptic membrane thickening) on dendritic spines (o) and symmetrical synapses on either soma or dendritic shafts (●). Six samples of 170 μm^2 were counted at each level. Vertical bars represent $\pm$ 1 s.d.

(Westrum and Blackstad, 1962) and 3 μm for the apical dendritic shaft, it appears that 60–90% of the total longitudinal resistance is to be found in the spine neck. All synapses thus may be electrotonically remote because the added resistance for the distal synapses could be of moderate significance only. However, this point requires further study.

2. *The effect of hyperpolarization on EPSP amplitude*

The sensitivity of the EPSP to hyperpolarizing current varies greatly from cell to cell. From the results in the previous section, it was concluded that most or all dendritic excitatory synapses are electrotonically remote from the soma. Accordingly, both proximal and distal EPSPs were almost insensitive to hyperpolarization. In later studies, however, we have sliced the hippocampus more obliquely or increased the slice thickness in order to obtain more powerful recurrent inhibition (see the Methods Section, p. 58). In these slices we commonly encounter EPSPs with voltage-sensitive amplitudes. In hippocampal pyramidal cells, the IPSP is produced by basket cells synapsing on the somata of pyramidal cells causing the IPSP to be easily manipulated by changes of the membrane potential (see Section on IPSP, p. 75). It therefore appears likely that the EPSP observed in response to orthodromic stimulation is a complex potential produced by the summation of an EPSP and an IPSP. Because the EPSP and the IPSP occur almost simultaneously, the IPSP may depress the EPSP by trying to clamp the membrane to the reversal potential for the IPSP.

A more direct test by Dingledine and Gjerstad (1979) was used to study this effect. The cell was activated alternately orthodromically (ortho) and antidromically (anti) to evoke a mixed EPSP/IPSP and an IPSP, respectively. Penicillin, which is known to block recurrent IPSPs and GABAnergic (γ-aminobutyric acid) action in the hippocampus (Dingledine and Gjerstad, 1980) was then added to the fluid superfusing the slice. Families of orthodromic (upper traces) and antidromic (lower traces) responses under penicillin application are shown in Fig. 3.6. In each group the lowest curves are control responses and the upper traces are the responses in the fully developed penicillin state. Penicillin blocked the IPSP, thereby unmasking the true EPSP.

The most likely mechanism for the increase in the apparent EPSP with increasing membrane potential observed in some cells, is, therefore, an interaction with an IPSP. When the soma membrane is hyperpolarized, the membrane potential approaches and may pass the reversal

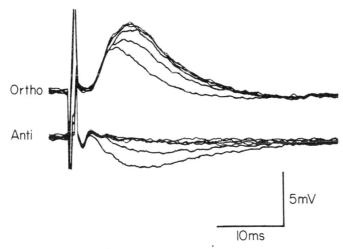

Fig. 3.6. Unmasking of EPSP (ortho) during IPSP (anti) blockade by penicillin (17 mM in a 1 nl droplet). Families of responses to orthodromic and antidromic stimulation just before and after application of a penicillin droplet. Each trace is the average of 10 consecutive stimuli. The lowest trace in each of the two responses (ortho and anti) was obtained during the control period, the middle trace in the period 0–30 s after penicillin and the upper traces in the four consecutive 30 s periods. (From Dingledine and Gjerstad, 1979.)

potential of the IPSP. The effect of the IPSP, which at the resting potential is to counteract the EPSP, is itself reduced and can even be reversed to a depolarizing effect.

The degree to which the apparent EPSP in a pyramidal cell is affected by the membrane potential is thus dependent on the slicing angle and the stimulus current required to evoke an EPSP. If the required stimulus current is high compared to the threshold current for neighbouring cells, the latter may induce recurrent inhibition, particularly when the slicing angle is oblique. In such situations, the IPSP will be well developed with a high voltage sensitivity of the EPSP. It should be pointed out, however, that the dendritic EPSP itself appears less affected by the membrane potential.

3. *EPSP–EPSP interaction*

A general problem is whether or not two neighbouring populations of synapses interfere with each other. In other words, do the EPSPs produced by two different synaptic populations sum linearly or non-linearly?

In order to study this problem, we activated two different inputs, one proximally and one distally in the stratum radiatum, or one input in stratum radiatum and one in stratum oriens (Langmoen and Andersen, 1980). The stimulus current was usually below 50 μA which activates fibres that synapse on a limited part of the dendritic tree only (Andersen *et al.*, 1980b). In some experiments the activated fibre population was limited further by lesions leaving 35 μm wide gates open for afferent input (see the Methods Section, p. 59). The two methods gave similar results.

The inputs were then activated sequentially: the first input alone, then the other alone, then both were delivered simultaneously, and then starting on the first one again. The signals were fed into a computer system that used a level discriminator to differentiate between the different input modes. In each series, 10–50 signals were averaged. The average responses to the two individual inputs were added algebraically and the sum was compared with the averaged response to double stimulation. Fig. 3.7, A and B show the average EPSP produced by proximal and distal synapses, respectively. C shows the results from simultaneous stimulation and D is the algebraic sum of A and B. C and D are expanded and superimposed in E. The cell was hyperpolarized by 3 mV while these data were sampled. The result of simultaneous stimulation of the two afferent inputs is practically identical to the algebraic sum of the individual responses. Similar results were found in more than

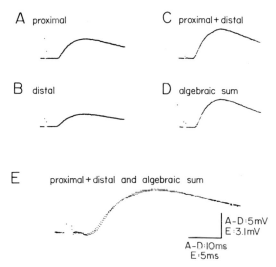

Fig. 3.7. EPSP due to stimulation of a proximal input (A), a distal input (B) and the proximal and distal input together (C) while the cell was hyperpolarized by 3 mV. All traces represent the average of 30 consecutive stimuli. The algebraic sum of the EPSP evoked by the proximal and distal input (A + B) is shown in part D. In part E the potentials in C and D are expanded and superimposed.

30 other cells. There seems, therefore, to be no interaction between neighbouring populations of synapses.

This algebraic summation took place only when the cells were hyperpolarized. At the resting potential, the EPSP produced by simultaneous stimulation of the two inputs declined faster and had a lower amplitude than the algebraic sum. This was probably due to a stronger IPSP which was expected to occur as a result of the amplifier effect the summation of two different inputs has on the field potential (see below).

As expected, the summed effect of two inputs gave rise to larger EPSPs intracellularly and, consequently, to a larger population spike in the extracellular record. The summation of the population spike was, however, non-linear. When submaximal stimulation of the proximal and distal input was used, the population spike due to simultaneous stimulation of the two inputs was several times larger than the algebraic sum. The summation of the EPSP to two afferent inputs thus can lead to a powerful amplification at the population spike level.

4. The IPSP

The IPSP in hippocampal pyramidal cells is produced by a recurrent pathway through basket cells synapsing on the somata of pyramidal cells (Ramon y Cajal, 1911; Lorente de Nó, 1934; Andersen *et al.*, 1964a, b). In addition, feed-forward inhibition probably exists, since IPSPs may occur in response to inputs too weak to elicit any population spikes. We have used the slice technique to study the mechanisms by which the IPSP inhibits cell discharges in hippocampal pyramidal cells and to describe the time course of the IPSP and its underlying conductance change (Dingledine and Langmoen, 1980).

The IPSP amplitude was readily manipulated by changing the soma membrane potential, probably reflecting the location of the synapses on or near the cell body. In different cells, the reversal potential was 7–12 mV more negative than the resting potential. Interestingly, the first part of the IPSP always reversed at a more positive potential than the later parts of the IPSP. This may suggest that the current associated with the IPSP is carried by two different ion species or that a proportion of the synapses is electrically remote from the soma. The *in vitro* IPSP was less intense with a shorter duration than its *in vivo* counterpart which lasts up to 600 ms (Kandel *et al.*, 1961; Andersen *et al.*, 1964a) and can be of sufficient intensity to drive the membrane potential nearly to the IPSP reversal potential (Andersen and Lømo, 1969; Allen *et al.*, 1977). In contrast, the *in vitro* IPSP lasts for only 50–150 ms and only attains 30–60% of its maximal possible amplitude.

The time course of the conductance change associated with the IPSP was studied using an AC bridge circuit (see the Methods Section, p. 62). In Fig. 3.8, the record in part C shows an average recurrent IPSP (35 trials). The AC bridge was switched on for alternate sweeps; superimposed sweeps are shown in part B. The changes in the amplitude of the potential envelope are proportional to the changes in the input impedance. The time course of the impedance changes during the IPSP can thus be plotted as the inverse ratio of the resting and experimental amplitude against time (Fig. 3.8B). There was a clear impedance change throughout the duration of the IPSP.

When recording extracellularly from the pyramidal layer, population spikes were triggered by orthodromic stimulation of radiatum fibres. A prior shock to the alveus (activating the recurrent inhibitory loop) reduced the size and increased the time to peak of the population spike for a period of up to 130 ms. This decreased probability of firing of the cell population was confirmed by intracellular recordings. The question arose whether this inhibition was caused by the removal of the membrane potential from the firing level by the hyperpolarization or whether the underlying conductance increase also interfers with the EPSP amplitude or the firing level.

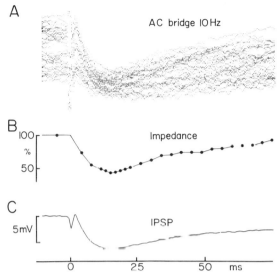

Fig. 3.8. The time course of the recurrent IPSP and its underlying conductance change. A, Alveus stimulation at 0.5 Hz evoked a recurrent IPSP. Simultaneously, a 0.15 nA, 10 Hz sine wave current was passed through the recording electrode: 35 sweeps of the combination of the IPSP and the sine wave were superimposed. B, The relative impedance throughout the IPSP calculated from part A by taking the amplitude of the voltage envelope as a percentage of the resting amplitude. C, The time course of the IPSP. The trace represents the average of 35 sweeps when the sine wave was not switched on. (From Langmoen and Dingledine, 1979.)

In order to study a possible EPSP/IPSP interaction, the membrane potential was held at the IPSP reversal potential while subthreshold EPSPs were elicited at various latencies before and after an antidromic stimulus. The EPSP was reduced during the period with the greatest impedance change (up to 25 ms) (Fig. 3.9). As expected, when comparing different cells the degree of shunting was correlated with the magnitude of the IPSP impedance change. Possible effects of the conductance increase on the firing level were investigated by eliciting spike discharge by delivering short depolarizing pulses after the antidromic conditioning stimulus. The firing level usually had a rather large standard deviation (1 mV) and was sensitive to small changes in the membrane potential. We did not find any significant change of the firing level, even during the peak of the conductance increase of the IPSP.

The major mechanism of the inhibitory action of the IPSP thus seems to be a change of the membrane potential towards the IPSP reversal

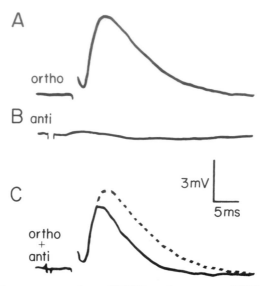

Fig. 3.9. Non-linear summation of EPSP and recurrent IPSP. The cell membrane was hyperpolarized to the reversal potential of the IPSP. The cell then received sequentially a subthreshold orthodromic activation (A), a subthreshold antidromic activation (B) and an orthodromic stimulus preceded by an antidromic volley at 5 ms delay (C, ——). – – – –, in part C represents the control EPSP (A).

potential. This inhibits cell discharges in two ways: first, the membrane potential is removed from the firing level, and second, the EPSP amplitude is reduced. This appears to be similar to the mechanisms for motoneuron IPSPs (Araki *et al.*, 1960; Cooms *et al.*, 1955; Curtis and Eccles, 1959). The EPSP is, however, only shunted during the period of largest conductance increase.

C. Properties of some putative transmitters

1. *GABA*

γ-Aminobutyric acid (GABA) may act as an inhibitory transmitter in the hippocampus. The evidence for this is that GABA depresses spontaneous and glutamate-induced cell discharges (Curtis *et al.*, 1970), the GABA metabolizing enzymes glutamic acid decarboxylase (GAD) and GABA transferase (GABA-T) are present in the hippocampus (Pohle and Matthies, 1970; Storm-Mathisen and Fonnum, 1971) and finally the presence of high-affinity GABA uptake (Hökfelt and Ljungdahl, 1971; Iversen and Bloom, 1972).

Iontophoretically delivered GABA effectively reduced the amplitude of the population spike and blocked spontaneous and induced unit activity (Andersen *et al.*, 1981). A strong effect was often observed even with weak ionotophoretic currents (3–6 nA).

The diffusion of GABA through the tissue is probably small since the response can usually be reduced by 50% by removing the electrode as little as 20–25 μm (in some cases only 10 μm) in either direction perpendicular to the dendritic tree. This observation allows mapping of GABA-sensitive areas in the dendritic tree. The effect of GABA was tested at the soma and in different locations along the dendritic tree. At an appropriate depth, GABA was found to be effective as far as 250 and 150 μm from the soma in the apical and basal direction, respectively. This corresponds to roughly 50% of the dendritic length in each of the two directions. GABA was also effective when applied in the pyramidal cell layer. On either side of this layer there was a 50 μm wide zone which showed less sensitivity to GABA. The best effect was observed at the depth of the recording electrode, probably because the dendritic trees run parallel to the slice surface.

The effect of the GABA application was, however, also dependent on the relative location of the stimulating and GABA-ejecting electrodes. GABA delivered in stratum radiatum effectively reduced the population spike elicited by radiatum fibres, whereas the oriens-induced population spike was enhanced. Conversely, an oriens-evoked population spike was inhibited by GABA application in this layer, but enhanced by delivery GABA in the stratum radiatum. The mechanism for these two responses was investigated by intracellular recording.

Intracellularly, two GABA responses were observed (Langmoen *et al.*, 1978; Andersen *et al.*, 1980a). Application of GABA to the cell body caused hyperpolarization of the impaled cell (Fig. 3.10A). This effect is similar to the effect of GABA in Deiters' nucleus (Obata *et al.*, 1967), on neocortical cells (Krnjević and Schwartz, 1967) and on motoneurons (Curtis *et al.*, 1968). In contrast, however, GABA could depolarize the cells when applied in their apical or basal dendritic trees (Fig. 3.10B). Both the hyperpolarizing and the depolarizing responses were associated with an increase in the cell input conductance. In some cases, the depolarizing response triggered cell discharges. This response, however, quickly vanished with larger GABA doses.

The hyperpolarizing response was associated with inhibition of spontaneous and evoked cellular discharges. EPSPs were reduced in amplitude when the GABA dose was strong enough to cause a considerable

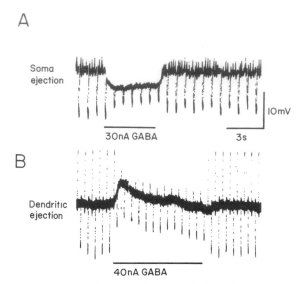

Fig. 3.10. The two separate GABA effects. A, When GABA was applied to the soma the cell hyperpolarized. B, When GABA was delivered into the dendritic tree the cell depolarized. Vertical lines are hyperpolarizing current pulses (60 ms, − 0.3 nA) for resistance measurement.

increase in the input conductance. Both the hyperpolarization and its associated conductance increase were augmented with increasing ionto-phoretic currents. Increasing doses decreased the latency of the response. The shortest observed latency was 80 ms. When the response was tested at different depths in the slice, the response was critically dependent on a position close to the cell body. A 50% reduction of the inhibitory effect required a removal of 30 μm only.

Dendritic application was remarkably effective provided that the electrode was located in a sensitive area. The best chance to find such sensitive areas was to aim at a point 100 to 150 μm away from the soma along a line perpendicular to the pyramidal cell layer. The best depth corresponded to the depth of the recording electrode. In contrast to glutamate, GABA only showed a single effective region in each track.

The depolarizing GABA effect was also associated with inhibition of cell discharges. The EPSP was reduced in amplitude only if the depolarization was associated with a considerable increase in input conductance.

The possibility that the two response types were mediated through interneurons was excluded by testing the GABA effects after blocking synaptic transmission. Calcium and magnesium concentrations of 0.2

and 10 mM, respectively, effectively blocked synaptic transmission, without changing either of the responses. GABA thus exerts its described effects directly on the pyramidal cell membrane.

The hyperpolarizing GABA effect was very sensitive to changes of the membrane potential. It increased with depolarization and decreased and reversed with hyperpolarization. The localization of the hyperpolarizing GABA response to the soma supports the idea that GABA acts as the transmitter of the basket cells which are known to synapse on the pyramidal cell somata (Ramon y Cajal, 1911; Lorente de Nó, 1934). This hypothesis was further supported by a similar reversal potential for the IPSP and the hyperpolarizing GABA response.

We then investigated whether local application of GABA to either the dendrites or the soma could selectively inhibit local excitatory potentials (Andersen *et al.*, 1979). The cell was given a suprathreshold orthodromic shock followed by a suprathreshold depolarizing pulse through the recording pipette after 120 ms (Fig. 3.11A). GABA application to the soma selectively blocked the response to direct activation in the soma (Fig. 3.11B). GABA to the site of afferent input to the dendrites blocked the spike elicited by the dendritic synaptic input (Fig. 3.11C) but either enhanced or did not affect the response due to direct stimulation of the soma. Since there was no gradual reduction of the spike and since the depolarizing potentials underlying the spike discharges were heavily shunted (Fig. 3.11D), the inhibitory effect was probably due to a clamping of the excitatory potentials rather than a direct effect on the spike-generating mechanism itself. GABA application can thus selectively inhibit the excitatory input to a localized part of the cell.

Bicuculline and picrotoxin, but also penicillin, are used to block GABA inhibition. Preliminary studies have shown that bicuculline and picrotoxin reduce both types of GABA responses and also suppress the IPSP. The effect was associated with the development of long trains of action potentials in response to orthodromic stimulation, suggesting that these blockers have other epileptogenic effects in addition to their action on GABA-mediated inhibition.

Dingledine and Gjerstad (1980) have studied the effect of penicillin on GABA responses and recurrent inhibition. Sodium benzyl penicillin (17 or 170 mM) was applied as a single droplet (1–2 nl) to the surface of the slice. Penicillin application did not produce significant changes in the membrane potential or input resistance. Both the hyperpolarizing and depolarizing GABA effect was reduced by penicillin. This applied to the conductance as well as the potential changes. Penicillin also decreased the amplitude of the IPSP. Studies of the IPSP/GABA reversal potential and conductance measurements demonstrated that this

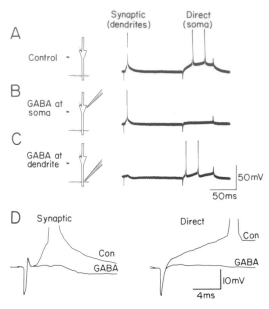

Fig. 3.11. Localized inhibitory effect of GABA. A, Control response. The cell first received an orthodromic volley (synaptic) which evoked an EPSP with a superimposed action potential. After 120 ms delay, it received a square-wave pulse through the intracellular electrode at the soma (direct) which evoked two action potentials. B, GABA application to the soma abolished the firing to the depolarizing pulse, whereas the orthodromic response being due to synaptic input in the dendrites was hardly changed. C, GABA application at the site of the afferent input in the apical dendrites abolished the synaptic response evoked at this site, but facilitated the response due to direct stimulation. D, In the left part the cell was orthodromically activated, the control response consisting of an EPSP with a superimposed action potential. GABA, applied to the site of afferent input in the dendrites "clamps" the EPSP and thereby abolished the spike discharge. To the right, the cell was given a depolarizing pulse through the intracellular pipette. The control response consisted of a membrane depolarization which reached the firing level and triggered an action potential. GABA applied to the soma "clamped" the depolarization and thereby abolished the spike discharge.

was mainly due to a reduction of the conductance increase normally associated with the IPSP. Application of penicillin also enhanced the orthodromic response. The time course was the same for the reduced effect of GABA, the reduced IPSP amplitude and the increased orthodromic response. This suggests that interference with recurrent inhibition is important in the development of the epileptic behaviour of pyramidal cells observed in the penicillin state.

Reduced extracellular or increased intracellular chloride concentration diminished the hyperpolarizing GABA response and the IPSP amplitude and increased the orthodromic response (P. Andersen, R. Dingledine, L. Gjerstad and I. A. Langmoen, unpublished observations). This supports the hypothesis that chloride currents mediate the IPSP in hippocampal pyramidal cells (Allen *et al.*, 1977).

2. *Glutamate*

Glutamate (2 M, pH 8.0) was ejected from an iontophoretic electrode (1–2 μm) (Schwartzkroin and Andersen, 1975). Small doses of glutamate (down to 1 nA for 100 ms) excited the cells with short latency (15–20 ms) provided the electrode was placed in a sensitive region. Sensitive spots were found inside an area that corresponded to the apical and basal dendritic tree of the pyramidal cells, having the shape of a double cone with tips pointing to the recording site (i.e. the cell body) (Dudar, 1974). The typical response consisted of a depolarization that evoked an initial burst of spike discharges followed by a pause, after which discharges were evoked with relatively constant frequency. With larger iontophoretic currents the action potentials were often inactivated. The following pause was related to the discharge frequency and the length of the glutamate ejection. With moderate doses, the discharge frequency was nearly constant throughout the pulse. Large doses of glutamate to highly sensitive spots often elicited responses similar to those observed in experimental epileptical foci (see Section III D2, p. 88). Repeated glutamate ejections did not reduce the sensitivity irrespective of the duration. The membrane potential repolarized within 0.1 to 1.5 seconds after the end of the glutamate current. The time required for repolarization was dose-dependent.

When the drug was ejected at different positions in the dendritic tree, a large variation in sensitivity was found. Moving the glutamate electrode perpendicularly to the dendritic axis, several separate areas of high sensitivity were found. With small glutamate doses, the distance from maximal to half-maximal response measured about 10 μm. A possible explanation is that the electrode passes different branches of the dendritic tree with their associated glutamate receptors. Using two glutamate electrodes located in the apical and basal dendritic tree, respectively, the effect of simultaneous ejection was compared with the effect evoked by each of the electrodes alone. The discharge frequency

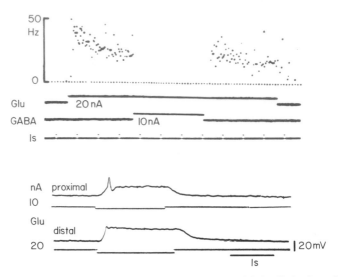

Fig. 3.12. Extracellular recording from a CA1 pyramidal cell during glutamate application (upper part); 20 nA glutamate drives the cell with a frequency which initially is near 50 Hz, but then gradually decreases. A 10 nA GABA application reversibly abolished this response. The lower part shows an intracellular recording from another CA1 cell during application of glutamate from two iontophoretic electrodes placed in the apical dendritic tree.

resulting from double ejection was higher than the algebraic sum of the discharge frequencies evoked by single ejections.

The data support the idea that glutamate may be the transmitter in CA1 excitatory synapses.

3. *Norepinephrine*

The hippocampus receives a noradrenergic input from the pontine locus coeruleus (Pickel *et al.*, 1974; Swanson and Hartman, 1975). Both activation of this pathway with locus coeruleus stimulation and local application of norepinephrine (NE; noradrenaline) inhibit hippocampal unit discharges (Segal and Bloom, 1974a, b). We decided to study the cellular mechanism for this action by intracellular recording (Langmoen *et al.*, 1981).

Droplets (1 nl) of Ringer's solution, containing NE in concentrations of 1 μM to 1 mM were applied to the slice surface, near the recording site. When carefully made, no mechanical artifacts were produced. Drop concentrations of 1 mM regularly gave good effects, whereas 1 μM

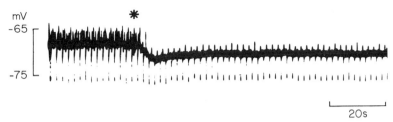

Fig. 3.13. Continuous recording of the membrane potential during NE application. The downward vertical deflections represent membrane responses to constant hyperpolarizing pulses (-0.3 nA), thus indicating cell input resistance. NE (1 mM) application (asterisk) hyperpolarized the cell and reduced the input resistance.

gave weak effects, if any. The typical effect was a hyperpolarization of 1–5 mV (mean -2.7 mV ± 1.7 s n) associated with decreased input resistance as measured by voltage deflections to hyperpolarizing current pulses (Fig. 3.13). The reduction in input resistance ranged from 10 to 30% (mean 22% ± 13). The time course for the development of the reduced resistance closely followed the increase in membrane potential. This suggests that the hyperpolarization was caused by increased conductance for a particular type of ion. Usually, both effects developed within 10 seconds but, occasionally, required up to 1 minute.

NE application caused a remarkable reduction in discharges due to depolarizing current pulses. This would be expected from the described effects on membrane potential and input resistance. Two observations, however, suggest that it may be due to an additional, separate phenomenon. First, whereas the mean reduction in membrane responses to hyperpolarizing pulses was roughly 20%, the depolarizing response was reduced by 50% (Fig. 3.14). Second, the effect on depolarizing responses developed faster than the effect on responses to hyperpolarizing pulses. NE was still effective after blocking synaptic transmission by increasing the magnesium and decreasing the calcium level in the incubation fluid, indicating that NE acts directly on the pyramidal cell membrane and not via interneurons.

It could be argued that the application of drugs as a drop to the slice surface could produce mechanical artifacts. We think this unlikely since mechanical distribution usually leads to depolarization and since Dingledine and Gjerstad (1979), who used the same method of application, saw no changes in membrane potential or input resistance. Finally, the observed effects were related to the drug concentration although the drop size was kept constant.

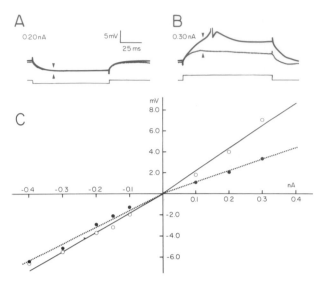

Fig. 3.14. A, Membrane responses to hyperpolarizing pulses before (——) and after (·······) NE application (1 mM, 1 nl drop). Each trace represents an average of 10 sweeps. B, Membrane responses to depolarizing current pulses before and after NE application. Note that the depolarizing response was reduced much more than the hyperpolarizing response. C, Membrane responses of different strength and polarity measured at 24 ms delay (arrowheads in A and B) plotted against respective current strength before (o) and after (●) NE application. Whereas the hyperpolarizing responses were only slightly changed, the depolarizing responses were reduced by 50%.

The most striking characteristic of the NE response was the stronger reduction of the responses to depolarizing than to hyperpolarizing current pulses. Tentatively, this can be explained by a reduction of the anomalous rectification which occurs in these cells when the membrane potential approaches the firing level (Hotson *et al.*, 1979). A mechanism operating through increased Cl⁻ or K⁺ conductances can, however, not be excluded.

4. Serotonin

The hippocampus receives an inhibitory input from the raphe nuclei which probably is serotonergic (Conrad *et al.*, 1974; Segal, 1975). In order to study the cellular mechanism of the serotonergic inhibition, serotonin (5-HT, 5-hydroxytryptamine) was applied either to the bath or in droplets during intracellular recording from pyramidal cells (I. Langmoen, unpublished observations).

When 5-HT was applied, the cells hyperpolarized. The hyperpolarization was associated with a decreased input resistance and reduction of spontaneous and evoked activity. When applied in droplets (1 1) 5-HT was less potent than NE (Section III C3, p. 84). A concentration of 1 mM was often not effective, whereas concentrations of 10 mM always produced the characteristic response. When applied through the incubation fluid 10 μM 5-HT produced the typical response, whereas concentrations of 0.1 mM were associated with an initial depolarization.

The effect on input resistance was usually too weak for shunting of the synaptic potentials. 5-HT, therefore, probably acts mainly through its effect on membrane potential.

D. The hippocampal slice as a model for studying general physiological and pathophysiological phenomena

1. *Long-lasting potentiation of synaptic transmission*

In vivo, relatively short-lasting tetanic stimulation of the perforant path/dentate granule cell synapse causes a long-lasting enhancement of synaptic transmission (Bliss and Lømo, 1973). A similar facilitation occurs for the Schaffer collateral/CA1 pyramidal cell synapses. A stimulation of 10 Hz for 15 seconds causes an increased response for several hours (Andersen *et al.*, 1973). This phenomenon also occurs in slices (Schwartzkroin and Wester, 1975) where a large increase in the CA1 population spike amplitude takes place after 3–50 Hz stimulation of radiatum afferents for 10–15 seconds. The slice preparation, however, also offers the possibility of a more direct investigation (Andersen *et al.*, 1977, 1980c; Wigström *et al.*, 1979). Using two afferent inputs, one in stratum radiatum and one in stratum oriens, it could be determined whether the long-lasting potentiation (LLP) was specific to the tetanized pathway, or whether the effect was more general. In addition, the presynaptic volley was recorded (see the Methods Section, p. 62) in order to detect any changes in the size of the afferent input.

A short tetanus to either input enhanced the population spike. In order to exclude the possibility that the changes were dependent on the size of the afferent input and to study parameters other than the presynaptic volley as the possible "independent variable", different stimulation strengths were studied for three different relationships: (i) afferent volley/population spike; (ii) afferent volley/field EPSP size and (iii) field EPSP/population spike amplitude. Facilitation of the first two relations occurred at all levels of stimulation in all 17 experiments, and in about half of the trials for the field EPSP/population spike relation. The latter observation is probably not due to a changed excitability of the pyramidal cell membrane since corresponding changes were not observed for the control input.

A short-lasting depression (2–4 minutes) was observed for the non-tetanized pathway. In cases with strong depression this was also observed for the tetanized pathway, apparently being superimposed on the potentiation. This phenomenon seems, therefore, not to be specific for any population of synapses, but is more likely of a generalized character.

Unit studies showed the LLP to be associated with an increased firing probability. This was confirmed by intracellular studies which also usually showed an increased EPSP amplitude following tetanization. The enhancement of the intracellular EPSP occurred, however, less often than the enhancement of the field EPSP and was also smaller. There was no change in membrane potential, input resistance and excitability as measured by intracellular current pulses.

The lack of parallel changes in the non-tetanized pathway and the stable membrane potential, input resistance and excitability indicate that long-lasting potentiation was specific for the tetanized pathway. The short-lasting depression was probably due to an ionic re-distribution since it occurred with similar time course for both pathways.

Since the major part of the potentiation occurs in the presynaptic volley/field EPSP relation, the phenomenon is probably largely pre-synaptic. A simple explanation would be a presynaptic loading of calcium during the tetanus.

An increased population spike was, however, also seen when the size of the field EPSP was kept constant. This suggests a postsynaptic mechanism as well. A change in the resistance of the dendritic spine neck, a change in receptor sensitivity or a locally increased membrane resistance in the region of the tetanized synapse are all theoretical possibilities.

2. *Studies on epileptogenesis*

In spite of much experimental work, the basic mechanism of epilepsy remains largely unknown. Although valuable information has been obtained from studies *in vivo*, these investigations have been hampered by the complexity of the organization of the tissue, the problems of controlling the experimental parameters and the stability, especially in intracellular recordings. The use of invertebrate preparations has overcome these obstacles, but observations from such preparations may not be compared directly with the situation in human epilepsy. The *in vitro* preparation of vertebrate brain tissue, however, combines the advantages of the simple preparations with the use of mammalian cortical neurons. Studies on this preparation, important for the understanding of epileptogenesis, have already appeared (Yamamoto and Kawai, 1967; Ogata, 1975; Schwartzkroin and Prince, 1977, 1978; Gjerstad *et al.*, 1978).

Using crystalline sodium benzyl penicillin as the epileptogenic agent, we first studied the pattern of the epileptiform activity in the hippocampus, and compared this with the pattern described from studies *in vivo* (Gjerstad *et al.*, 1981a).

Orthodromic stimulation of CA1 afferent fibres normally gave one population spike when recorded extracellularly in the pyramidal cell layer. The intracellular counterpart was a single action potential superimposed on an EPSP. The addition of 3.4 mM benzyl penicillin induced a hyperactivity which occurred both spontaneously and in response to orthodromic stimulation. Extracellular responses to orthodromic stimulation in the penicillin state were characterized by 3 to 10 population spikes. Simultaneous intracellular recording showed a similar number of action potentials superimposed on a pronounced depolarization (Figs. 3.15 and 3.16). The rising phase of the EPSP and the first action potential were not changed, whereas later spikes, riding on the slow depolarization, had a longer duration and lower amplitude. The later part of the depolarization, with its superimposed spikes, could occur in an all-or-none pattern.

In addition to this evoked activity, spontaneous epileptiform activity also occurred (Fig. 3.16). The spontaneous burst consisted of 6 to 12 population spikes, occurring in an all-or-none fashion. It was suppressed by orthodromic stimulation at more than 1 Hz and disappeared at temperatures below 34°C. It was not dependent on the penicillin concentration above a certain concentration. Simultaneous recordings from CA1 and CA3 showed that the spontaneous burst in CA1 always followed a burst in CA3 at a certain latency. When the connection to CA3 was cut, the spontaneous CA1 burst disappeared, while the CA3 burst remained. In response to orthodromic stimulation, bursts occurred both in CA1 and in CA3.

Both the CA1 and the CA3 thus have the ability to generate bursts of population spikes in response to stimulation in the presence of penicillin. In addition, the CA3 can generate spontaneous bursts which subsequently may trigger bursts in the CA1 region.

The intracellular burst response described here is similar to the

A   B

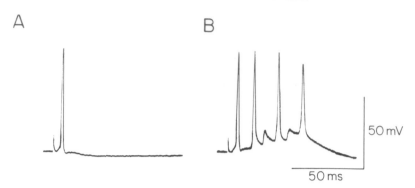

Fig. 3.15. Effect of penicillin on the orthodromic response. The control response (A) obtained before penicillin application consisted of a single spike, but 1 min after penicillin application (B) an orthodromic shock of the same intensity elicited a burst of four spikes (R. Dingledine and L. Gjerstad, unpublished observations).

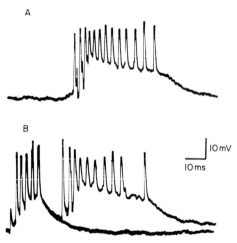

Fig. 3.16. Spontaneously (A) and orthodromically evoked (B) penicillin responses recorded intracellularly from a CA1 pyramidal cell. The first part of the burst in B, consisting of five spikes superimposed on an underlying depolarization, persisted after cutting the fibres from CA3 and was probably developed by an intrinsic mechanism in the CA1 region. The later part of the burst in B and the burst in A were abolished when the afferents from CA3 were cut and were probably triggered from the CA3 region. (From Gjerstad et al., 1981a.)

observation from experimental foci *in vivo* (Matsumoto and Ajmone Marsan, 1964; Prince, 1968; Matsumoto *et al.*, 1969; Dichter and Spencer, 1969). Both the spontaneous burst of spikes superimposed on a large depolarization and the synchronized activation of a large group of cells, are in accordance with these findings. It therefore seems justified that the transverse hippocampal slice is a favourable experimental tool in studying basic mechanisms of epilepsy.

(i) *Factors affecting the epileptiform activity* (L. Gjerstad, unpublished observations). The efficient concentration of penicillin was established by varying the concentration from 0.85 to 17 mM. Some hyperactivity was present already at 0.85 mM. It did not increase with doses above 3.0 mM. Accordingly, a concentration of 3.4 mM (2000 I.U./ml) was used in later studies.

The epileptiform activity was enhanced by increasing the temperature. The number of spikes increased and their latency decreased. The response was reduced by lowering the temperature. Spontaneous bursts were not observed at temperatures below 34°C.

Stronger epileptiform activity was also observed when the K^+ concentration was increased from 3.25 through 6.25 to 9.25 mM. It was also enhanced by lowering the Ca^{2+} level, whereas it was reduced by increased Ca^{2+} concentration.

(ii) *Afferent input and synaptic transmission.* Among the possible mechanisms for the development of epileptiform activity are enhancement of excitatory and reduction of inhibitory synaptic transmission (cf. Gjerstad *et al.*, 1981b; Dingledine and Gjerstad, 1980). Neither the presynaptic volley nor the field EPSP changed in spite of the development of a marked epileptiform activity. Plots of stimulus-current strength versus prevolley size and prevolley size versus field EPSP confirmed this observation for a broad range of stimulus intensities (Fig. 3.17). The rising phase of the intracellular EPSP was also unchanged. However, the EPSP showed a moderate increase in its peak amplitude, time to peak and half width, whereas the IPSP was reduced. The enhanced and prolonged EPSP which followed penicillin application is probably explained by a reduction of the recurrent IPSP and not by an increased synaptic input. First, the increase in the apparent EPSP was paralleled by a decrease of the antidromically activated recurrent IPSP. Second, the time courses for the development of these two effects were similar. Third, the initial part of the field EPSP, an index of the synaptic current, was unchanged. Our data, therefore, suggest that the afferent input and the excitatory synaptic transmission are unchanged in the penicillin state.

(iii) *Membrane properties.* We systematically compared input resistance

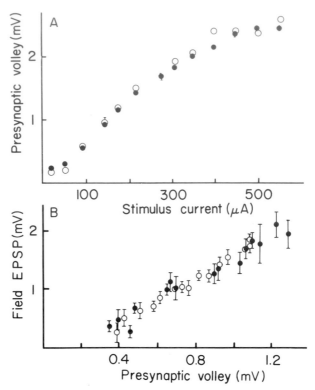

Fig. 3.17. The presynaptic volley and field EPSP before (o) and after (●) penicillin application. A, The size of the presynaptic volley plotted against the stimulus current. Standard deviations are not shown if they are smaller than the symbol size. B, Amplitude of field EPSP plotted against the amplitude of the presynaptic volley. Vertical lines indicate ± 1 s.d. (Modified from Gjerstad *et al.*, 1981b.)

and membrane potential before and after penicillin application without being able to detect any significant changes (Gjerstad *et al.*, 1981a). Penicillin gave the same result whether it was applied through the incubation fluid (3.4 mM) or dissolved in droplets delivered to the surface (1–2 nl, 17 mM).

In order to study possible changes in ionic currents underlying the action potential, directly and orthodromically evoked spikes were differentiated. The initial spike due to afferent stimulation and all spikes due to depolarizing current pulses were unchanged. The later spikes in the orthodromic response in the penicillin state showed prolonged rising and falling phases. This was, however, probably due to partial sodium inactivation because the spikes took off from a long-lasting depolarization.

Other aspects of the response to intracellular depolarizing-current pulses were also unchanged in most cells. The number of spikes and their latency from pulse onset were unaltered. In 5/143 cells, however, depolarizing pulses elicited burst responses. It is, however, questionable whether this response represented a true epileptiform burst, since it occurred independent of orthodromically evoked hyperactivity.

In conclusion, the epileptiform activity seen in the penicillin state does not seem to be explained by changes in membrane potential, input resistance or in the ionic currents underlying the soma action potential.

(iv) *Two types of bursts.* The burst observed with intracellular recording during penicillin application belonged to two different classes (P. Andersen, L. Gjerstad, J. J. Hablitz and I. A. Langmoen, unpublished observations). The first occurred typically in an all-or-none fashion. It was usually not associated with a field potential discharge and was easily blocked by artificial hyperpolarization of the membrane. The last observation suggests that the mechanism generating this type of burst is located electrotonically near the soma. This type of burst was also occasionally observed in slices not being subject to penicillin treatment. Such cells were usually slightly depolarized but did not otherwise show obvious signs of injury.

The other type of burst had a form that varied from trial to trial. Increasing synaptic input increased the number of spikes and the underlying depolarization. This burst was always associated with a synchronous field potential consisting of population spikes which were well timed to the intracellularly observed spikes. At threshold, a burst either did or did not occur, both in the field potential and in the cellular record. One contributor to this type of burst is probably the depolarizing after-potential (DAP) following the action potential. These DAPs increase with number of spikes in a burst and contribute to the underlying depolarization by summation when the spikes occur at high frequency. A second contributor is a depolarizing wave which may occur independently of the DAPs when the cell is triggered orthodromically at high frequencies of stimulation. At stimulation rates of 5–20 Hz, the first spike (which was unchanged from the control situation) persisted, whereas the spikes superimposed on the long-lasting depolarization were inactivated and disappeared. The long-lasting depolarization itself, however, persisted, suggesting that it was partly independent of the DAP following spike discharges.

3. *Physiological changes in the aged dentate area*

During the last few years a variety of anatomical and physiological changes have been described in the aged brain, including an altered ability to perform different behavioural tasks. The complexity of the mammalian nervous system complicates a comparison between physiological and anatomical data. The dentate area has a relatively simple structure. Because the slice technique offers the possibility of highly stable intracellular recordings, the dentate area *in vitro* appeared as a suitable model system for investigation of physiological changes due to ageing and their correlation to anatomical alterations.

In the senescent fascia dentata there is a significant loss of afferent synaptic contacts. According to Geinisman and Bondareff (1976) the number of synaptic contacts within the termination zone of the perforant path is 27% less in old than in young rats. Recently, Barnes and McNaughton (1980), working in this laboratory, have asked the question whether this loss of afferent synaptic contacts are compensated by any physiological mechanism. In addition, they investigated whether any physiological changes due to ageing occurred at the cellular level.

Comparisons of synaptic transmission and membrane properties were made between 96 young (10 months) and 94 old (31 months) granule cells. In the old animals, a given field EPSP alway gave rise to a larger population spike. However, the stimulus intensity versus fibre potential relation was steeper in the young rats, showing that a given stimulus pulse triggered a larger number of fibres in these animals. The afferent fibres were, however, more effective in the old rats as the same fibre response gave rise to a larger field EPSP than in the young. This was confirmed by intracellular studies where a given fibre response triggered EPSPs of higher amplitude in senescent than in the young rats. A direct comparison between stimulus intensity and field EPSP showed that the stimulus current in the old group was less effective in generating a synaptic current. Interestingly, however, the EPSP at threshold was lower (20.0 versus 24.1 mV) and the latency of the spike at threshold was shorter (3.40 to 3.92 ms) in the old animals. No significant differences were observed for cell parameters like resting potentials, amplitude and width of the spike, EPSP latency and half-width, rheobasic current and whole-cell input resistance, and time constant.

The smaller field EPSP and fibre response observed to a given stimulus intensity in senescent rats support the observation that the number of synaptic contacts is reduced and suggest that this is associated with a decreased number of afferent fibres. The afferent fibres in old rats seem more efficient, however, since the ratio of the EPSP to fibre potential was increased. The greater efficiency of the field EPSP in evoking the population spike in the old rats is probably explained by the observed decrease in discharge threshold of the dentate granule cells in this group.

IV. Discussion

A. Are brain slices physiological?

Are the results obtained from brain slices comparable to those obtained from intact preparations or have pathological changes significantly altered the behaviour of the cells?

A partial answer to this question can be obtained by comparing the morphology of the slices with intact preparations. Further information can be gained by comparing the behaviour of neurons in the slices with their counterparts in intact preparations.

Studies made by Garthwaite *et al.* (1979) and in our own laboratory have shown that brain slices often contain clear pathological changes. These consist of swellings of axonal and dendritic profiles, formation of water-filled spaces between neurons, particularly near the soma, and darkening of neuronal profiles. These changes appear to be time dependent and related to the physiological status of the slice. The pathological changes are most developed in the first hour after cutting. With increasing incubation time, there is a gradual improvement of the pathology. However, even in slices with successful physiological responses, we have always seen some degree of dendritic swelling and a certain amount of clear spaces around the cell bodies. These changes are probably related to the ionic-pump failures seen in newly prepared brain slices (Tower, 1960).

There is a clear correlation between the physiological status of the slice and the histopathology. When large amplitude responses can be obtained with weak stimulation (less than 50 μA) the electron micrographs show a relatively well-prepared tissue. On the other hand, when strong stimulation is needed, slices often appear clearly pathological.

When the membrane parameters of neurons are compared, the cells from brain slices appear to have similar data, if not better, than the corresponding cells of intact preparations. Unfortunately, no comprehensive studies are available for the membrane properties of CA1 pyramids in intact preparations. When comparing the data from cat CA3/CA2 neurons (Kandel *et al.*, 1961) with CA1 neurons from guinea pig slices, the membrane potential of the latter appears to be higher (63 versus 52 mV). The action potentials were similar (71 versus 70 mV), whereas the input resistance was considerably higher (31 versus 8 MΩ). Although the input resistance of the large CA3 pyramids is expected to be lower than for the smaller CA1 pyramids, the reported data seem to indicate that the properties of the cells of the slices are equal to, or better than, those in intact preparations.

The shape and amplitude of the excitatory synaptic potentials also compare favourably with those in intact preparations. However, the amplitude of the IPSP is lower in the slice. The amplitude discrepancy is partly explained by the higher membrane potential and input resistance of the cells in the slices. The duration of the IPSP in slices is also significantly shorter than in intact preparations. The duration of the IPSP in intact preparations is, however, often increased due to the use of anaesthetics, particularly barbiturates.

A contributary factor for a smaller percentage of IPSPs in slices is the sectioning procedure. Since basket cells are an important element in IPSP production, the amputation of basket-cell axonal branching might be significant. The basket cell axons tend to have a longitudinal distribution which will be severely cut by the transverse sectioning. This interpretation was partly corroborated by the fact that more obliquely cut slices contained more cells that showed clear IPSPs.

B. For what purposes can slices be used?

On the basis of our experience during the last 8 years, we feel that brain slices are best suited for certain specific types of experiments. First, they are suitable for studies of cellular neurobiological problems. In particular, the methods seem useful for experiments where stability is important (intracellular recording from small cells and for long periods and long-term studies of cellular events). The technique may also be of importance for experiments in which changes of the environmental parameters are desirable. Here, brain sub-systems can be forced far outside the ranges they normally encounter. Clearly this calls for some care in interpretation. Nevertheless, a large number of interesting experiments can be performed in this way.

The second area where slices seem to offer good opportunities are experiments in which anatomical and/or biochemical data are to be compared with physiological observations from a certain region. The ease with which slices may be transferred to further histological or biochemical studies makes this possible. The third area involves experiments in which visual control during the experiment is important. Fourthly, experiments in which multiple electrodes are needed for recording or monitoring the activity of a single cell or a small cell group can benefit from the use of slice preparations. Examples are multiple inputs to a given cell or iontophoretic application of drugs to different parts of the cell. Finally, experiments that need to be done without anaesthesia can easily be performed in slices.

C. Necessary precautions with slice experiments

As with all methods, a faulty technique may give misleading results. This applies to slice work as well. In particular, it is important to make sure that the slice is in an adequate physiological condition. In particular, well-developed postsynaptic responses should be obtainable with relatively weak stimulation. Experiments in which only fibre response or direct activation of the cells are obtainable may well give misleading results.

It is hardly necessary to stress that all studies require that the behaviour of the neurons in the slice should be compared with the response of the corresponding cells in intact preparations. Usually slice experiments can only detail the general behaviour of a small part of the system, whereas studies on larger integrated systems must rely on the use of intact preparations.

D. New information on the hippocampal system provided by brain slice work

If we first focus on knowledge acquired about individual hippocampal neurons, slice studies have contributed a long series of interesting results. First, the high input resistance of hippocampal neurons has been disvered. This is of great importance when comparisons are made about the capability of cells in intact preparations. Such data are also important for the evaluation of synaptic influence on the cell. Second, slice work has given us knowledge about calcium spikes in CA1 neurons (Schwartzkroin and Slawsky, 1977). The necessary substitution experiments with barium and blocking experiments with cobolt would have been extremely difficult, if not impossible, to perform in intact preparations. Important new data on the efficiency of various synapses and the mechanism of potentiation could not have been obtained in intact preparations because of the difficulties with the experimental manipulation and the necessary long-lasting recordings. Research using slices has also indicated the existence of two types of GABA receptors and the mechanisms involved in the effects of NE and 5-HT. The interesting studies on epileptiform behaviour are new and potentially important for understanding the epileptogenic process and therapy for epilepsy. Finally, the compensatory mechanisms apparently existing in old neurons of the hippocampal formation have been found by work on slices. In none of these cases are these data self-supporting. They all have to be compared with reactions obtained from the intact animal.

When we turn our interest to acquired knowledge about the hippocampus as a system, the data obtained from slices are more scarce. The slice work has clearly supported the idea about the lamellar arrangement of the hippocampus. Otherwise it has not contributed much to the understanding of the physiology of the hippocampus as a full system. This probably stresses the fact that the slice work is more suitable for studies of general biological problems at the cellular level than for a study of systems as a whole.

E. Future challenges

There are a number of possible future developments in brain slice work. One possibility is to improve the duration of the period in which slices can give meaningful and interesting results. By taking precaution against infection and by the addition of metabolic requirements, which at the moment are not added to artificial cerebrospinal fluid, it is possible that the life span of the slices may be extended. If slices could be made to work for several days, many interesting manipulations of the system *in vitro* could be performed. In such a case, studies on degeneration and regeneration after lesions and on development may be possible. Another important avenue would be to develop the microsurgery. By removing those parts of the slices that are without interest for a given experimental group, one could probably minimize the difficulties of interpretation. For example, it should be possible to greatly restrict the amount of information that can reach a certain cell group in order to reduce the number of donor and receiving cells to manageable levels. It should also be possible to manipulate the brain by surgery and unbend previously curved structures so that new parts of the central nervous system could be made available for slice cutting. It might also be possible to use the technique on other areas of the central nervous system. With various new ways of cutting (slower, slicing under water, slicing in cooled condition, slicing on preoperated structures) it should be possible to make sections of areas that have hitherto resisted this treatment. In particular, heavily myelinated structures have appeared difficult because of the toughness of the myelinated areas. Some of these regions might, however, yield to new cutting techniques.

Another interesting avenue would be trials to stain the cells while still alive. So far, experiments with intravital staining have been difficult, since most cell dyes appear to be toxic, even in small doses. However, with injections of inert particles, dark-field illumination, ultra-thin slices and possibly other techniques, new avenues might be opened in the future. Finally, advances may be obtained by cooling the brain part during sectioning or during the whole preparation procedure. As with all techniques, only further experimentation will determine the range of the possibilities and give firm data for distinguishing interesting physiological data from concomitant pathological reactions.

V. Conclusions

1. The apparatus and procedures used for the preparation and incubation of hippocampal slices *in vitro* are described.

2. A brief survey of methods for stimulation and recording is given, including methods for restriction of the afferent input and for its recording.

3. A good preparation shows population spikes with an amplitude of several millivolts to stimulation at 50 μA (50 μs, 1 Hz). The cell layer is dark and distinctly different from the grey surroundings. Repetitive firing, dry surface, or translucent, mottled or whitish slices are usually signs of deterioration.

4. A combined identification and evaluation of hippocampal pyramidal cells include: (a) antidromic activation; (b) typical responses to orthodromic and direct activation; (c) spike amplitude above 60 mV, membrane potential above 55 mV and input resistance above 15 MΩ.

5. The two most important factors for experimental success are the gas flow and careful handling during the preparation procedure.

6. Based on a study of 166 CA1 pyramidal cells, their spike amplitude were 71 ± 9 mV, the membrane potential 63 ± 6 mV, the input resistance 31 ± 12 MΩ, and the spike width of 1.6–1.9 ms. Stimulation of afferent fibres gave rise to an EPSP, which, at a sufficient strength, had a single superimposed action potential. Antidromic stimulation elicited a single action potential which took off from the baseline level. Subthreshold antidromic stimulation could give rise to a negative potential deflection identified as an IPSP. Depolarizing current pulses elicited spikes that decreased in latency and increased in number with increasing current strength.

7. The efficiency of distal versus proximal synapses has been compared and was judged to be similar by the shape indices of the corresponding EPSPs.

8. The apparent EPSP is a complex potential produced by summation of the real EPSP and the IPSP. Its voltage sensitivity is explained by IPSP

contamination. There is a linear summation of the EPSPs when the effect of the IPSP is counteracted by hyperpolarization.

9. The IPSP is associated with a reduced input impedance with a time course similar to the IPSP itself. The IPSP was associated with an increased chloride permeability.

10. GABA application gave two different responses. When applied at the soma, it hyperpolarized the cell, whereas it depolarized the cell when applied among the dendrites. Bicuculline blocked both responses.

11. Glutamate excites the pyramidal cells. Sensitive points are found at well-localized spots within the distribution of the dendritic tree.

12. NE hyperpolarized the pyramidal cells and decreased the input resistance. In addition it had a selective suppressive effect on membrane depolarization.

13. 5-HT hyperpolarized the cells and decreased the input resistance.

14. Three examples showing the value of the slice technique in model studies are described. They include studies on long-lasting potentiation, epileptogenesis and ageing. Long-lasting potentiation is homosynaptic and associated with an increased EPSP without detectable postsynaptic excitability changes.

15. Both CA1 and CA3 regions are able to generate bursts of population spikes in response to stimulation in the presence of penicillin. The afferent input and excitatory transmission are unchanged.

16. Epileptiform activity seen after application of penicillin is not explained by changes in membrane potential, input resistance or in the ionic currents underlying the soma action potential.

17. Two different types of bursting activity are observed with intracellular recordings during penicillin application.

18. Older (31-month-old) rats showed a larger population spike, in response to a given field EPSP, and a larger field EPSP with the same fibre response than younger 10-month-old rats. There is a decreased number of afferent fibres in old rats, but the remaining fibres seem to be more efficient.

References

Allen, G. I., Eccles, J., Nicoll, R. A., Oshima, T. and Rubia, F. J. (1977). The ionic mechanisms concerned in generating the i.p.s.ps of hippocampal pyramidal cells. *Proc. Roy. Soc. Lond. Ser. B.* **198**, 363–384.

Ames, A., Sakanove, M. and Endo, S. (1964). Na, K, Ca, Mg, and Cl concentrations in choroid plexus fluid and cisternal fluid compared with plasma ultrafiltrate. *J. Neurophysiol.* **27**, 672–681.

Andersen, P. and Lømo, T. (1969). Organization and frequency dependence of hippocampal inhibition. *In* "Basic Mechanisms of the Epilepsies" (H. H. Jasper, A. A. Ward and A. Pope, eds.), pp. 604–609. Little Brown and Co., Boston.

Andersen, P., Eccles, J. C. and Løyning, Y. (1964a). Location of postsynaptic inhibitory synapses of hippocampal pyramids. *J. Neurophysiol.* **27**, 592–607.

Andersen, P., Eccles, J. C. and Løyning, Y. (1964b). Pathway of postsynaptic inhibition in the hippocampus. *J. Neurophysiol.* **27**, 608–619.

Andersen, P., Bliss, T. V. P. and Skrede, K. K. (1971). Unit analysis of hippocampal population spikes. *Exp. Brain Res.* **13**, 208–221.

Andersen, P., Teyler, T. and Wester, K. (1973). Long-lasting change of synaptic transmission in a specialized cortical pathway. *Acta physiol. scand.,* Suppl. **396**, 34.

Andersen, P., Sundberg, S. H., Sveen O. and Wigström, H. (1977). Specific long-lasting potentiation of synaptic transmission in hippocampal slices. *Nature (Lond.)* **266**, 736–737.

Andersen, P., Silfvenius, H., Sundberg, S. H., Sveen O. and Wigström, H. (1978). Functional characteristics of unmyelinated fibres in the hippocampal cortex. *Brain Res.* **144**, 11 18.

Andersen, P., Gjerstad, L. and Langmoen, I. A. (1979). Localized GABA application selectively blocks excitation from various parts of hippocampal pyramidal cells *in vitro*. *Neurosci. Letters, Suppl.* **3**, S219.

Andersen, P., Dingledine, R., Gjerstad, L., Langmoen, I. A. and Mosfeldt Laursen, A. (1980a). Two different responses of hippocampal pyramidal cells to application of gamma-aminobutyric acid (GABA). *J. Physiol. (Lond.)*, **305**, 279–296.

Andersen, P., Silfvenius, H., Sundberg, S. H. and Sveen, O. (1980b). A comparison of distal and proximal dendritic synapses on CA1 pyramids in hippocampal slices *in vitro*. *J. Physiol.*, **307**, 273–299.

Andersen, P., Sundberg, S. H., Sveen, O., Swann, J. W. and Wigström, H. (1980c). Possible mechanisms for long-lasting potentiation of synaptic transmission in hippocampal slices from guinea pigs. *J. Physiol.*, **302**, 463–482.

Andersen, P., Bie, B. and Ganes, T. (1981). Distribution of GABA sensitive areas on hippocampal pyramidal cells. *Exp. Brain Res.*, in press.

Araki, T., Eccles, J. C. and Ito, M. (1960). Correlation of the inhibitory postsynaptic potential of motoneurons with the latency and time course of inhibition of monosynaptic reflexes. *J. Physiol (Lond.).* **154**, 354–377.

Barnes, C. A. and McNaughton, B. L. (1980). Physiological compensation for loss of afferent synapses in rat hippocampal granule cells during senescene. *J. Physiol. (Lond.).* **309**, 473–485.

Blackstad, T. W. (1956). Commissural connections of the hippocampal region in the rat, with special reference to their mode of termination. *J. comp. Neurol.* **105**, 417–536.

Bliss, T. V. P. and Lømo, T. (1973). Long-lasting potentiation of synaptic transmission in the dentate area of the anesthetized rabbit following stimulation of the perforant path. *J. Physiol. (Lond.)* **232**, 331–356.

Conrad, L. C. A., Leonard, C. M. and Pfaff, D. W. (1974). Connections of the median and dorsal raphe nuclei in the rat: An autoradiographic and degeneration study. *J. Comp. Neurol.* **156**, 179–206.

Coombs, J. S., Eccles, J. C. and Fatt, P. (1955). The electrical properties of the motoneurone membrane. *J. Physiol. (Lond.)* **130**, 291–235.

Curtis, D. R. and Eccles, J. C. (1959). The time courses of excitatory and inhibitory synaptic actions. *J. Physiol. (Lond.)* **145**, 529–546.

Curtis, D. R., Hösli, L. and Johnston, G. A. R. (1968). A pharmacological study of the depression of spinal neurones by glycine and related amino acids. *Exp. Brain Res.* **6**, 1–18.

Curtis, D. R., Felix, D. and McLennan, H. (1970). GABA and hippocampal inhibition. *Br. J. Pharmacol.* **40**, 881–883.

Davson, H. (1967). "Physiology of the Cerebrospinal Fluid." Churchill, London.

Dichter, M. and Spencer, W. A. (1969). Penicillin-induced interictal discharges from the cat hippocampus. I. Characteristics and topographical features. *J. Neurophysiol.* **32**, 649–662.

Dingledine, R. and Gjerstad, L. (1979). Penicillin blocks hippocampal IPSPs, unmasking prolonged EPSPs. *Brain Res.* **168**, 205–209.

Dingledine, R. and Gjerstad, L. (1980). Reduced inhibition during epileptiform activity in the *in vitro* hippocampal slice. *J. Physiol. (Lond.)*, **305**, 297–313.

Dingledine, R. and Langmoen, I. A. (1980). Conductance changes and inhibitory actions of hippocampal recurrent IPSPs. *Brain Res.*, **185**, 277–287.

Dudar, J. D. (1974). *In vitro* excitation of hippocampal pyramidal cell dendrites by glutamic acid. *Neuropharmacol.* **13**, 1083–1089.

Garthwaite, J., Woodhams, P. L., Collins, M. J. and Balazs, R. (1979). On the preparation of brain slices: morphology and cyclic nucleotides. *Brain Res.* **173**, 373–377.

Geinisman, Y. and Bondareff, W. (1976). Decrease in the number of synapses in the senescent brain: a quantitative electron microscopic analysis of the dentate gyrus molecular layer in the rat. *Mech. Age. Dev.* **5**, 11–23.

Gjerstad, L., Langmoen, I. A. and Andersen, P. (1978). Factors affecting epileptiform pyramidal cell discharges *in vitro*. In "Advances in Epileptology—1977" (H. Meinardi and A. J. Rowan, eds.), pp. 443–449. Swets and Zeitlinger, Amsterdam.

Gjerstad, L., Andersen, P., Langmoen, I. A., Lundervold, A. and Hablitz, J. (1981a). Synaptic triggering of epileptiform discharges in CA1 pyramidal cells *in vitro*. *Acta Physiol. scand.*, in press.

Gjerstad, L., Langmoen, I. A. and Andersen, P. (1981b). Monosynaptic transmission during epileptiform activity induced by penicillin in hippocampal slices *in vitro*. *Acta Physiol. scand.* in press.

Gray, E. G. (1959). Axo-somatic and axo-dendritic synapses of the cerebral cortex: an electron microscopic study. *J. Anat.* **93**, 420–433.

Haas, H. L., Schaerer, B. and Vosmansky, M. (1979). A simple perfusion chamber for the study of nervous tissue slices *in vitro*. *J. Neurosci. Methods* **1**, 323–325.

Hagiwara, S. and Tasaki, I. (1958). A study of the mechanism of impulse transmission across the giant synapse of the squid. *J. Physiol. (Lond.)* **143**, 114–137.

Hotson, J. R., Prince, D. A. and Schwartzkroin, P. A. (1979). Anomalous inward rectification in hippocampal neurons. *J. Neurophysiol.* **42**, 889–895.

Hökfelt, T. and Ljungdahl, Å. (1971). Uptake of [^{3}H]noradrenaline and γ-[^{3}H]aminobutyric acid in isolated tissues of rat: An autoradiographic and fluorescence microscopic study. *Progr. Brain Res.* **34**, 87–102.

Iversen, L. L. and Bloom, F. E. (1972). Studies on the uptake of [^{3}H]-GABA and [^{3}H]glycine in slices and homogenates of rat brain and spinal cord by electron microscopic autoradiography. *Brain Res.* **41**, 131–143.

Kandel, E. R., Spencer, W. A. and Brinley, F. J. (1961). Electrophysiology of hippocampal neurons. I. Sequential invasion and synaptic organization. *J. Neurophysiol.* **24**, 225–242.

Katz, B. (1942). Impedance changes in frog's muscle associated with electrotonic and 'endplate' potentials. *J. Neurophysiol.* **5**, 169–184.

Krnjević, K. and Schwartz, S. (1967). The action of γ-aminobutyric acid on cortical neurones. *Exp. Brain Res.* **3**, 320–336.

Langmoen, I. A. and Andersen, P. (1980). Summation of dendritic synaptic potentials in hippocampal pyramidal cells. *Acta physiol. scand.*, **109**, 13A.

Langmoen, I. A. and Dingledine, R. (1979). On the time-course of recurrent inhibition in hippocampal pyramidal cells *in vitro*. *Acta Physiol. scand.* **105**, 40–41A.

Langmoen, I. A., Andersen, P., Gjerstad, L., Mosfeldt Laursen, A. and Ganes, T. (1978). Two separate effects of GABA on hippocampal pyramidal cells *in vitro*. *Acta Physiol. scand.* **102**, 28–29A.

Langmoen, I. A., Segal, M. and Andersen, P. (1981). Mechanisms of norepinephrine actions on hippocampal pyramidal cells *in vitro*. *Brain Res.* **208**, in press.

Li, C.-L. and McIlwain, H. (1957). Maintenance of resting membrane potentials in slices of mammalian cerebral cortex and other tissues *in vitro*. *J. Physiol. (Lond.)* **139**, 178–190.

Lorente de Nó, R. (1934). Studies on the structure of the cerebral cortex. II. Continuation of the study of the Ammonic system. *J. Psychol. Neurol. (Lpz.)* **46**, 113–177.

Matsumoto, H. and Ajmone Marsan, C. (1964). Cortical cellular phenomena in experimental epilepsy: ictal manifestations. *Exp. Neurol.* **9**, 305–326.

Matsumoto, H., Ayala, G. F. and Gumnit, R. J. (1969). Neuronal behaviour

104   I. A. LANGMOEN AND P. ANDERSEN

and triggering mechanism in cortical epileptic focus. *J. Neurophysiol.* **32,** 688–703.

Obata, K., Ito, M., Ochi, R. and Sato, N. (1967). Pharmacological properties of the postsynaptic inhibition by Purkinje cell axons and the action of γ-aminobutyric acid on Deiter's neurones. *Exp. Brain Res.* **4,** 43–57.

Ogata, N. (1975). Ionic mechanisms of the depolarization shift in thin hippocampal slices. *Exp. Neurol.* **46,** 147–155.

Pickel, V. M., Segal, M. and Bloom, F. E. (1974). A radioautographic study of the efferent pathways of the nucleus locus coeruleus. *J. Comp. Neurol.* **155,** 15–42.

Pohle, W. and Matthies, H. (1970). Die Topohistochemie von Transmitter-Systemen in Kortex und Hippokampus des Kaninchens. *Acta biol. med. germ.* **25,** 447–454.

Prince, D. A. (1968). The depolarization shift in 'epileptic' neurons. *Exp. Neurol.* **21,** 467–485.

Ramon y Cajal, S. (1911). Histologie du Systéme nerveux de l'Homme et des Vertébrés. Vol. 2, 993 pp. A. Maloine, Paris.

Richards, C. D. and Sercombe, R. (1970). Calcium, magnesium and the electrical activity of the guinea-pig olfactory cortex *in vitro. J. Physiol. (Lond.)* **211,** 571–584.

Schwartzkroin, P. A. (1975). Characteristics of CA1 neurons recorded intracellularly in the hippocampal *in vitro* slice preparation. *Brain Res.* **85,** 423–436.

Schwartzkroin, P. A. (1977) Further characteristics of hippocampal CA1 cells *in vitro. Brain Res.* **128,** 53–68.

Schwartzkroin, P. A. and Andersen, P. (1975). Glutamic acid sensitivity of dendrites in hippocampal slices *in vitro. In* "Advances in Neurology" (G. W. Kreutzberg, ed.), Vol. 12, pp. 45–51. Raven Press, New York.

Schwartzkroin, P.A. and Prince, D. A. (1977). Penicillin-induced epileptiform activity in the hippocampal *in vitro* preparation. *Ann. Neurol.* **1,** 463–469.

Schwartzkroin, P. A. and Prince, D. A. (1978). Cellular and field potential properties of epileptogenic hippocampal slices. *Brain Res.* **147,** 117–130.

Schwartzkroin, P. and Slawsky, M. (1977). Probable calcium spikes in hippocampal neurons. *Brain Res.* **135,** 157–161.

Schwartzkroin, P. and Wester, K. (1975). Long-lasting facilitation of a synaptic potential following tetanization in the *in vitro* hippocampal slice. *Brain Res.* **89,** 107–119.

Segal, M. (1975). Physiological and pharmacological evidence for a serotonergic projection to the hippocampus. *Brain Res.* **94,** 115–131.

Segal, M. and Bloom, F. E. (1974a). The action of norepinephrine in the rat hippocampus. I. Iontophoretic studies. *Brain Res.* **72,** 79–97.

Segal, M. and Bloom, F. E. (1974b). The action of norepinephrine in the rat hippocampus. II. Activation of the input pathway. *Brain Res.* **72,** 99–114.

Skrede, K. K. and Westgaard, R. H. (1971). The transverse hippocampal slice: a well-defined cortical structure maintained *in vitro. Brain Res.* **35,** 589–593.

Storm-Mathisen, J. and Fonnum, F. (1971). Quantitative histochemistry of glutamate decarboxylase in the rat hippocampal region. *J. Neurochem.* **18,** 1105–1111.

Swanson, L. W. and Hartman, B. K. (1975). The central adrenergic system. An immunofluorescence study of the location of cell bodies and their efferent connections in the rat utilizing dopamine-beta-hydroxylase as a marker. *J. comp. Neurol.* **163,** 467–505.

Tower, D. B. (1960). "Neurochemistry of Epilepsy". Charles C. Thomas, Springfield.

Westrum, L. E. and Blackstad, T. W. (1962). An electron microscopic study of the stratum radiatum of the rat hippocampus (regio superior, CA1) with particular emphasis on synaptology. *J. comp. Neurol.* **119,** 281–292.

Wigström, H., Swann, J. W. and Andersen, P. (1979). Calcium dependency of synaptic long-lasting potentiation in the hippocampal slice. *Acta physiol. scand.* **105,** 126–128.

Yamamoto, C. (1972). Intracellular study of seizure-like afterdischarges elicited in thin hippocampal sections *in vitro*. *Exp. Neurol.* **35,** 154–164.

Yamamoto, C. and Kawai, N. (1967). Seizure discharges evoked *in vitro* in thin section from guinea pig hippocampus. *Science* **155,** 341–342.

Yamamoto, C. and McIlwain, H. (1966). Electrical activities in thin sections from the mammalian brain maintained in chemically-defined media *in vitro*. *J. Neurochem.* **13,** 1333–1343.

4

The Preparation of Brain Tissue Slices for Electrophysiological Studies

C. D. RICHARDS

Department of Physiology, Royal Free Hospital School of Medicine, Pond Street, London NW3 2QG, U.K.

I. Introduction

Brain tissue slices have been used for biochemical studies of the nervous system ever since their introduction by Warburg in the 1920s. Though many authors have subsequently used brain slices for neurochemical studies, the work of McIlwain and his associates stands out for its

107

attempts to explain neurophysiological and neuropharmacological phenomena in biochemical terms (see McIlwain and Bachelard, 1976). Initially there was much scepticism about the possibility of maintaining brain tissue alive *in vitro* and McIlwain realized the importance of establishing electrophysiological criteria of the viability of brain slices. Early attempts were unsuccessful, but Li and McIlwain (1957) reported that steady potentials analogous to membrane potentials could be recorded from slices of neocortex maintained *in vitro* under suitable conditions. Shortly afterwards Hillman *et al.* (1963) showed that neuronal injury discharges could be recorded as electrodes penetrated slices of neocortex. These observations suggested that slices of brain contained living neurons and were cause for optimism that it would ultimately be possible to find a way of maintaining a neuronal population electrically excitable *in vitro*. This was achieved by Yamamoto in 1966 while he was visiting McIlwain's laboratory. He found that stimulation of the lateral olfactory tract in slices of guinea pig olfactory cortex elicited evoked potentials that could be recorded from the adjacent cortex (Yamamoto and McIlwain, 1966a, b). These potentials had many of the characteristics expected of synaptic potentials and were later shown to be the result of synchronous excitation of the superficial neurons of the olfactory cortex (Richards and Sercombe, 1968a, b; 1970). Soon after, Richards and McIlwain (1967) showed that slices of neocortex could also produce evoked potentials in response to direct stimulation of the pial surface.

These advances were not immediately exploited by other groups. Indeed, for several years only two groups had any real interest in using brain slices for electrophysiological studies: Yamamoto's in Japan and my own in the United Kingdom. However, since the early 1970s, interest in the field has steadily grown to the point where it is opportune to review the practical aspects of maintaining brain slices *in vitro* for electrophysiological study for those wishing to enter the field.

II. General Considerations

The primary advantage of the technique is that of control over the environment of the neurons that make up the preparation. This includes control over the composition of the bathing medium, its pH, temperature and degree of oxygenation. Other advantages include the absence of respiratory and arterial pulsations, excellent visual control over the placement of electrodes, particularly with structures such as the hippocampus, which have clearly defined cell layers, and the isolation of one population of neurons from the influence of other populations. This last provides a degree of control over the electrical activity of the cells which is virtually unattainable *in vivo*.

The primary disadvantage of the technique is that a compromise must be struck between the needs of the tissue for oxygen and the experimental requirement of intact neural pathways. A secondary worry concerns the physiological status of isolated preparations, although this can largely be countered by comparing various measures of tissue viability from intact animals with those obtained in brain slices. Examples are the tissue levels of ATP, sodium and potassium; the characteristics of the evoked potentials in both situations and the ultra-structure revealed by electron microscopy. It would be foolish to pretend that the results *in vivo* and *in vitro* are identical, but the measures obtained in healthy slices compare well with those obtained from intact tissue (for a review of this literature, see Richards *et al.*, 1976).

III. Choice of Section and Thickness of Slices

The ideal slice preparation has at least one clearly defined afferent pathway making synaptic contact with a discrete population of neurons. Because of the practical constraints imposed by the oxygen requirements of the tissue, both the afferent pathway and the postsynaptic cells need to lie within a single plane if one wishes to study synaptic events. It is no accident that the majority of studies on brain slices *in vitro* have been carried out on layered structures such as the olfactory cortex and hippocampus. However, as Brown and Halliwell (1979) have shown, it is possible to cut slices for such studies from less well-defined areas by the use of a suitable template to guide the cutting edge. No such restriction need apply where the properties of single neurons are the object of study.

Brain tissue slices contain living cells which, like all mammalian cells, need a constant supply of O_2. However, if the metabolism is to be completely aerobic, the rate at which O_2 diffuses from the two surfaces of a slice must balance the O_2 consumption of the cells throughout the slice. The limiting thickness for full aerobic metabolism of a slice is given by the relation (Dixon, 1951):

$$d' = \sqrt{8\,C_o\frac{D}{A}} \tag{1}$$

where d' is the limiting thickness (in cm) of a slice in which the O_2 tension reaches zero at its centre; C_o is the O_2 tension in the atmospheres; D is the diffusion coefficient of O_2 through the tissue (14 nl O_2 cm^{-2} min^{-1} at 1 atm cm^{-1} gradient); A is the rate of O_2 consumption in ml O_2 g tissue^{-1}.

Yamamoto and McIlwain (1966b) have measured the O_2 consumption of slices of olfactory cortex under various conditions (see Table 4.1) from which we may calculate the theoretical maximum limiting thickness. As an illustration, let us consider the limiting thickness for a slice undergoing stimulation at 100 Hz. The O_2 consumption is 39.2 μl O_2 g^{-1} min^{-1}; the O_2 tension would be 0.95 atm maximum when a mixture of 95% O_2/5% CO_2 is used for gassing. However, measurements of the O_2 tension of bathing media in slice-incubation chambers give values slightly lower than this (about 0.85–0.9 atm) because of the back diffusion of atmospheric air. These values entered into eqn. (1) give d' as 510 μm. To allow for compression of the tissue during cutting and to give some margin for error, nominal slice thicknesses of about 420 μm are used. Harvey et al. (1974) have measured the actual thickness of slices of olfactory cortex and compared them with the nominal thickness set by the cutting guide. As expected, they found that the actual thickness of a slice tended to be greater than the nominal thickness (see Table 4.2).

If it is necessary to use thicker slices than those calculated from eqn. (1), it is possible to work at lower temperatures to reduce the O_2 consumption. Several authors have adopted this approach (e.g. Harvey et al., 1974; Pickles and Simmonds, 1978; Gilbey and Wooster, 1979), but it is not without its perils. The chief problem lies in the comparison of data obtained from slices incubated at 25–30° C with that obtained from intact mammals whose core temperatures are 36–39° C (Spector, 1956). Most mammals are unable to continue an active existence with a core temperature below 33–34° C. Thus, unless the chief object of an experi-

Table 4.1. The O_2 consumption and limiting thickness of olfactory cortex slices under various conditions at 37°C. (Data from Yamamoto and McIlwain, 1966b.)

	O_2 uptake (μl g^{-1} min^{-1})	Calculated limiting thickness (mm)
Unstimulated	29.6	0.58
Olfactory tract stimulated at 100 Hz	39.2	0.51
Whole cortex stimulated at 100 Hz	51.0	0.45

Table 4.2. Comparison between the thickness of slices and that of the slicing guides. (Data from Harvey et al., 1974.)

Guide thickness (mm)	Thickness of tissue (mm)		
	Average	Range	n
0.25	0.31	0.28–0.34	2
0.38	0.41	0.37–0.46	3
0.42	0.47	0.46–0.48	2
0.50	0.52	0.46–0.60	8

ment is the study of the temperature-dependence of a particular process, it is unwise to work with brain slices at low temperatures. Obviously such considerations do not apply to poikilotherms.

The alternative of increasing the O_2 tension of the bathing medium might be practicable if a pressure chamber is available, but the possible toxic effects of high O_2 tensions would need to be carefully evaluated.

IV. Methods of Preparation

Obviously, it is not possible to cover all the variations of technique that have evolved over the past few years. In the following pages I shall describe how to remove the brain of a guinea pig and prepare blocks of hippocampus and olfactory cortex for slicing. The various methods of slicing will then be discussed.

A. Removal of the brain

Guinea pigs of 300–450 g are best for dissection. Smaller animals have fragile skulls which tend to split erratically and large animals have strongly calcified skulls that are difficult to work with. The animal is first killed either by decapitation with a guillotine or by a sharp blow from a heavy bar low down on the neck followed by section of the spinal column and great vessels of the neck with a large pair of scissors. The scalp is removed and the skull is then split open by inserting the point of one blade of a stout pair of scissors through the foramen magnum and, keeping the point against the top of the inside of the skull to avoid damage to the brain, the scissors are closed. The skull will split along the central suture when the scissors are closed and the bones of the back of the roof of the skull can then be prised away. The frontal bones are removed in the same way as far as the olfactory bulbs. If care is exercised, the dura remains intact and may be cut sagitally, with the point of a scalpel or with fine scissors, and reflected. The forebrain is removed by first sectioning the olfactory bulbs and then inserting a spatula between the cerebellum and cerebral hemispheres and pushing downwards and forwards. The brain should be lifted carefully and immediately transferred to a closed petri dish containing one or two pieces of filter paper thoroughly moistened with an artificial cerebro-spinal fluid (ACSF). The cerebellum and brainstem can be removed in a similar fashion if desired. It is essential that the brain be kept moist after removal from the body. Great speed is not required but the procedure outlined takes just over a minute when fully mastered.

To prepare blocks of tissue for slicing either the olfactory cortex or the hippocampus, the brain is divided along the midline with a large half-round scalpel blade (e.g. Type A). The half not in use should be replaced in the closed petri dish and the other transferred to a small cutting table (see below) which has previously been covered with filter paper moistened with ACSF. Hardened, acid-washed papers such as Whatman No. 50 (10 cm diameter) are best.

The olfactory cortex is exposed by lying the half brain on its medial (cut) surface. The olfactory cortex can then be readily identified by the large white band of the lateral olfactory tract on the anterior ventro-lateral aspect. A slab of tissue some 3–4 mm thick is cut from the brain following the contours of the surface. The slab prepared in this way should contain the lateral olfactory tract with about 3 mm of cortex on either side and 5 mm of cortex posterior to the tract. The block can then be readily sliced by hand (see below).

The hippocampus is exposed by placing the half brain on its dorsal surface. A small spatula about 5 mm wide is inserted through the third ventricle to remove the lower structures of the brain, including the thalamus and hypothalamus. The ventral surface of the hippocampus is thereby exposed. The spatula is then inserted into the lateral ventricle between the cortex and the dorsal surface of the hippocampus and the hippocampus is freed from the remainder of the brain by sectioning the septum and entorhinal cortex with the spatula. Two different pre-parations can be made. First, a longitudinal slice of the dentate gyrus can be prepared by placing the hippocampus on its dorsal surface and slicing the gyrus along its length by hand (see Richards and White, 1975). Second, a transverse slice may be made to include parts of all three of the main layers of the hippocampus, i.e. the granule cell layer and the CA3 and CA1 pyramidal cell layers (see Skrede and Westgaard, 1971). This preparation is made by placing the hippocampus on its ventral surface (taking care to keep it moist with ACSF) and cutting slices along the hippocampal lamellae. The exact orientation of the slice with respect to the midline varies according to the pathway under study (see Andersen *et al.*, 1971). For the perforant path or antidromic invasion of the CA1 cells the slices are cut to include an angle of about 65–70° between the long axis of the hippocampus and the midline. Such slices can be cut from suitably oriented small blocks of hippocampus by hand, but it is easier to use a Vibratome or tissue chopper (see below).

B. Methods of slicing

There are three methods of preparing slices for electrophysiological studies, each with its own advantages and disadvantages. These are: (1) cutting by hand with a razor strip and glass guide; (2) cutting slices from small blocks of tissue with the aid of a vibratome; (3) preparing slices from larger blocks of tissue with the aid of a tissue chopper.

1. *Cutting slices by hand*

This method is of greatest value for tissues such as neocortex or olfactory cortex where large areas tangential to a surface need to be cut. It is also possible to compensate for the curvature of a surface with this method. The required equipment is simple and inexpensive.

The cutting instrument is usually a razor strip 3 in long by $\frac{1}{2}$ in broad, which has a honed edge on one side only. They can be obtained in the U.K. from Gilette Surgical Ltd., Isleworth, Middlesex (Valet 3 blade strips 4–5/6 in × $\frac{3}{4}$ in). These blades are supplied individually wrapped and protected by a silicone preparation. Before use it is advisable to clean the blade with chloroform and then with ethanol, taking care not to damage the cutting edge. Rinse with distilled water and dry before use. With care one blade will last for several experiments without a noticeable deterioration in performance. Immediately after use the blade should be cleaned by rinsing with distilled water, dried and replaced in its wrapper.

The thickness of the tissue is determined by a glass template or guide which is constructed from glass microscope slides (76 mm × 36 mm) and strips of coverslips of No. 1, 2 or 3 thickness (22 mm × 50 mm), the thickness of the slice being determined by the depth of the recess (see McIlwain and Rodnight, 1966). The glass is first cleaned in nitric or chromic acid (precleaned slides do not need such drastic treatment).

They are then washed and dried. The cover slips are sorted according to their thickness which is determined with a micrometer. No. 1 coverslips are about 0.12 mm, No. 2 0.15 mm and No. 3 0.20 mm. The coverslips are scored along their long axis with a diamond pencil to make strips 3 mm × 50 mm. Two strips of coverslip are fixed at one end of the large microscope slide on opposite edges with a suitable mountant such as Canada Balsam or DPX. Press the coverslips gently against the slide along their length, remove as much surplus mountant as possible, and allow to dry. Repeat the procedure to mount a second strip over each of those already fixed. This should give a guide of thickness (about 320–400 μm) according to the thickness of coverslip selected. The exact depth of recess should be determined at three or four points on each side of the guide and should be uniform within ± 0.01 mm. It is useful to have a selection of guides of various thicknesses available. Once constructed they last indefinitely—as long as they are not dropped!

The cutting table consists of a heavy base about 2 in (50 mm) high and 2 in (50 mm) diameter to which a platform of perspex 3 in (75 mm) square and $\frac{1}{4}$ in (6 mm) thick is fixed. The base is usually of brass or aluminium. The top can be fixed rigidly by araldite or a similar adhesive. The table provides a firm platform clear of the working bench on which the slicing takes place.

After cutting, the slices are floated off the cutting blade into a petri dish filled with ACSF. A small right-angled hook made from thick aluminium wire is used to handle the slices. The horizontal arm is beaten flat to facilitate manoeuvring the slices.

To cut a slice of olfactory cortex, a slab of frontal cortex is prepared as described previously. The surface of the cutting table is covered by two hardened filter papers soaked in ACSF and a small mound of the filter paper is placed across one corner and moistened. The filter papers help to prevent any movement of the slab during the cutting process and the mound compensates for the pronounced curvature of the olfactory cortex at the junction of the prepiriform cortex and the piriform lobe. The slab is placed over the mound with the posterior region towards the experimenter. The guide and the blade are moistened with ACSF and excess fluid is shaken off. The guide is then laid gently over the surface of the slab and the surface layer cut by a slow sawing action while the blade is moved steadily forward. Cut through all tissue before attempting to remove either the blade or the guide. The slice usually adheres to the blade and can be floated off in a petri dish containing oxygenated ACSF before it is transferred to the incubation chamber. *Do not press on the guide* during cutting or damage may result. If thick slices are needed, use guides of the appropriate thickness. In the early stages it is helpful to have some visual indication of the condition of a slice. This can be effected by placing the petri dish containing the slice in ACSF over a piece of black perspex or black card. The slice should be translucent and be even in appearance without excessively thick (opaque) or thin (transparent) parts. It may help to try a few sections of neocortex for practice before cutting a slice of olfactory cortex.

Once the techniques are mastered, it is possible to remove the brain, cut a slice and place it in an incubation chamber in less than 3 minutes; although speed is not essential, a good dissection is. Further advice on slice techniques can be found in McIlwain and Rodnight (1966).

2. Cutting slices with a vibratome

The vibratome is an instrument especially designed for cutting fresh or lightly fixed tissue. It consists of a platform which can be adjusted for height and a vibrating cutting edge which can be advanced at different speeds. The tissue is mounted on the platform which is in a well that can be filled with a suitable bathing solution. Such an instrument is supplied by Oxford Laboratories, Foster City, California, U.S.A. (U.K. agents, A.R. Horwell, London NW6). As supplied, I have found that the amplitude of vibration is too small but if the core of the solenoid used to impart the vibrating action is replaced by one of higher magnetic permeability, the movement can be substantially increased (increased amplitude gives rise to greater cutting action as the frequency of vibration is constant). The cutting edge is made from ordinary safety razor blades broken in half. Gilette Platinum or 7 O'Clock blades are very satisfactory as they have a single bevel on their cutting edge. The blades are mounted in the Vibratome at an angle of 17° which can be checked by the protractor supplied by Oxford Laboratories.

Before use, the central well of the instrument is filled with an oxygenated ACSF to a depth sufficient to cover the cutting blade. The tissue is mounted on a small aluminium block (about 1 cm^3) by smearing the top surface of the block with a very thin coat of tissue glue (e.g., IS 12, Locktite Ltd.) and *lightly* pressing the tissue into position. The aluminium block is then transferred to the vibratome and mounted in the small vice that forms the base of the cutting platform. The tissue block is raised until its upper surface is level with the cutting edge and then it is trimmed carefully by successive passes of the cutting blade until a uniform surface is obtained. The trimming is best carried out with small increments of the height of the cutting platform (~ 100 μm) and the speed of the cutting and amplitude of vibration can be optimized during this process. Once trimming is complete, the height of the block is advanced by an amount equal to the required thickness of the slice and the slice is obtained by advancing the cutting edge as before. Once cut, the slice usually floats free though if it hangs by a few threads of tissue these may be cut by fine scissors. The slice is picked up into a small glass tube 3–5 mm diameter by suction and transferred to a petri dish full of ACSF before mounting in an experimental chamber. Several slices may be obtained from one block of tissue.

For reliable cutting it is best to use a small block of tissue 1.5–2 mm in height to avoid distortion of the tissue block. Moderate to slow speeds of advance of the cutting blade together with large amplitudes of vibration are preferred. The whole process should be watched carefully through a binocular dissecting microscope. The magnifier supplied with the vibratome is quite inadequate.

The vibratome can produce excellent slices of tissues such as the hippocampus where the structures required lie within a single plane. The disadvantages are the cost of the instrument and the slowness with which individual slices are obtained.

3. *Preparing slices with tissue choppers*

Both the preceding methods require considerable manual skill and produce few slices from a given block of tissue. Mechanical devices such as the automatic tissue chopper devised by McIlwain (1961) or the simpler machine devised by Duffy and Teyler (1975) have been used to provide slices for electrophysiological recording. In these machines, the tissue block is placed on a platform under a blade which can be made to fall very rapidly in a chopping action. Following the first cut, the tissue block is moved sideways by an amount corresponding to the desired thickness of the slice and a second cut made. This procedure is repeated as often as necessary; the tissue is then transferred to a petri dish and the slices are floated apart before being transferred to an incubation chamber.

The advantages of this method are obvious. It is quick, produces a large number of slices from a given block of tissue, and requires little manual skill. The disadvantages are cost (compared to manual slicing), lack of control over the cutting process, and possible tissue damage due to the blunting of the cutting edge on each stroke of the blade as it hits the platform. The seriousness of this last objection I cannot assess as I have no first-hand experience of this method.

C. Incubation media

It has been traditional in biochemical studies to use Krebs–Henseleit solutions as incubation media for brain tissue slices. This is certainly a mistake if one accepts that one should mimic the cellular environment *in vivo*. It is now clear that the extracellular environment of nerve cells corresponds not to an ultrafiltrate of plasma but to cerebrospinal fluid (CSF). The chief differences are that the pH, K^+ and Ca^{2+} concentrations are lower in CSF than in plasma. The composition of the extracellular fluid of the brain substance measured with ion-sensitive electrodes corresponds to that of the CSF measured by more traditional means (Prince *et al.*, 1973; Heinemann *et al.*, 1977). These differences are important as changes in the levels of calcium and potassium modify the electrical excitability of neurons; furthermore, the level of calcium in the bathing medium has a powerful influence on the magnitude of synaptic potentials (Richards and Scrcombe, 1970). The composition of two ACSF solutions suitable for maintaining slices of brain tissue for electrical recording is given in Table 4.3 together with the composition of CSF. The first is the solution I have routinely used over the past 10 years; the second solution provides a more accurate mimic of CSF, particularly in respect of the K^+ concentration (Gardner-Medwyn and Nicholson, 1978). To provide a solution of pH 7.35 with the standard gas mixture of 95% O_2/5% CO_2, it is necessary to reduce the bicarbonate to 16 mM. The concentration of glucose in the artificial solutions is higher than that in the CSF to ensure that sufficient glucose is available for all cells throughout the thickness of the slice. Physiological levels would be appropriate for slices superfused by ACSF on both surfaces, but slices

Table 4.3. Composition (mmol/l) of CSF and ACSF solutions used for maintaining electrical excitability in brain slices. All solutions to be equilibrated with 95% O_2/5% CO_2 before use to give a pH 7.35 at 37° C.

	Solution 1	Solution 2	CSF
NaCl	134	137	131
KCl	5	2	2
KH_2PO_4	1.25	1.25	1
$MgSO_4$	2.0	1.1	1.1
$CaCl_2$	1.0	1.1	1.15
$NaHCO_3$	16.0	16.0	23
Glucose	10.0	10.0	3

incubated half-immersed in such solutions (i.e. those that are superfused over one surface only) tend to show instability of their electrical activity if the glucose concentration is lowered below 6 mM: 10 mM glucose provides an adequate margin of safety.

The artificial solutions should be fully saturated with a 95% O_2/5% CO_2 gas mixture before use. At low temperatures the pH will be below 7.2, but as the solutions are brought to 37° C the pH falls within the range 7.3–7.4.

D. Incubation chambers

These may be more or less complicated depending on the needs of the experimenter. Both of the chambers I shall describe were designed for mammalian tissues and have thermostatically controlled water jackets. Such elaboration is not necessary for brain tissue obtained from poikilotherms.

Two main methods of incubation have been employed. In the first, the tissue is incubated partially submersed in ACSF with its upper surface exposed to a humidified atmosphere of 95% O_2/5% CO_2. In the second, the tissue is totally immersed and superfused on both surfaces by oxygenated ACSF.

The advantages of the first are that it is easy to provide full oxygenation of the tissue, the recording of extracellular field potentials is facilitated (particularly at the liquid–gas interface) and problems due to stimulus artifacts are much reduced. The main disadvantages are that the tissue may be disturbed by minor surface tension forces (e.g. bubbles breaking or sudden changes in the flow of liquid), the tissue may rapidly dry out unless adequate precautions are taken to keep the atmosphere very moist, and rapid changes in the composition of the bathing solution are not so readily achieved as with the second method.

The principal advantage of the second method is that the composition of the bathing solution may be completely changed within 2–3 minutes. Additionally, movement artifacts are minimized and the problems of tissue dessication are avoided. The disadvantages are that field potentials are less easy to record because of the shunting of current via the bathing solution, stimulus artifacts may prove troublesome, and care must be taken to ensure adequate oxygenation of the bathing solution.

1. *Chamber for maintaining brain slices at the gas–liquid interface*

The chamber described here is that of Doré and Richards (1974). The following account refers to Fig. 4.1. The apparatus consists of an inner assembly in which the slice is incubated and which is maintained at the desired temperature by a thermostatically controlled water jacket. The tissue slice rests on a grid of platinum gauze (4) (1 mm mesh size, Johnson Matthey Ltd., London) in a inner chamber (7) made of Delrin or Perspex. The inner chamber is fixed in an outer chamber (6) made of Perspex by the tubing connectors (13). These connectors are so constructed that they pull against three studs set in the base of the inner chamber. This preserves a gap beneath the chamber which facilitates the draining of ACSF. Solutions reach the tissue slice by way of silicone rubber tubes (not shown in Fig. 4.1) which are attached to connectors on the base of the Perspex outer chamber (13) and on the wall or lid of the water jacket (17). After reaching the inner chamber the solutions pass through non-return valves (8) made from $\frac{1}{8}$ inch-diameter ceramic balls (Insley Industrial Ltd., Bracknell, Berks., U.K.) and a seating inserted in the base of the inner chamber. Although six connections are available, only one solution is permitted to flow at a time and its rate of flow is governed by the head of pressure applied (i.e. to the height of the reservoir above the chamber). The pressure head is constant for all solution lines and a rotary switch clamps five of the six tubes connecting the reservoirs of saline to the incubation chamber. The pressure head is set up by a Marriotte bottle arrangement. When the saline has reached the tissue it flows away by gravity into the outer chamber (6) and thence to waste via a sintered phosphor bronze disk (16) (to avoid air locks forming in the outflow tube) and silicone rubber tubing fixed to simple connectors.

The whole inner assembly rests on a shelf formed by extensions of the outer chamber base (10) which in turn rests on supports fixed to the sides of the water jacket. The water jacket (5) is filled with distilled water to the level of the inlet tubes (17) and its temperature is kept at $38 \pm 0.2°$ C by an immersion heating element (15), the output of which is continuously regulated by a thermistor-controlled feedback circuit. A gas mixture of 95% O_2/5% CO_2 is passed through the water of the water bath (to warm and moisten the gas) and then is diverted over the tissue via radially disposed drillways (4) and a removable baffle plate (1).

The thermistor (20) is a glass encapsulated type 3in long supplied by ITT Electronic Services, Harlow, Essex, U.K. (Type F53D, 5KΩ at 20° C), and it is mounted on a Perspex holder (19) that can be pushed into place through an insert screwed in the wall of the water jacket. The heater element consists of a core of PTFE around which a length of fabric-covered nichrome wire has been wound. The core is then mounted in a close-fitting stainless-steel case. The nichrome wire is attached to a specially adapted 3-way Din plug via screw terminals.

Soldering the nichrome wire to the plug is not satisfactory. The resistance of the heater coil is about 12Ω and the current is supplied from a 30V DC supply with low ripple (< 1mV r.m.s.).

2. *Superfusion chamber*

The chamber described here is that of Richards and Tegg (1977), but

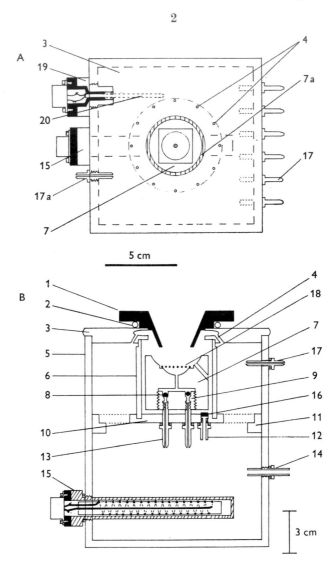

2

the principles of construction can readily be adapted. The apparatus consists of a small recessed chamber (3 cm × 2 cm × 1.5 cm deep) mounted on top of a thermostatically controlled water bath (see Fig. 4.2). The details of the water bath heater and thermistor are similar to those just described.

The tissue slice rests on a grid of platinum gauze just above (1mm) the bottom of the recess and is prevented from floating by a nylon mesh which is placed over the slice. The nylon mesh is stuck at its edges to a Perspex square that fits precisely into the recess. The friction between the Perspex and the wall of the chamber prevents the nylon mesh floating on top of the fluid. The tissue is superfused by solutions emanating from a small hole at the bottom of one end of the recess. The level of fluid above the slice is maintained by a suction pipette mounted on a micromanipulator.

Solutions are prewarmed in tubes made of silicone rubber or nylon which are fixed to connectors on the top plate of the incubation assembly and to connectors at the base of the recessed chamber. To reach the tissue the solutions pass through non-return valves made from $\frac{1}{8}$ inch-diameter ceramic balls which seat on the top of the inner connectors. Only one solution at a time is permitted to flow, the remainder being clamped by a rotary switch. The rate of flow is governed by the head of pressure applied as with the chamber of Dore and Richards (see above). A gas mixture of 95% O_2/5% CO_2 is bubbled through the water bath before being diverted over the surface of the solution in the recessed chamber by drillways and a simple baffle plate.

Fig. 4.1. Incubation chamber assembly. A, Top view with baffle plate, O-ring and platinum grid removed. B, Sectional view: all interconnecting tubes are omitted for clarity. Parts are: 1, baffle plate (perspex); 2, O-ring (synthetic rubber); 3, top plate (friction fit; perspex); 4, airways (twelve, radially disposed); 5, water jacket (perspex); 6, outer chamber (perspex); 7, inner chamber (Delrin or perspex); 7a, flat top of inner chamber; 8, non-return valves (ceramic balls); 9, inner chamber base (Teflon); 10, base plate of outer chamber (perspex); 11, supporting bracket for outer chamber extension; 12, drain tube connector (connects to 14; Teflon); 13, feed tube connector (connects to 17; 6 off, Teflon); 14, drain tube connector; 15, immersion heating element (see text); 16, sintered phosphor bronze disk; 17, inlet tube connector (6 off, Teflon); 17a, gas inlet tube (1 off, connects to a porous store block (e.g. aquarium aerator) which is immersed in the water bath (not shown); 18, grid of platinum gauze (1mm mesh); 19, thermistor mounting assembly; 20, thermistor, glass encapsulated. All screw inserts are rendered watertight by the use of rubber O-rings wherever necessary. The sketches are to scale. (From Doré and Richards, 1974.)

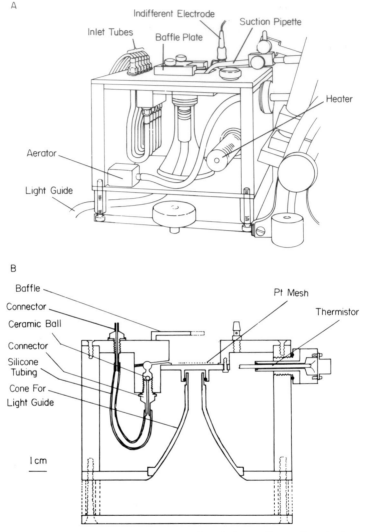

Fig. 4.2. Superfusion chamber. A, Simplified three-dimensional view of the apparatus to show general arrangement. The small parts such as tubing connectors are similar to those used in Fig. 4.1. B, A cross-sectional view of the chamber to show the detail of the non-return valve arrangement and the cone for the light guide. (From Richards and Tegg, 1977.)

One particular advantage of this arrangement is that it is possible to illuminate the preparation from below. I have found a fibre optics light source (LS10 Barr and Stroud, Glasgow, U.K.) to be very satisfactory for this and the bath has been adapted to permit the introduction of a light pipe from below.

E. Methods of recording

The problems of recording potentials from brain tissue slices are essentially similar to those of recording from the intact brain, except that the problems of pulsation and movement are largely avoided. For most purposes simple saline-filled micropipettes of 2 4 μm tip can be used. They are connected to a suitable low-noise voltage follower or preamplifier (e.g. Neurolog NL102 DC preamplifier, Digitimer Ltd., Tewin Road, Welwyn Garden City, U.K.). The indifferent or reference electrode can be either a silver–silver chloride wire dipped in the bathing medium (which I have found satisfactory for many years) or a more elaborate half-cell connected to the saline via an agar bridge.

Stimulating electrodes can be made from insulated wires (preferably silver or platinum to avoid polarization during high-frequency stimulation) or from glass micropipettes filled with Woods metal and plated at the tip with gold and platinum (Dowben and Rose, 1953). The problems of intracellular recording will be dealt with elsewhere in this volume.

F. Analysis and measurement of field potentials

In areas such as the olfactory cortex or hippocampus the afferent nerve fibres make synaptic contacts with the postsynaptic cells in a very restricted part of the dendritic tree. Such areas are ideal for recording evoked field potentials because stimulation of the afferent nerve fibres results in synchronous activation of the postsynaptic cells. Although the currents passing through a cell membrane during the synaptic activation of a single cell are too small to be detected by an extracellular electrode, the synchronous activation of a population of cells, all of which have similar connectivity, results in the summing of the potentials which can then be recorded from the extracellular fluid. The magnitude of such potentials depends on the degree of activation of individual cells and on the number of cells activated. Furthermore, it also depends on the location of the recording electrode with respect to the neuronal elements generating the signal. With so many variables one may be forgiven for wondering whether anything can be deduced from field potential recordings. In fact, a surprising amount of information can be obtained provided care is taken with interpretation as the following examples will show.

Let us consider the perforant path and the granule cells of the dentate gyrus of the hippocampus. We will, for simplicity, consider the longitudinal slice preparation described by Bliss and Richards (1971) and by Richards and White (1975). The slice is mounted pial surface uppermost. The recording and stimulating electrode are placed on the surface of the slice about 1mm apart with the stimulating electrode near to the edge formed by the hippocampal fissure (see Fig. 4.3). Stimulation of the

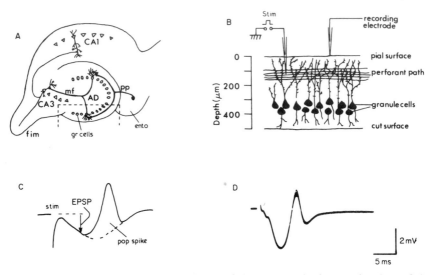

Fig. 4.3. Diagrammatic representation of the anatomical organization of the dentate gyrus together with examples of the potentials elicited by perforant path stimulation. A, Schematic drawing of a transverse section through the hippocampus to show the area from which preparations of the dentate gyrus are made (indicated by the broken lines). Abbreviations: CA1, CA3, pyramidal cells of the hippocampus: fim, fimbria; mf, mossy fibres; AD, area dentata; gr cells, granule cells; pp, medial perforant path; ento, entorhinal area. B, Schematic drawing of a section through a preparation of the dentate gyrus to show the innervation of the granule cells by the medial perforant path and the placement of stimulating and recording electrodes. C, Drawing of the characteristic potential evoked by perforant path stimulation labelled to show the population EPSP (excitatory post synaptic potentials) and population (pop) spike. Abbreviations: Stim, stimulus artifact. D, Photograph of ten faint superimposed sweeps to show the small variation in the responses evoked by stimulation of the afferent fibres at 1 Hz. Calibration vertical bar 2 mV (positive upwards) horizontal bar 5 ms. (From Richards and White, 1975.)

preparation results in a negative wave being recorded after a brief latency. If the stimulating electrode is now pushed through the cortex in 20 μm steps (assuming a fine microelectrode is used for stimulating) the negative potential grows in size even though the stimulus voltage and duration are kept constant. The response reaches a maximum at about 100 μm below the pia and then declines as the electrode is moved deeper (Fig. 4.4A). This simple experiment shows that the perforant path is located about 100 μm below the pial surface and is in agreement with the anatomical studies. For the rest of the experiment the stimulating electrode is positioned where it evoked the maximum response, i.e. about 100 μm below the pia. The stimulus strength and duration are kept constant (strictly, the stimulating current should be monitored but in practice this is seldom necessary). If the recording electrode is now advanced through the cortex the evoked negative wave is first found to grow and then to decline in amplitude and finally reverse in polarity.

A profile of the potential against depth is seen in Fig. 4.4B. The first

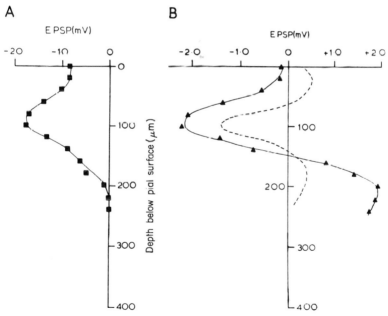

Fig. 4.4. The changes in the amplitude of the population EPSP of the dentate gyrus with different positions of the stimulating electrode (A) and recording electrode (B). In B, the second spatial derivative of the population EPSP is also plotted (− − − −) to show an area of maximum inward current density in the region innervated by the perforant path (about 100 μm below the pial surface).

point to note is that a potential can be recorded at most depths through-out the preparation (except at the reversal point). The second is that the maxima tend to be broad and ill-defined. Humphrey (1968) has shown that the areas generating the current can be located by plotting the second spatial derivative of the potential against depth. Positive values for d^2V/dx^2 indicated areas in which current is leaving nerve cells—'source' zones—and negative values indicate areas in which current enters the cells—'sink' zones. By itself such analysis cannot discriminate a passive source from an active one but consideration of the whole situation usually resolves matters. In the present case, the maximum amplitude is recorded in the vicinity of the perforant path and the perforant path is known to have an excitatory influence on the granule cells, so the 'sink' recorded in this region is due to activation of excitatory synapses. Note the passive source areas above and below the activated zone. The sink zone corresponds to the zone of the granule cell dendrites that is innervated by the perforant path.

In this simple example, field potential analysis has been used to deduce the nature of an evoked potential and locate the activated region. However, for such an analysis to be valid two criteria must be satisfied. First, the degree of activation of the cells must remain constant (i.e. there must be no change in the number of cells contributing or of the average size of their individual contributions). Second, the potentials recorded at any point should be constant throughout the procedure; in practice this is met by the criterion of reversability—if an electrode records the same potential from the same point at the beginning and end of the analysis then constancy is assumed. These criteria are especially important as the analysis requires that repeated stimuli elicit the same potential so that effectively one analyses the *instantaneous* value of the potential at a series of points. To achieve this the potential must be measured at a fixed latency from the stimulus. If there is a drift of potential with time for any reason then the analysis is not valid.

Other problems associated with the use of field potentials are best appreciated by illustration with specific examples. Field potentials have been used to follow changes in neuronal activity as various manoeuvres are performed, e.g. repetitive high-frequency stimulation or depression by anaesthetic drugs. For such studies to be amenable to analysis, it is essential that both the stimulating and recording electrodes are kept fixed in position throughout the experiment. A fundamental premise is that under standard conditions the recorded amplitude is proportional to the number of active elements contributing. This may be readily verified by recording the response of a peripheral nerve to a stimuli of different strengths. As the amplitude of the potential reflects the local extracellular current flow during activity, a decrease in amplitude must result from a smaller average contribution from a given population of cells. This situation can arise because: (i) fewer cells contribute to the current, each cell contributing the same amount as before; (ii) there is a smaller contribution from each of the cells of the population; (iii) there is a change in the synchrony with which the cells are activated (especially important with cell discharge—see below); (iv) there is a decrease in the average membrane potential of the cells making up the population (so the driving potential for current flow is reduced); (v) there is a decrease in the resistivity of the bathing medium.

A similar set of alternatives can be considered for increases in the amplitude of field potentials. To illustrate how these possibilities are considered, let us examine the action of halothane on the evoked potentials of the olfactory cortex (Richards, 1973).

The appearance of the evoked potentials is shown in Fig. 4.5. The potential recorded from the cortical surface consists of an initial biphasic wave the lateral olfactory tract (LOT) compound action potential (conducted response) followed by a negative wave on which one or more positive peaks are superimposed. The negative wave has been identified as the extracellular field potential arising from synchronous activation of the superficial synapses made between the axons of the LOT and the pyramidal cells of the cortex. The positive peaks reflect the synchronous discharge of the pyramidal cells in response to the synchronous activation. The negative wave is therefore called a 'population EPSP' and the positive peaks 'population spikes'. Further details of the interpretation may be found in the papers of Richards and Sercombe (1968b, 1970) and Richards and ter Keurs (1971).

The amplitude of the LOT compound action potential was measured peak to peak. The amplitude of the EPSP field potential was measured at a fixed latency from the stimulus artifact on the falling phase and so estimates the rate of growth of the potential. The exact latency chosen (2–4 ms) was dependent on the conduction time of the LOT fibres and was adjusted so that the EPSP was measured at about 80% of its maximum value. The potential is measured in this way as it is often difficult to be sure where the LOT action potential ends and the EPSP begins. For some situations, however, it is necessary to estimate the rate of growth of the EPSP in such a way that it is independent of the LOT conduction time, e.g. when local anaesthetics are applied. Under these circumstances, it is necessary to measure the EPSP from its onset. Population spikes were measured by their area. This was achieved by projecting the waveform onto squared paper and then counting the squares between the observed waveform and that which would have

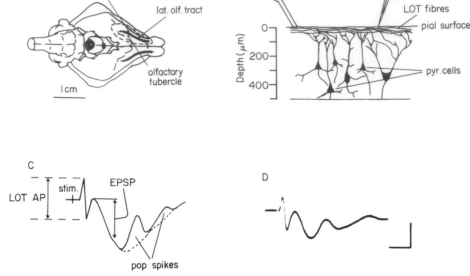

Fig. 4.5. Diagrammatic representation of the anatomical organization of the olfactory cortex together with examples of the evoked field potentials elicited by LOT stimulation. A, The ventral surface of the guinea-pig brain to show the area from which slices are taken (– – – –). B, Schematic drawing of a section through the prepiriform cortex to illustrate the innervation by the lateral olfactory tract and the electrode placements. C, A drawing of the characteristic potential evoked by LOT stimulation and recorded from the surface of the prepiriform cortex adjacent to the LOT. The components of the potential have been labelled to show the nomenclature and the measurements used. Abbreviations: LOT AP, compound action potential of the lateral olfactory tract; stim, the stimulus artifact; EPSP, the population EPSP; pop spikes, two population spikes with their area estimated by the difference between the recorded potential and the estimated decay of the population EPSP (– – – –). D, An example of an actual recording, consisting of ten faint superimposed sweeps to show the very small variation in the size and shape of the evoked potentials. The preparation was stimulated at 1 Hz throughout. Calibrations: vertical bar 2 mV, positive upwards, horizontal bar 2 ms. (From Richards et al., 1975.)

been expected had no population spike been present. The accuracy of such an estimate is greatest when the EPSP and population spike are smallest as the time course of the EPSP decay is more easily predicted.

As Fig. 4.6 shows, halothane decreased the amplitude of the population EPSP and population spike but did not alter the amplitude of the LOT compound action potential. Of the five possibilities listed earlier,

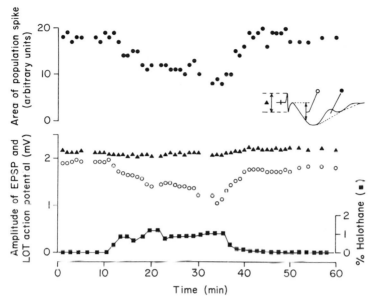

Fig. 4.6. The quantitative changes in the evoked field potentials during exposure to 0.8–1% halothane. The top line shows the area of the first population
spike plotted against time (●). The amplitude of the LOT compound action
potential (▲) and of the EPSP field potential (at 3–5 ms from the stimulus
artifact) (○) are shown in the lower graph (see insert figure). The halothane
tension (■) in the superfusing gas stream is shown at the bottom of the figure
(ordinate on the right). (From Richards, 1973.)

(iii) and (v) can now be dismissed. There was no change in the synchrony of the afferent volley; if there had been, the amplitude and duration
of the LOT action potential would have been affected. Similarly, there
has been no change in the resistivity of the bathing medium because the
effects on the EPSP and population spike are selective (if the change had
been caused by a change in the resistivity of the bathing medium, all
three components should have been reduced). As the LOT compound
action potential has not been affected, it is probable that the number of
active synapses has not fallen (other experiments comparing the action
of tetrodotoxin with that of halothane were needed to resolve this point
completely; see Richards, 1980). Changes in membrane potential cannot be totally excluded but recordings of individual cell discharges failed
to show any consistent change in the amplitude of the evoked spikes
when halothane was applied, so this possibility is also unlikely. The most
probable cause of the depression of the EPSP is, therefore, a smaller

contribution from each cell to the total extracellular current, i.e. to a decrease in the synaptic current occurring in each cell and so to a decreased EPSP.

The area of the population spike was reduced so that the total number of cells contributing must have fallen—neurons discharge in an all-or-none manner and the amplitude of individual cell discharges did not fall. If the height of the population spikes had been used as a measure of cell discharge it would have shown a greater change; as the synchrony of cell discharge decreases the population spike broadens and the overall amplitude falls. This decrease in amplitude can occur even though the total number of cells contributing to the population spike remains constant (measured by the total area of the population spike).

Similar lines of argument can be used to deduce the reasons for a potentiation of evoked field potentials. Note, however, that field potential analysis by itself can only give a partial answer and much more detailed analysis is needed to establish the underlying mechanisms of any change.

For a thorough treatment of the theory of field potentials, the reader is referred to the articles by Humphrey (1968), Rall and Shepherd (1968) and Nicholson and Llinas (1971).

V. Conclusions

1. The use of brain tissue slices for electrophysiological studies has been a major technical development in neurobiology.

2. In conjunction with field potential analysis, it has been possible to analyse how various ions and drugs affect central synaptic transmission.

3. As field potential analysis gives information about the average behaviour of a neuronal population, while single-cell studies give detailed information about the state of a particular cell, it is to be expected that in the future these techniques will be used in combination.

4. In addition to this analytical role, it is likely that brain slices will be used as bioassay systems in the search for new neuroactive materials.

References

Andersen, P., Bliss, T. V. P. and Skrede, K. K. (1971). Lamellar organisation of hippocampal excitatory pathways. *Exp. Brain Res.* **13,** 222–238.

Bliss, T. V. P. and Richards, C. D. (1971). Some experiments with *in vitro* hippocampal slices. *J. Physiol. (Lond.)* **214**, 7P–9P.

Brown, D. A. and Haliwell, J. V. (1979). Neuronal responses from the rat interpeduncular nucleus *in vitro*. *J. Physiol (Lond.)* **292**, 9P–10P.

Dixon, M. (1951). *Manometric Methods*, 3rd Edn. Cambridge University Press, London.

Doré, C. F. and Richards, C. D. (1974). An improved chamber for maintaining mammalian brain tissue slices for electrical recording. *J. Physiol (Lond.)* **239**, 83P–85P.

Dowben, R. M. and Rose, J. E. (1953). A metal-filled microelectrode. *Science* **118**, 22–24.

Duffy, C. J. and Teyler, T. J. (1975). A simple tissue slicer. *Physiol. Behav.* **14**, 525–526.

Gardner-Medwyn, A. R. and Nicholson, C. (1978). The measurement of extracellular potassium and calcium concentration during passage of current across the surface of the brain. *J. Physiol. (Lond.)* **275**, 66P–67P.

Gilbey, M. P. and Wooster, M. J. (1979). Mono and multisynaptic origin of the early surface negative wave recorded from guinea-pig olfactory cortex *in vitro*. *J. Physiol. (Lond.)* **293**, 153–172.

Harvey, J. A., Scholfield, C. N. and Brown, D. A. (1974). Evoked surface-positive potentials in isolated mammalian olfactory cortex. *Brain Res.* **76**, 235–245.

Heinemann, U., Lux, H. D. and Gutrick (1977). Extracellular free calcium and potassium during paroxysmal activity in the cerebral cortex of the cat. *Exp. Brain Res.* **27**, 237–243.

Hillman, H. H., Campbell, W. J. and McIlwain, H. (1963). Membrane potential in isolated and electrically stimulated mammalian cerebral cortex: effects of chlorpromazine, cocaine, phenobarbitone and protamine on the tissue's electrical and chemical responses to stimulation. *J. Neurochem.* **10**, 325–339.

Humphrey, D. R. (1968). Re-analysis of the antidromic cortical response: II. On the contribution of cell discharge and PSPs to the evoked potentials. *Electroenceph. Clin. Neurophysiol.* **25**, 421–442.

Li, C.-L., and McIlwain, H. (1957). Maintenance of resting membrane potentials in slices of mammalian cerebral cortex and other tissues *in vitro*. *J. Physiol. (Lond.)* **139**, 178–190.

McIlwain, H. (1961). Techniques in tissue metabolism: 5. Chopping and slicing tissue samples. *Biochem. J.* **27**, 213–218.

McIlwain, H. and Bachelard, H. S. (1976). *Biochemistry and the Central Nervous System*, 4th edn. Churchill–Livingstone, London.

McIlwain, H. and Rodnight, R. (1966). "Practical Neurochemistry", Ch. 6. Churchill, London.

Nicholson, C. and Llinas, R. (1971). Field potentials in the alligator cerebellum and theory of their relationship to Purkinje cell dendritic spikes. *J. Neurophysiol.* **34**, 209–531.

Pickles, H. G. and Simmonds, M. A. (1978). Field potentials, inhibition and the effect of pentobarbitone in the rat olfactory cortex slice. *J. Physiol. (Lond.)* **275,** 135–148.

Prince, D. A., Lux, H. D. and Neher, E. (1973). Measurement of extracellular potassium activity in the cat cortex. *Brain Res.* **50,** 489–495.

Rall, W. and Shepherd, G. M. (1968). Theoretical reconstruction of field potentials and dendrodendritic synaptic interactions in olfactory bulb. *J. Neurophysiol.* **31,** 884–914.

Richards, C. D. (1973). On the mechanism of halothane anaesthesia. *J. Physiol. (Lond.)* **233,** 439–456.

Richards, C. D. (1980). The input–output relations of a cortical synapse studied by progressive tetrodotoxin poisoning. *J. Physiol. (Lond.).,* **305,** 93P.

Richards, C. D. and McIlwain, H. (1967). Electrical responses in brain samples. *Nature (Lond.)* **215,** 704–707.

Richards, C. D. and Sercombe, R. (1968a). Electrical activity observed in guinea-pig olfactory cortex maintained *in vitro* and its modification by changes in the ionic composition of the bathing medium. *J. Physiol. (Lond.)* **196,** 94P–95P.

Richards, C. D. and Sercombe, R. (1968b). Electrical activity observed in guinea-pig olfactory cortex maintained *in vitro*. *J. Physiol. (Lond.)* **197,** 667–683.

Richards, C. D. and Sercombe, R. (1970). Calcium, magnesium and the electrical activity of guinea-pig olfactory cortex *in vitro*. *J. Physiol. (Lond.)* **211,** 571–584.

Richards, C. D. and Tegg, W. J. B. (1977). A superfusion chamber suitable for maintaining mammalian brain tissue slices for electrical recording. *Br. J. Pharmacol.* **59,** 526P.

Richards, C. D. and ter Keurs, W. J. (1971). The effects of tetrodotoxin on the evoked potentials of the guinea-pig prepiriform cortex. *Brain Res.* **26,** 446–449.

Richards, C. D. and White, A. E. (1975). The actions of volatile anaesthetics on synaptic transmission in the dentate gyrus. *J. Physiol. (Lond.)* **252,** 241–257.

Richards, C. D., Russell, W. J. and Smaje, J. C. (1975). The action of ether and methoxyflurane on synaptic transmission in isolated preparations of the mammalian cortex. *J. Physiol. (Lond.)* **248,** 121–142.

Richards, C. D., Smaje, J. C. and White, A. E. (1976). The use of laminated cortical structures for neurophysiological studies *in vitro*. *Exp. Brain Res., Suppl.* **1,** 207–212.

Skrede, K. K. and Westgaard, R. H. (1971). The transverse hippocampal slice: a well-defined cortical structure maintained *in vitro*. *Brain Res.* **35,** 589–593.

Spector, W. S. (ed.) (1956). "Handbook of Biological Data." W. B. Saunders Company, Philadelphia.

Yamamoto, C. and McIlwain, H. (1966a). Potentials evoked *in vitro* in preparations from the mammalian brain. *Nature (Lond.)* **210,** 1055–1056.

Yamamoto, C. and McIlwain, H. (1966b). Electrical activities in thin sections from the mammalian brain maintained in chemically-defined media *in vitro*. *J. Neurochem.* **13,** 1333–1343.

5

Intracellular and Extracellular Recordings in the Isolated Olfactory Cortex Slice and Some Problems Associated with Assessing Drug Action

C. N. SCHOLFIELD

Department of Physiology, Medical Biology Centre,
The Queens University of Belfast, 97 Lisburn Road,
Belfast, Northern Ireland BT9 7BL, U.K.

I. Introduction

This chapter is concerned mostly with the technical details of obtaining stable electrical recordings from slices of olfactory cortex and some of the short comings of using such a tissue *in vitro*. The results of most experiments using this preparation have been published elsewhere and are referred to below.

II. The Composition of the Bathing Solution

The composition of the Krebs solution used was one which produced potentials mostly closely resembling those recorded *in vivo* (Harvey *et al.*, 1974). The composition of the Krebs solution used is (mM): Na^+, 144; K^+, 5.9; Ca^{2+}, 2.5; Mg^{2+}, 1.3; Cl^-, 123; HCO_3^-, 25; SO_4^{2-}, 1.3; HPO_4^-, 1.2; D-glucose, 11. This solution contains higher K^+ and Ca^{2+} concentrations than those in the cerebrospinal fluid (CSF) and extracellular solutions of mammalian brain (Katzman, 1976; McIlwain and Bachelard, 1971). The higher K^+ concentration helps the tissue to maintain an intracellular K^+ concentration nearer that *in vivo* (McIlwain and Bachelard, 1971). This increased K^+ causes the membrane to be depolarized by a few mV (Scholfield, 1978a). The Ca^{2+} concentration is important in synaptic transmission (Richards and Sercombe, 1970). Ca^{2+} also appears to be important in the integrity of the tissue (see below).

In my experiments, the temperature of the bathing solution (25°C) also differs substantially from that *in vivo*. Using a lower temperature has the advantage of experimental convenience and enables one to use much thicker tissues than at 38°C and still maintain adequate oxygenation of cells in the middle of the slice (Harvey *et al.*, 1974). A thicker slice ($< 600 \mu m$) contains elements generating much larger inhibitory potentials and P-waves. Set against these advantages is the possibility that temperature might influence the activity of cells, of drugs, etc. Reduced temperature produces a slowing of the action potential and synaptic responses but does not otherwise change electrophysiological functions (Scholfield, 1978a, b) and the action of pentobarbitone (Scholfield, 1978c). However, the membrane lipids change phase (approx. 21°C in mammalian kidney; Grisham and Barnett, 1973) which might change the action of exogenous agents. So temperatures below this should be avoided.

Reduced temperature probably has a stabilizing action on the neuron membrane. Thus, the temperature change from 25° to 37°C has no effect on the amplitude of the various extracellular potentials recorded from the prepyriform cortex, whereas the destabilizing action of reduced Ca^{2+} and Mg^{2+} is greater at 37°C than at 25°C (see below).

III. Slice Preparation

Pial surface slices of about 600 μm thick are prepared with a bow cutter and guide (McIlwain, 1975). A new cutter will cut about 5–50 slices after which it appears to still cut but the electrical responses, particularly the "P-wave", are smaller.

The bow cutter is made from a stainless-steel wire folded into the shape of a miniature hack saw. The blade is prepared from gillette "valet" blades, 12 cm long. A strip of marking tape is placed along the entire length of each side so that only 1.6 mm wide border along the sharpened edge of the blade is left uncovered. The unmasked blade edge is placed between two steel parallels, taking care not to damage the cutting edge, and clamped into a vice. The edge of a steel block, 1 cm × 2 cm × 10 cm) is pressed into the angle between one parallel and the unclamped part of the valet blade. The blade is draped in a cloth to catch flying splinters and the edge of the steel block opposite the blade is walloped so that the valet blade is sheared off leaving the cutting edge remaining clamped between the parallels. The cutting edge is removed and the ends silver-soldered to the bow cutter frame.

It might be expected that the removal of the brain from the animal should be as rapid as possible. To see whether this assumption is true, a series of experiments were performed where the animal was decapitated and the head left at room temperature (approx. 22°C) for varying periods before removal of the brain and preparation of the slices (Fig. 5.1). Surprisingly, the rate of disappearance of electrical activity is slow, such that a substantial amount of activity is still present after 60 min.

Guinea pigs of either sex and weighing between 250 and 500 g are beheaded and the brain removed within 1.5–2.0 minutes, less the olfactory bulbs. A large section of brain incorporating the amygadaloid and olfactory cortices is cut and placed on a moistened circle of filter paper on a raised block. The olfactory cortex surface is made flat by placing strips of filter paper under the brain section. The slice is cut using a bow cutter using the edge of Gilette valet blades and a guide milled from a sheet of Perspex. Cutting of brain slices appears to be an acquired skill, since the slices cut by novices may look good but give poor electrical responses. This particularly applies to the longer latency activity. In cutting slices, consideration should be made of the sharpness of the blade, the slight amount of pressure placed on the brain, rate of cutting and the amplitude of the cuts. The hand holding the guide and that holding the cutter should be steadied on the side of the raised block and on the bench to reduce the effect of hand tremor (Fig. 5.2). Just enough pressure should be applied to the brain surface with the guide to cause

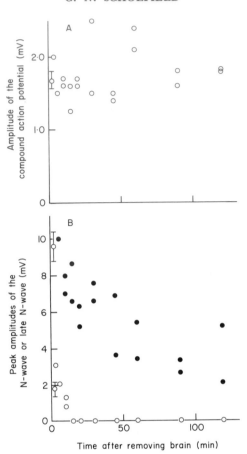

Fig. 5.1. The effect of delaying removal of the guinea pig brain after decapitation on the amplitude of electrical activity recorded from olfactory cortex slices *in vitro*. The slices were prepared as in the text and incubated in Krebs solution for 3 hours and then the electrical activity to single, supramaximal LOT stimuli were recorded with extracellular, pial surface electrodes. A is the peak amplitude of the negative phase of the compound action potential and B is the maximum amplitudes of the synaptic N-wave ($\bullet$) and the late-N-wave (o) (Harvey *et al.*, 1976; Pickles and Simmons, 1978). The abscissa is the time (min) between decapitating the animal and removal of the brain.

the moisture on the olfactory cortex area to wet the guide. In cutting, the edge of the blade is at 90° to the direction of cut. Cutting is done with a sawing motion using about 2 mm strokes and advancing 1–2 mm with each stroke, the whole slice taking about 4 seconds to cut. Fig. 5.3 shows the area of brain used to make the slice.

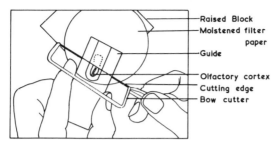

Fig. 5.2. A drawing made of the bow cutter and guide cutting a slice of olfactory cortex.

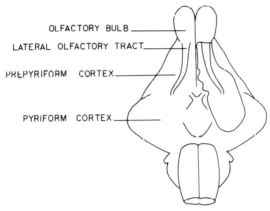

Fig. 5.3. A ventral view of the guinea-pig brain. On the right side an outline of a slice of olfactory cortex has been drawn and is superimposed onto the brain from which it was cut.

Slices are placed in a holder suspended in Krebs solution bubbled with 95% O_2/5% CO_2. The holders consist of a "Perspex" ring with nylon mesh (2 mm mesh) cemented onto the bottom of this ring with a solution of "Perspex" in chloroform. The two ends of a stainless-steel wire loop are fixed through two holes opposite each other with which the holder may be supported. The slice is placed on this nylon mesh and a similar nylon mesh covers the slice. The upper mesh is held in place by a stainless-steel wire circlip held down by the shape of the wire loop (Fig. 5.4A).

IV. Recording Bath

The baths used for extracellular and for intracellular are essentially the same (Fig. 5.4B). They are constructed of two Perspex blocks. The lower

A

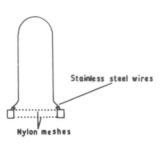

B

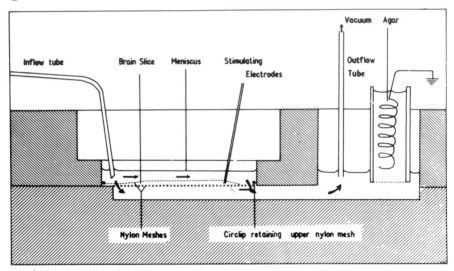

Fig. 5.4. A, The holder for preincubating the slices. B, The bath for recording intracellular or extracellular electrical activity. The arrows indicate the direction of flow of Krebs solution as observed by the addition of drugs.

one is attached to a solid base and contains a trough about 2 mm deep and a little larger (8 mm × 15 mm) than a slice of olfactory cortex and connected to an auxiliary trough via a channel. A fine nylon mesh (from an aquarium fishing net) is laid over these troughs and another Perspex block bolted onto the top of the low block and the mesh pulled taut. The joints are made water-tight with silicone grease. The nylon mesh should be presoaked in distilled water since it stretches on hydration.

The auxiliary trough contains a pipe through which Krebs solution is withdrawn and a large Ag/AgCl earth electrode contained in a saline/ agar-filled cylinder. The height of the bathing solution is then governed by the solution suck-off tube. There are two suck-off tubes for extracellular recording; one to maintain about 2 mm of solution over the slice and one to momentarily lower the meniscus to the level of the slice during recording.

The brain slice is floated onto the nylon mesh in the recording bath and the solution lowered so that the slice is in the middle of the mesh. Another nylon mesh (course mesh) having the same area as the upper part of the recording bath is placed onto the slice and held in position by a U-shaped stainless-steel wire. This circlip is held in position by its tension pushing against the vertical walls of the recording bath. Take care not to squash the tissue. The flow of Krebs solution is resumed. Perspex slides are placed over the bath to maintain an O_2/CO_2-rich atmosphere.

The inflow tube is positioned such that the equal proportions of solution pass underneath and over the slice. The solution originates from a reservoir through which 95% O_2/5% CO_2 passes. The flow rate through the bath is about 2 ml min^{-1} and the bath volume is about 1 ml. This type of bath has two advantages over earlier bulky baths: (i) rapid changes in solution composition and (ii) no need to have additional oxygenation in the recording bath.

(i) When recording from single surface neurons, responses to GABA are maximized in about 5 seconds (Brown and Scholfield, 1979) indicating a rapid change in composition. (ii) Earlier reports had stated that to obtain good electrical responses, reduced swelling etc., it was necessary to provide vigorous oxygenation in the bath containing the tissue. Since this amount of oxygen provides an enormous excess over the tissue's metabolic needs, the functions of the bubbling is more likely one of providing agitation of the solution immediately next to the slice. Without agitation, the stationary layer of solution next to the slice imposes a greater diffusion distance (see Cuthbert and Dunant, 1970). In the present bath design, the small bath and high flow rate minimizes stationary layers.

Present experiments are usually performed at 25°C. The temperature is maintained by delivering the solution from a water-jacketed reservoir at 25°C. Thus the solution is equilibrated at this temperature with O_2 and CO_2. The solution passes into the bath via a 1 mm stainless-steel tube surrounded by a 4 mm Perspex water jacket containing circulated water at 25°C. The amount of cooling in the recording bath is small. Thus with the temperature at 38°C in the reservoir, the temperature in the recording bath at the end furthermost from the inflow tube is 37°C.

For intracellular recording, the recording bath is mounted on a steel plate 17 mm thick × 35 cm × 35 cm. The micromanipulators to hold the electrodes are also attached to this plate. The plate rests on the small rubber bungs on a 12 mm steel plate 60 mm × 40 mm and reinforced underneath with 10 cm-deep diagonal ribs. Laboratory scaffolding poles are threaded into the plate to carry support apparatus. This plate is mounted on the bench on four large rubber bungs. Thus the recording bath is mounted on these plates to reduce vibration. Probably lighter supports would suffice since the high rates of fluid flow across the slice must produce some vibration.

The bath and tubing is made from either disposable apparatus or made such that it can be completely disassembled for cleaning. This

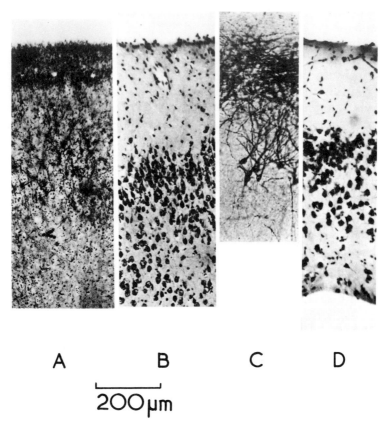

A   B   C   D

200 μm

Fig. 5.5. Transverse sections through the prepyriform cortex. A, is from an animal which had the ipsilateral olfactory bulb removed 3 days previously. The brain was fixed by perfusion with formol-saline and sections were cut at 20 μm with a freezing microtome and stained with Finke–Heimer stain for degenerating axons. B, As A but stained with Kliver–Barrara stain. C, A Golgi-lox preparation of a normal guinea pig. D, A section through a slice of olfactory cortex incubated in a recording bath for 18 hours and prepared as in B.

prevents fungus from accumulating in the chambers containing Krebs solution.

V. Histology of Olfactory Cortex Slices

Experiments have been conducted to observe the distribution of axons from the LOT through the tissue. Fig. 5.5B is a transverse section through the guinea pig brain at the level of the olfactory cortex. It shows a non-nuclear molecular layer within the first 200 μm from the pia. Next to this is a dense layer of neuron somas. The dendrites from these neurons project mainly into the molecular layer as shown in the Golgi-stained preparation (Fig. 5.5C) (see also Halliwell, 1976). The section in Fig. 5.5A is from an animal that had its ipsilateral olfactory bulb removed 3 days previously and stained for degenerating axons and nerve terminals. It shows that the LOT presynaptic elements are confined to a 100 μm band next to the pia. Also noteworthy are some stained neurons in Fig. 5.5A. These are not seen in other brain areas nor in unoperated contralateral olfactory cortex and probably result from transneuronal degeneration (Scholfield, 1979b).

Another series of experiments were done to observe the action of incubation on cell structure in slices of olfactory cortex. Fig. 5.5D shows a section of a slice of olfactory cortex. Thus such sections contain all the molecular layer and the dense underlying soma layer. A new water-miscible plastic mounting medium (JB-4, Polysciences Inc., Pennsylvania, U.S.A.) has been used to observe finer structures under the light microscope. In the dense soma layer and in the molecular layer there is little difference between fresh tissue and incubated tissue. But within a band of about 100 μm from the cut surface, the smaller elements appear swollen and the nuclei and cytoplasm of the neurons is condensed and deeply staining suggesting that these neurons are dying. This explains why it is much more difficult to record from neurons near to the cut surface (see below). In Ca^{2+}-free solution, however, there is also gross swelling of the LOT axons which is in agreement with the electrophysiological observations (see below).

VI. Recording Apparatus

For extracellular recording, a 28-gauge Ag/AgCl wire is placed on the pial surface of the slice. This is connected via an AC amplifier (time constant approx. 10 s), containing cheap low-noise amplifier, and a storage oscilloscope. Alternatively, saline-filled microelectrodes may be used (Halliwell, 1976; Pickles and Simmonds, 1978) to record from different depths through the slice. Recordings can be made whilst the slice is submerged but are reduced compared with those with the upper surface of the slice exposed to air. The compound action potential recorded from the lateral olfactory tract is particularly vulnerable to the shorting action of the bathing solution. With shorting, the separation of the N-wave and the action potential (see Fig. 5.7) is poor, unless the electrode is pushed slightly into the tissue as with glass microelectrodes. When the fluid level is lowered there is still some shorting as indicated by the fact that the action potential doubles in amplitude over about 20 seconds while the solution drains off.

For intracellular recording, single-barrelled microelectrodes are used and pulled, just before using, on a gear-type of puller (Scholfield, 1978a) using 1 mm glass tubing containing a filament (Clarke Electromedical). The electrode resistance varies between 70 and 400 MΩ. About 150 MΩ is the best compromise between electrode noise and frequency response and ease of impalement. The shape of the tip is the most important factor in obtaining stable recordings. A constant taper of about 1 in 10–20 over about 20 μm and without any sudden shoulders is desirable. The electrodes are filled with 4 M potassium acetate, pH 7.0 aided by capillarity; filling is faster at about 40°C. In this tissue, potassium acetate-filled electrodes do not appear to affect inhibitory potentials (Scholfield, 1978b) but in other tissues potassium citrate electrodes are necessary (Kelly *et al.*, 1969). The rectification produced by these electrodes is citrate > acetate > chloride.

The microelectrode is connected to an amplifier with capacity compensation. It also incorporates circuitry to pass constant current into the electrode and to compensate for the electrode resistance. The circuit for this amplifier was derived by Colburn and Schwartz (1972) but using cheap integrated circuit amplifiers. Current is applied to the electrode via a 10^9 Ω resistor. A high input resistance is maintained by the output voltage of the current-pumping amplifier following the input voltage. The amplifier-input offset current can be nulled by adjusting the voltage bias to the current-pump amplifier with the input on open circuit. When the input current is balanced, the input resistances can be measured; typically 10^{11}–10^{12} Ω. This measurement is done with the input connected to an open-circuited microelectrode in the micromanipulator to check for current leakages.

VII. Results

Fig. 5.6 shows sample intracellular recordings from a neuron in the prepyriform and of the olfactory cortex slice. Action potentials produced by direct current injection (Fig. 5.6D) or orthodromic stimulation are rather longer in most other mammalian CNS tissues. The olfactory cortex action potential is about 3–4 ms at the base compared to hippocampal (~ 2 ms) and chick colliculus (~ 1 ms) at the same temperature.

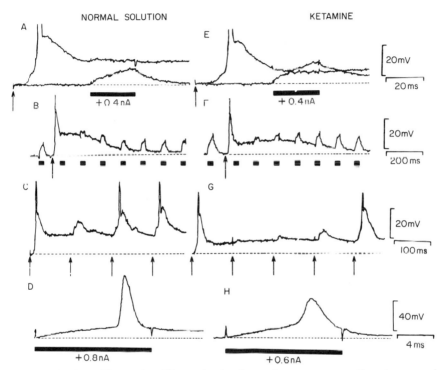

Fig. 5.6. Action of ketamine. Traces in the first column are recordings in normal solution. The second column shows records from the same cell in 0.5 mM ketamine. A, The slice was superfused with normal solution. Current (0.4 nA) was injected into the resting cell during the period marked by the filled bar. This gave rise to a current-induced depolarization. Following this, the LOT was stimulated at the point marked by the arrow. Current was again injected into the cell during the period marked by the bar. The traces are superimposed. B, Current pulses (0.4 nA) were injected into the cell at regular intervals during the periods marked by the bars. At the arrow, the LOT was stimulated. C, A train of stimuli were delivered to the LOT (marked by the arrows). The dashed line is the resting E_m. D, An action potential produced in another cell generated by passing $+0.8$ nA of current into the cell during the period marked by the filled bar. E–G, Recordings were made in the same way as A–C but after superfusing with 0.5 mM ketamine for 30 minutes. H, as D, but after superfusing with 2.0 mM ketamine for 30 minutes.

The conductance increase following the action potential is relatively small and its reversal potential is depolarizing (Scholfield, 1978a).

LOT stimulation produces an excitatory potential (i.e. a depolarization giving rise to an action potential) which is rapid in onset (Fig. 5.6A). This is followed by a depolarizing tail which is accompanied by an increased conductance (Fig. 5.6A, B). This is due to an admixture of a slightly depolarizing inhibitory potential and the tail of the excitatory potential (Scholfield, 1978b). Trains of LOT stimuli produce a reduced second response due to it coinciding with the inhibitory potential of the first response. However, subsequent responses are large, because inhibition is more frequency-dependent than the excitatory potential (Fig. 5.6C).

Ketamine produces a small potentiation of the inhibitory conductance (Fig. 5.6F, E and G) and also a small depression of the excitatory potential (Fig. 5.6E) which is probably a local anaesthetic-like action since the action potential is also depressed (Fig. 5.6H). A much wider range of general anaesthetics have also been tested and they all appear to prolong inhibition (Scholfield, 1980).

A. Impalement of neurons

The microelectrode is advanced through the cut surface (uppermost) of the slice at about 5 $\mu m\ s^{-1}$. The manipulator is gently tapped downwards. Every 10 μm, the capacity compensation is overadjusted for about 0.1 second so that the circuit oscillates. This aids penetration of the electrode into the cell. Since the duration and degree of oscillation is difficult to control with the single potentiometer controlling capacity compensation, another circuit was constructed to deliver a preset period of square-wave oscillation to the amplifier input via a low-capacitance delay. But this appeared to present no particular advantage compared to the overadjustment of the capacity-compensation method of impalement at its present stage of development. On first impalement, the recorded potential jumps to a value of up to − 75 mV, then subsides over a few seconds to a few minutes. With good initial impalement, the membrane potential gradually recovers to a steady value of about − 75 mV over several minutes (Scholfield, 1978a). Often, coinciding with the restoration of the membrane potential, the membrane resistance also increases. With other impalements showing a poor initial negativity, the membrane potential continues to decline. Often during prolonged periods of recording the electrode resistance increases and becomes noisy, particularly during the passage of current. This can usually be reversed by overadjustment of the capacity compensation but can destroy the cell in the process. In experiments with γ-aminobutyric acid (GABA), the likelihood of this is reduced by applying a high concentration of this amino acid, thus producing a shunt for the high current during oscillation. Typical potentials recorded with these techniques are described by Scholfield (1978a, b).

The neurons in the olfactory cortex have input resistances of 9–500 MΩ. These high input resistances have also been measured by the author in the dentate gyrus. The cells in these brain areas have somas of about 12 μm in diameter. The majority of neurons in the mammalian brain are of about this size whereas previous *in vivo* intracellular recordings have been performed on a small population of large neurons. Thus these more numerous cells appear to be more successfully studied by using the present *in vitro* technique.

Since these cells are small, it is more difficult to obtain stable recordings, perseverance is required, sometimes several hours being required, other times, a stable recording can be obtained immediately. It is easier to obtain stable recordings from neurons deep in the tissue than cells near to the cut surface. Perhaps the greatest difficulty is in producing microelectrodes with the appropriate electrical characteristics (see above). The low-resistance bevelled electrodes have not worked in this system. Perhaps now that mechanically more stable *in vitro* systems are available, more research might be devoted to the production of microelectrodes, and particularly to the composition of the hydrated glass at the tip to obtain microelectrodes with fine tips, low noise, high resistance and good frequency response and good sealing with the cell membrane.

With extracellular recordings, slow potential changes can also be obtained by applying various solutions to one surface but not to the other. Thus by such a differential application of KCl (150 mM) or glutamate (10 mM), depolarizations of up to 10 mV may be obtained. This approach has been refined to study the action of GABA (Brown and Galvan, 1979).

Intracellular recordings are useful in determining synaptic events at the soma and pial surface; extracellular recordings are useful in determining events at the superficial dendrites. However, other potentials are recorded from this tissue (Harvey *et al.*, 1974; Simmonds and Pickles, 1978). To distinguish the origin of these potentials, it is necessary to record at different levels through the slice (Haberly and Shepherd, 1973; Halliwell, 1976; Pickles and Simmonds, 1978).

The results in Fig. 5.1 show that slices prepared under the same conditions show potentials which have similar sequence and amplitudes. Thus this preparation may be used to assess effects of experimental manipulations conducted *in vivo*. Changes in electrical activity after removal of an olfactory bulb have been followed (Scholfield, 1979b). The guinea pigs were bulbectomized under ether anaesthesia and allowed to recover. The animals were killed at various times, slices prepared and set up for extracellular recording. No electrical activity was seen in slices from animals more than 1.8 days postoperated. Before 1.6 days, the electrical activity was like that from the unoperated controls.

B. Some problems associated with drug applications

When brain slices are first cut, no electrical activity can be evoked (see Harvey *et al.*, 1974) and up to 3 hours of incubation are required for their full development. When the electrical activity of the tissue is fully developed, electrical activity can be reduced or lost by the slice being slightly damaged, which might coincide with drug applications or electrode position being changed. In chloride-free solution, or in a high concentration of bicuculline, activity also can disappear, particularly following LOT stimulation. The loss of activity is preceded by a long seizure discharge and a neuronal depolarization. Probably this situation arises because of loss of inhibition, which can produce either excitation or a conversion of inhibition into excitation in low chloride followed by a loss of excitability. Thus here, loss of excitability is an indirect consequence of the procedure.

With intracellular recording, the membrane potential measurements are fairly constant. Membrane resistance sometimes "drifts" in either direction by up to 50% over several hours. Membrane resistance decreases with hyperpolarization. Thus, although a hyperpolarization might be accompanied by a decreased R_m, it does not necessarily indicate an increased K or inhibitory conductance. There are a number of procedures which cause a progressive depolarization without any change in R_m. After about a 20–25 mV depolarization, the depolarization accelerates and R_m decreases and is sometimes accompanied by spike generation. Such agents are glutamate or ouabain addition, substitution of Na^+ by Li^+ and omission of Ca^{2+}, etc. The point at which depolarization accelerates corresponds to the threshold for spike generation (Scholfield, 1978a), and is probably due to increased voltage-dependent Na^+ and K^+ conductance increases.

C. Accessibility of drugs to the neurons

Brain consists of a fairly tightly packed cellular structure with little extracellular space and one might expect a somewhat restricted access of drug to the recording site. This problem has been studied with GABA. It was observed that neurons near to the cut surface of the slice showed a much greater sensitivity to the action of GABA than deeper neurons (Galvan and Scholfield, 1978). The most likely explanation of this finding is that the GABA diffusing through the extracellular space is rapidly removed by cellular uptake, leaving little to pass deeper into the slice and act on deeper neurons. This is borne out by autoradiographic accumulation of GABA at the cut surfaces and the avid uptake of GABA into this tissue (Brown et al., 1980). Thus, these differing actions of GABA reflect the uptake of GABA rather than any differences in neuronal sensitivity. For GABA, there are consequences of this on the interpretation of its effects: (a) the potency of GABA is much greater than would be supposed from the higher dose of GABA necessary to produce an effect on deeper neurons; (b) the weak GABA action, i.e. that manifested on deep neurons, appears to be mediated through different receptor–channel complexes than the more potent action. The evidence for this is that the more sensitive "receptor" is antagonized by leptazol (Scholfield, 1979a) whereas the weak "receptor" is not, and that reversal potential for the former is more negative (Brown and Scholfield, 1979). Since leptazol also antagonizes inhibition in this tissue, it would seem likely that this receptor mediates inhibition. Thus, in this instance, observations on the less sensitive GABA "receptor" lead to erroneous conclusions about the nature of the inhibitory receptor.

D. Action of changing ionic composition

Any study on the ionic species involved in the generation of a particular potential requires alteration of the concentration of that particular ions. As in in vivo studies, the intracellular concentration may be increased by including the ion in the electrolyte of the microelectrode. With in vitro preparations, the extracellular ionic composition may also be changed.

The extracellular Na^+ concentration can be reduced to 20 mM: a reduced membrane resistance and potential is produced. Further reduction of Na^+ produces a complete dissipation of the membrane potential. However, the smaller reductions in Na^+ that can be achieved do produce a complete abolition of the action potential. The reduced membrane resistance and potential might in part be due to the failure of the Na^+-dependent reuptake of synaptically active amino acids, etc.

Reduction of extracellular Cl^- to 30 mM has little action on membrane properties but inhibitory potentials are more depolarizing. A single stimulus to the LOT can sometimes cause a long seizure discharge terminating in complete depolarization and loss of synaptic activity, recorded both intracellularly and extracellularly. It presumably reflects a massive discharge excitatory synaptic transmission unchecked by inhibition. Further reduction of extracellular Cl^- in the absence of synaptic activation also produces a depolarization which is not prevented by adding 20 mM Mg.

The concentrations of PO_4^{2-}, SO_4^{2-} and HCO_3^- have not been manipulated in my laboratory.

By using the directly activated action potential in the lateral olfactory tract (LOT) the stabilizing action of Ca^{2+} and Mg^{2+} have been assessed. If the Ca^{2+} of the bathing solution is removed, all or most electrical activity is lost (both synaptic and non-synaptic). The LOT action potential may be completely restored by adding 10 mM Mg^{2+}, while synaptic activity is, of course, still blocked (Fig. 5.7). Intracellular recordings show that the neurons in the olfactory cortex depolarize in Ca^{2+} concentrations of less than 0.5 mM by 20–70 mV. This depolarization is confirmed with extracellular recordings where a postsynaptic population spikes rides on the excitatory postsynaptic potential (EPSP). With short exposures to low Ca^{2+} (approx. 5 minutes) the EPSP greatly decreases in amplitude without any diminution in the population spike (Fig. 5.7B). The only reasonable explanation of this is that the postsynaptic membrane is near to the threshold in low Ca^{2+} making the smaller EPSP more effective in generating spikes. This stabilizing action of Mg^{2+} has been quantified, and it was found that 1.7 mM Mg^{2+} produces a 50% activation of the LOT action potential. Likewise, Ca^{2+}

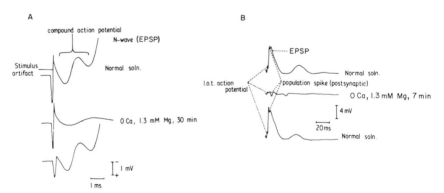

Fig. 5.7. A, Recordings from the lateral olfactory tract LOT after a single stimulus. The positive–negative transient is the summed extracellular activity of the axons in the LOT—the compound action potential. This is followed by the postsynaptically generated N-wave. In the middle trace, the bath was perfused with Ca^{2+}-free solution for 30 minutes and then returned to normal solution (lower trace). B, A similar recording to A but with a lower voltage sensitivity, slower sweep speed and a 7 minute exposure to Ca^{2+}-free solution.

in Mg^{2+}-free solution produced a 50% activation at 0.16 mM, i.e. a 10-fold greater potency than Mg^{2+} (Fig. 5.8).

These experiments have great significance from the point of view of experiments studying the Ca^{2+}-dependent release of neurotransmitter. Thus, simply removing the Ca^{2+} causes presynaptic and also postsynaptic depolarization and loss of membrane function as well as reducing transmitter release. A 95% reduction in the EPSP can be achieved by reducing Ca^{2+} to 0.5 mM and increasing Mg^{2+} to 10 mM, without changing the membrane potential. Addition of a Ca^{2+}-chelating agent would doubtless produce a more rapid depolarization (see Brown et al., 1972).

With decreased K^+ concentration, there is a hyperpolarization and an increase in membrane resistance in spite of the anomalous rectification. With ketamine there is also an increased membrane resistance but with no change in membrane potential (Scholfield, 1980). If, however, there is a leakage around the intracellular recording electrode the membrane potential measurement is reduced (Adams and Brown, 1975). Then an increased membrane resistance produces an apparent depolarization because the shunt is proportionately greater, thus leading to erroneous results.

Some erroneous interpretations might also be produced by the accumulation of extracellular K^+ where a large number of cells are

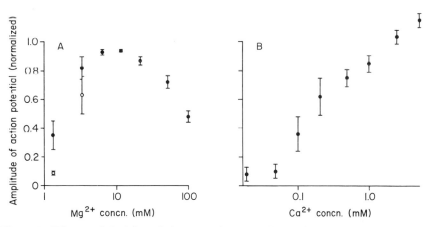

Fig. 5.8. The peak height of the negative transient of the compound action potential was plotted against (A) the Mg^{2+} concentration (in Ca^{2+}-free solution) or (B) the Ca^{2+} concentration (in Mg^{2+}-free solution) (absissae, mM). In A, experiments in 1.3 and 3 mM Mg were repeated at 37°C (o). Points are means of 3–6 slices and bars are the s.e.m.

synchronously activated. General anaesthetics substantially increase the duration and intensity of the inhibitory conductance (Scholfield, 1980). Accompanying this, there is a depolarizing shift in the reversal potential during the time-course of the inhibitory conductance. A parallel depolarizing shift is also seen in glial cells, but without the initial EPSP. A similar shift is also seen with high GABA concentrations. This suggests that the depolarization is due to shift in the ionic gradients, probably K^+. A slow depolarization in glial cells is also seen with convulsants where there is prolonged and intense excitatory activity.

Another series of confusing results have been obtained with applications of aspartic or glutamic acid. Initial experiments indicated that the reversal potential for the depolarization produced by these amino acids were similar to that of GABA, whereas very small glutamate-induced depolarizations produced proportionately much smaller decreases in membrane resistance. However, in the presence of tetrodotoxin (TTX; 1 μM) and tetraethylammonium (TEA; 10 mM), the reversal potential is much more depolarizing, indicating that the action of these amino acids alone is a mixture of the receptor activated effect and voltage dependent Na^+ and K^+ conductance increases. However, even with TEA and TTX, the current–voltage relationship in the depolarizing direction still shows considerable alinearity, probably due to residual voltage-dependent K^+ conductance and perhaps extracellular K^+ accumulation.

VIII. Conclusions

1. The guinea-pig olfactory cortex slice is a reliable tissue for electro-pharmacological and electrophysiological studies.

2. Both intracellular and extracellular recordings are more convenient, longer lived and more stable than *in vivo*.

3. The ionic composition of the medium may be raised and drugs applied directly to the bathing solution.

4. However, there are several factors which complicate the interpretation of the results from these experiments: (a) the restricted extracellular space which is sensitive to change in ion composition, particularly K^+; (b) the higher surface area and volume ratio of the neurons again causing rapid changes in ionic gradients; (c) active uptake of drugs reducing the effective extracellular concentration; (d) the synaptic in-

tcractions between neurons; (e) the uncertainty of hormonal influences, etc. normally present *in vivo*.

References

Adams, P. R. and Brown, D. A. (1975). Actions of γ-aminobutyric acid on sympathetic ganglion cells. *J. Physiol. (Lond.)* **250,** 85–120.

Brown, D. A. and Galvan, M. (1979). Responses of the guinea-pig isolated olfactory cortex slice to GABA recorded with extracellular electrodes. *Br. J. Pharmac.* **65,** 347–353.

Brown, D. A. and Scholfield, C. N. (1979). Depolarisation of neurones in the isolated olfactory cortex of the guinea-pig by γ-aminobutyric acid. *Br. J. Pharmac.* **65,** 339–345.

Brown, D. A., Brownstein, M.J. and Scholfield, C. N. (1972). Origin of the hyperpolarisation that follows removal of depolarising agents from the isolated superior cervical ganglion of the rat. *Br. J. Pharmac.* **44,** 651–671.

Brown, D. A., Collins, G. G. S. and Galvan, M. (1980). Influence of cellular transport on the interaction of amino acids with GABA-receptors in the isolated guinea-pig olfactory cortex. *Br. J. Pharmac.* **68,** 251–262.

Colburn, T. R. and Schwartz, E. A. (1972). Linear voltage control of current passed through a microelectrode with a variable resistance. *Med. Biol. Engng.* **10,** 504–509.

Cuthbert, A. W. and Dunant, Y. (1970). Diffusion of drugs through stationary water layers as the rate limiting process in their action at membrane receptors. *Br. J. Pharmac.* **40,** 508–521.

Galvan, M. and Scholfield, C. N. (1978). Cellular uptake of γ-aminobutyric acid influences its potency on neurones of olfactory cortex *in vitro. J. Physiol. (Lond.)* **284,** 129P–130P.

Grisham, C. M. and Barnett, R. E. (1973). The role of lipid-phase transition in the regulation of the (sodium + potassium) adenosine triphosphatase. *Biochemistry* **12,** 2635–2637.

Haberly, L. B. and Shepherd, G. M. (1973). Current-density analysis of summed evoked potentials in opossum prepyriform cortex. *J. Neurophysiol.* **36,** 789–802.

Halliwell, J. V. (1976). Synaptic interactions underlying piriform evoked responses studied *in vitro. Exp. Brain. Res., Suppl.,* **1,** 223–228.

Harvey, J. A., Scholfield, C. N. and Brown, D. A. (1974). Evoked surface-positive potentials in isolated mammalian olfactory cortex. *Brain Res.* **76,** 235–245.

Katzman, R. (1976). Maintenance of a constant brain extracellular potassium. *Fed. Proc.* **35,** 1244–1247.

Kelly, J. S., Krnjevic, K., Morris, M. E. and Yim, G. K. W. (1969). Anion permeability of cortical neurones. *Ex. Brain Res.* **7,** 11–31.

McIlwain, H. (1975). *In* "Practical Neurochemistry", (H. McIlwain ed.), pp. 117–119. Churchill–Livingstone, London.

McIlwain, H. and Bachelard, H. S. (1971). *Biochemistry and the Central Nervous System.* 4th edn., p. 88. Churchill–Livingstone, London.

Pickles, H. G. and Simmonds, M. A. (1978). Field potentials, inhibition and the effect of pentobarbitone in the rat olfactory cortex slice. *J. Physiol. (Lond.)* **275,** 135–148.

Richards, C. D. and Sercombe, R. (1970). Calcium, magnesium and the electrical activity of guinea-pig olfactory cortex *in vitro. J. Physiol. (Lond.)* **211,** 571–581.

Scholfield, C. N. (1978a). Electrical properties of neurones in the olfactory cortex slice in vitro. *J. Physiol. (Lond.)* **275,** 547–557.

Scholfield, C. N. (1978b). A depolarising inhibitory potential in neurones of the olfactory cortex *in vitro. J. Physiol. (Lond.)* **275,** 547–557.

Scholfield, C. N. (1978c). A barbiturate-induced intensification of the inhibitory potential in slices of guinea-pig olfactory cortex. *J. Physiol. (Lond.)* **275,** 559–566.

Scholfield, C. N. (1979a) Leptazol antagonises the post-synaptic actions of γ-aminobutyric acid. *Br. J. Pharmac.* **67,** 443P–444P.

Scholfield, C. N. (1979b). Electrical activity in the isolated olfactory cortex following afferent denervation. *Ir. J. med. Sci.* **148,** 245.

Scholfield, C. N. (1980). Potentiation of inhibition by general anaesthetics in neurones of the olfactory cortex *in vitro. Pflugers Arch.* **383,** 249.

6

Electrophysiology of the Rat Dentate Gyrus *In Vitro*

S. Y. ASSAF*, V. CRUNELLI† AND J. S. KELLY‡

* *MRC Neurochemical Pharmacology Unit, Department of Pharmacology, Medical School, Hills Road, Cambridge CB2 2QD, U.K.,*
† *Institute Mario Negri, Via Eriterea 62, 20157 Milano, Italy and*
‡ *Department of Pharmacology, St. Georges Hospital Medical School, Cranmer Terrace, London SW17 0RE, U.K.*

I. Introduction

Intracellular recording from the small-bodied granule cells $(8–10\mu m)$ has proved difficult, and detailed analysis of the various synaptic inputs and the ionic mechanisms underlying their actions have not been carried out *in vivo*. However, the clear spatial segregation of granule cell bodies and their apically oriented dendrites has facilitated electrophysiological analysis of their input–output relationships. In particular, Lømo (1971) has shown that in anaesthetized rabbits, stimulation of the cortical input in the perforant pathway results in extracellularly recorded field potentials with laminar profiles that can be interpreted on the basis of excitatory synapses on granule cell dendrites. The dentate gyrus also receives subcortical inputs which are relayed via the mossy fibre axons of granule cells to hippocampal pyramidal neurons (Andersen *et al.*, 1971; Assaf and Miller, 1978a).

In the transverse hippocampal slice described by Skrede and Westgaard (1971), we have been able to obtain particularly stable intracellular recordings from granule cells and thus evaluate their responses to both orthograde and retrograde stimulation and during the microiontophoretic application of putative neurotransmitters. In this chapter, we describe our most recent work on the neurophysiological characterization of granule cells, their responses to stimulation of the perforant path and the actions of γ-aminobutyric acid (GABA), 5-hydroxytryptamine (5-HT; serotonin) and glutamate.

II. Methods

A. Preparation of the tissue slice

Hippocampal slices $200–350\mu m$ thick were obtained from decapitated rats weighing 120–180g by techniques described in detail elsewhere (Dingledine *et al.*, 1980). Within 2–3 minutes of death, the brain is gently removed and a block of tissue containing the hippocampus is cut with a razor blade. The overlying neocortex is removed to reveal the alvear side of the hippocampus which is positioned towards the Oxford Vibratome® blade set at an angle of 15–20°C. In the Vibratome, the tissue is submerged in warmed (37°C) continuously oxygenated $(95\%O_2/5\%CO_2)$ medium. Under the microscope, slices are cut at the slowest speed and largest amplitude settings available on the Vibratome. With this procedure, crushing of the tissue by the blade can be eliminated. Our attempts to modify the primitive circuitry of the Vibratome with more sophisticated signal generators have not improved its performance. By cutting the fimbria, the hippocampus, including the subiculum, float free from the lateral ventricals and no further manipulation is required. One or two slices are removed to the recording chamber since we rarely need to record from more than one slice. The medium used consisted of 134mM NaCl, 5mM KCl, 1.25mM KH_2PO_4, 2mM $MgSO_4$, 2mM $CaCl_2$, 16mM $NaHCO_3$ and 10mM glucose.

B. Perfusion in the slice chamber

The chamber is a modification of that originally designed by Yamamoto and McIlwain (1957) and described by Schwartzkroin (1975) and is identical to that used in the laboratory of Professor Andersen at the Institute of Neurophysiology, Oslo. The slices are gently laid on a single layer of Kodak lens tissue which covers a nylon net suspended on a small reservoir. The lens tissue dips into a circular efflux reservoir which serves as a suction chamber thereby enhancing flow. The flow is adjusted to about 1–3ml min^{-1} and the slices are kept in an interface of liquid and humidified gas. The thinnest possible film of fluid covers the top surface of the slice. Not only does this appear to be essential for long-lasting intracellular recordings but it also facilitates the visual positioning of electrodes and reduces the traverses of our high-resistance recording micropipettes.

C. Recording procedures

Recording electrodes are made by pulling omega dot thin-walled tubing (Clarke Instruments, OD = 1.2mm) to a long fine tip on a vertical puller (Narishige Model PE2). The tips are not bevelled and when filled with 1M potassium acetate the electrodes have a resistance of 150–300MΩ. We are currently using a WPI 707 probe modified so that a simple push button causes the capacitance feedback to oscilliate and causes the electrode to vibrate its way into the cell. Once inside a cell, the bridge is balanced by minimizing the near-instantaneous voltage transients evoked by a hyperpolarizing rectangular pulse of current. As was proposed by Nelson and Frank (1967) and explained by Engel *et al.* (1972), the fastest components of the voltage response originate across the resistance ascribed to the electrode tip and the slower components across the neuronal membrane. For purely aesthetic reasons, we tend to favour a second method in which bridge balance is optimized by matching the peak amplitude of action potentials on and off pulses of current passed through the bridge (Martin and Pilar, 1963). Ideally both methods should give identical balance (see Fig. 6.6). However, when electrode resistance is high (300MΩ) we favour the method that matches the amplitudes of action potentials since the misbalance is easy to compensate for during the analysis of results.

D. Stimulating procedures

Pathways were stimulated using monopolar electrodes. We have tried several types of electrodes including glass pipettes with tips as large as $50\mu m$ and containing 1M NaCl. Unfortunately, NaCl diffuses out of these electrodes and appears to damage the slice. However, pipettes filled with the perfusing medium appear to minimize damage to the tissue and, if the layer of fluid on the slice is kept to a minimum, evoke only a small stimulus artifact (Fig. 6.1).

E. Iontophoresis

Under visual control using a Narishige 3-axis hydraulic micromanipulator (Model MO-102R) an independently mounted four barrel microelectrode (total tip diameter 4–$7\mu m$) was positioned extracellularly near the soma of the impaled neurons. The damped hydraulic mechanism allow the iontophoretic electrode to be repositioned several times during a single impalement. The iontophoretic barrels were filled with GABA (0·2M, pH 4.5), 5-HT (0.05M, pH 3.5), sodium glutamate (0.2M, pH 7.0) and NaCl (sometimes the pH was adjusted to 3.5) for current balancing. The main features of the custom-built iontophoresis unit (Linton Laboratories) have been described elsewhere (Kelly et al., 1975). The ejecting currents (1–320nA) were monitored on a pen recorder and recorded on one channel of a Racal Store 4D. After each period of intracellular recordings, the actions of the drugs were reexamined immediately after withdrawal of the recording electrode from the impaled cell. This allowed direct evaluation of the records for electrical coupling between the iontophoretic electrode and the recording electrode. This proved more reliable than the much publicized method of current balancing (Bloom, 1974). Even with independently mounted electrodes, the coupling could be as great as 5mV. The direction of coupling was $+$ve for GABA and 5-HT and $-$ve for glutamate. However, all records in which a significant degree of coupling became apparent during the subsequent analysis on a PDP-12 computer were discarded.

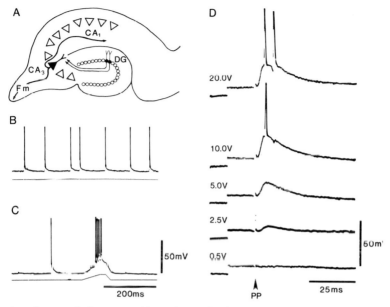

Fig. 6.1. Intracellular recordings from the hippocampal slice to show spontaneous and evoked responses of granule cells. A, Schematic drawing of the transverse hippocampal slice to show the projection from the dentate gyrus (DG) to CA3 neurons which in turn project to extrahippocampal sites via the fimbria (Fm). B, Voltage records made on moving film to show a train of spontaneously occurring action potentials. C, Action potentials initiated by the intracellular injection of a ramp of current through the recording electrode. D, shows the graded membrane depolarizations evoked by stimulating the perforant path (PP) with intensities indicated on the left of each trace. The calibration pulse at the beginning of these and subsequent records is 10mV 10ms[-1].

III. Results

Stable intracellular recordings for greater than 15 minutes and as for as long as 5 hours have been made from over 200 granule cells. Iontophoretic experiments were performed only on cells with resting membrane potentials between 50–70mV. The average input resistance determined from the slopes of linear-regression lines drawn through current–voltage plots as shown in Fig. 6.2 was 57.2MΩ. If one assumes that the somata of granula cells is 10μm in diameter and that the site of current injection is primarily into the soma, the resistance per unit membrane would be approximately 1.0×10^{14} ohm cM^{-2}. This value is similar to the value calculated for cat motoneurons (Coombs and Eccles, 1955; Frank and Fourtes, 1956; Burke and Bruggencater, 1971).

Several types of voltage perturbations occur spontaneously in granule cells with resting potentials in excess of 50mV and include action potentials with overshoots (e.g. Figs. 6.1 and 6.6) and thus amplitudes between 70–120mV, smaller amplitude spikes of less than 15mV similar to the so-called "dendritic spikes" described in pyramidal cells (Spencer and Kandel, 1961; Schwartzkroin, 1975) and depolarizing postsynaptic potentials which may elicit a full blown action potential.

A. Action potentials

With very few exceptions, the large action potentials (70–120mV, baseline to peak) overshoot zero potential. Since the majority of granule cells did not fire spontaneously, the desired number of spikes was evoked by the injection of a depolarizing ramp of current (0.05–0.5nA) as illustrated in Figs. 6.1, 6.6, 6.12 and 6.13. As described below, action potentials could also be initiated by antidromic invasion from the mossy fibres (Fig. 6.3) and orthodromic stimulation of the perforant path (see Fig. 6.5) and were thus used to identify the impaled neuron as granule cells. Although spontaneous bursts similar to those described by Prince and his colleagues in the CA3 region of the guinea pig slice were uncommon, rhythmical bursts of full-blown action potentials (see Fig. 6.7) occurred in five granule cells. In those cases, the number of spikes occurring during a burst could be increased or decreased by the passage of depolarizing or hyperpolarizing current through the recording electrode. Full-blown action potentials were completely blocked by adding TTX (0.1–1μM; tetrodotoxin) to the perfusate and could be recorded in low Ca^{2+} (0.1mM) and high Mg^{2+} (8mM) media suggesting that they were associated with TTX-sensitive Na^+ channels.

Depolarizing after-potentials 15–20ms in duration were a conspicuous feature of action potentials recorded in granule cells having resting membrane potentials greater than 50mV. Depolarizing after-potentials were associated with spontaneously occurring action potentials (e.g. Figs. 6.1 and 6.7) and action potentials evoked either by the intracellular passage of current, antidromic invasion from the mossy fibres (Figs. 6.1 and 6.2) or orthodromic activation from the perforant path (see Fig. 6.7). Fig. 6.2 shows how the properties of the after-potentials were investigated in an antidromically driven granule cell. As far as we know, after-potentials never occur in the absence of full-blown action potentials. When an intracellular depolarizing pulse current is just of sufficient

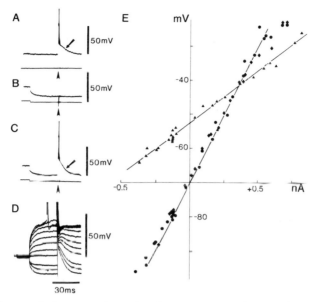

Fig. 6.2. Intracellular records and voltage current plots to show the "all or none" nature of the depolarizing after-potential and its reversal by the intracellular injection of current. A, B and C show the presence and absence of antidromic action potentials and associated after-potentials (arrows) evoked by stimulation in the CA3 region near threshold for antidromic invasion (5V, 50μs). In C, the amplitude of the after-potential is clearly increased by the passage of hyperpolarizing current. D, shows a family of superimposed voltage recordings during the passage of hyperpolarizing and depolarizing current pulses. Note that the reversal of the antidromically evoked after-potential coincides with the peak of the hyperpolarizing after-potential of a spike evoked by depolarizing current. In E, voltage current plots show the input resistance to be 80MΩ at rest (●, ◆) and 30.7MΩ at the peak of the after-potential (▲) with a reversal level of − 41mV. In A–C and in subsequent figures stimulus artifacts are indicated by arrow heads. (Assaf and Kelly (1980). *J. Physiol. (Lond.)* **296**, 68P.)

magnitude to initiate a full-blown action potential the after-potential of the directly evoked spike is identical to that which follows the antidromic spike (Fig. 6.2D, top trace). A family of superimposed voltage recordings such as those shown in Fig. 6.2D and their corresponding current records (not shown) can be used to obtain the voltage–current relationships similar to those in Fig. 6.2E. Clearly, there is a significant decrease in input resistance during the depolarizing after-potential. The reversal of the after-potential is the point of intersection of linear regressions computed from current–voltage relationships measured on and off the after-potential (see also Fig. 6.8). Table 6.1 summarizes the data obtained from five granule cells analysed in this manner.

The decrease in resistance and the clear-cut reversal of the depolarizing after-potential must result from the opening of ionic channels. The reversal levels of depolarizing after-potential range from -40 to -46mV and are depolarizing with respect to the resting potential. Thus after-potentials of granule cells are unlike the hyperpolarizing potential of cat motoneurons (Krnjeviĉ et al., 1979b) and similar to depolarizing after-potentials of hippocampal pyramidal cells (Schwartzkroin, 1975). Others have suggested that the after-potential of the motoneuron and the CA1 pyramidal neuron (Traub and Llinas, 1979) are heterogeneous and indeed some granule cell after-potentials appeared to consist of at least two components. In those cells, a depolarizing hump was observed to follow the spike during the early phase of the after-potential which thus resembled the delayed depolarization of a motoneuron. We have not yet identified the ionic species that underly the after-potential. Since the reversal is well above the resting membrane potential, the ionic flux can not be due to K^+ alone and increases in Cl^-, Na^+ or Ca^{2+} conductance is more than likely. Our preliminary experiments indicate that lowering Ca^{2+} to a level where synaptic potentials are eliminated reduces the amplitude of the after-potential. On both experimental (Wong and Prince, 1978) and theoretical (Traub and Llinas, 1979) grounds, others have postulated that an increased Ca^{2+} conductance

Table 6.1. Reversal levels of the depolarizing after-potentials determined in five granule cells in the hippocampal slice. Abbreviations: V_m, resting membrane potential; R_m, resting input resistance; R_{ap}, input resistance measured during the after-potential which followed an antidromic action potential; V_{ap}, the reversal potential of the after-potential determined as shown in Fig. 6.2.

V_m (mV)	R_m (MΩ)	R_{ap} (MΩ)	V_{ap} (mV)
-70	80	30	-41
-61	48	29	-46
-62	72	33	-44
-70	80	23	-40
-62	65	28	-42

in part underlies the depolarizing after-potential in hippocampal pyramidal cells.

The physiological role and the anatomical origin of depolarizing after-potentials are not yet clear. As shown in Fig. 6.3 the larger depolarizing after-potentials are associated with full-blown action potentials in the soma–dendritic region of the cell. During antidromic invasion, only stimuli that invade the soma and initiate a full-blown action potential produce appreciable depolarizing after-potentials (Fig. 6.3B). In the case of CA3 neurons, which also have depolarizing after-potentials, Wong and Prince (1978) have suggested that the depolariz-

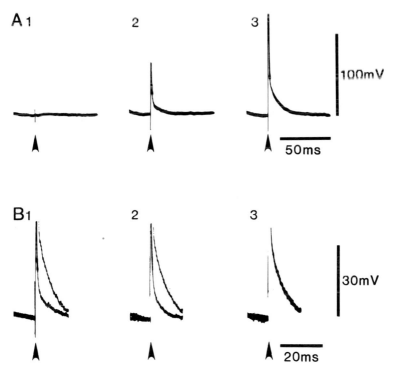

Fig. 6.3. The after-potential resulting from antidromic invasion of a granule cell. A, shows three responses evoked by stimulation in the CA3 region below (A1) and just above (A3) threshold for antidromic invasion (7V, 50 s). In A, the stimulus does not invade the soma and only a small axonal spike and a distinctly smaller after-potential is evoked. B, After-potentials from the same cell shown at high gain. In records B1 and B2, full-blown action potentials are superimposed on axonal spikes. In B3, several records of full-blown action potentials are superimposed in order to emphasize that the changes in the after-potentials in B1 and B2 are not due to spontaneous fluctuations.

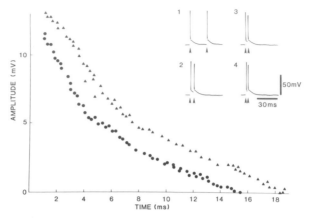

Fig. 6.4. Summation of depolarizing after-potentials is insufficient to sustain bursting discharge. Inserts 1–4 show pairs of antidromic action potentials evoked by pairs of identical stimuli (5V) at successively shorter inter-spike intervals. The graph shows the amplitudes of the first (●) and second (▲) after-potentials measured at the indicated latencies after the peak of their preceding spikes. Even at inter-spike intervals as short as 3ms, the second after-potential is only minimally larger than that following a single spike and is insufficient to sustain bursting.

ation associated with the depolarizing after-potentials causes the bursting discharge. This is unlikely in granule cells. Bursts of action potentials are rarely associated with the after-potential and depolarizations of the same amplitude as the depolarizing after-potential do not cause bursting. Moreover, the depolarizing after-potential is just beginning to reverse at the point where the passage of depolarizing current initiates an action potential (Fig. 6.2D, top trace). As shown in Fig. 6.4, depolarizing after-potentials from two adjacent spikes do not summate to any appreciable degree and are thus unlikely to initiate a sustained burst. This observation is unlike those of Schwartzkroin (1975) on CA1 pyramidal cells but was predicted on theoretical grounds by Traub and Llinas (1979). However, *in vivo* depolarizing after-potentials could initiate bursting either by invading the dendrites and thus causing active propagation of "dendritic spikes" (see Figs. 6.9 and 6.10) or by interacting with incoming synaptic events (Fig. 6.7, see below).

B. Synaptic potentials

As previously reported by Dudek *et al.* (1976), electrical stimulation of the incoming perforant-path fibres results in graded membrane depolarizations which are recorded by electrodes placed in the somatic region of granule cells (Figs. 6.1, 6.5, 6.6 and 6.7). Unlike depolarizing after-potentials, the amplitude and duration of these potentials was dependent on stimulus intensity and the larger depolarization gave rise to single or multiple spikes. Single spikes were more common. In some cases all-or-none small amplitude (SA) (15mV) spikes (described in the next Section) occurred on the perforant path-evoked depolarizations at lower thresholds than the full-blown action potentials (Figs. 6.1 and 6.5). Since the perforant path-evoked depolarizations are readily blocked by lowering Ca^{2+} to 0.1mM and increasing Mg^{2+} to 8mM in the perfusing media, they are in all probability excitatory postsynaptic potentials (EPSPs) mediated by the release of transmitter from perforant path terminals. However, as we will describe in detail elsewhere, the EPSPs may be contaminated by currents other than those resulting from synaptic events and may include "dendritic spikes".

The input resistance of the cell is reduced, presumably due to an increase in the permeability to some depolarizing ion. However, the resistance change during the EPSP was not as great as that seen during the depolarizing after-potential in the same cell (Fig. 6.8). In addition, EPSPs were not as sensitive to current injection as the depolarizing after-potential and in only 5 out of 38 attempts was the EPSP reversed (Fig. 6.5). In those cases the reversal potentials of the EPSP computed in a similar manner to that described for depolarizing after-potentials (see Fig. 6.2) ranged between -40 and -15mV. However, these reversal levels can only be approximations since the perforant path synapses may be located as much as 150–200μm from the soma (Lømo, 1971; Assaf and Miller, 1978a). It is therefore not surprising that we encountered difficulties during attempts to reverse EPSPs by the injection of current into the soma. During our attempts to reverse the EPSP, the injection of the depolarizing current into the soma often initiated small amplitude spikes (Section III C, p. 166) or full-blown action potentials. Presumably the initiation of these spikes is dependent on the summation of the synaptic potential and the depolarizing pulse. Of course, this summation would not occur if the EPSP was generated in the soma and mediated by simple increase in ionic conductance. More strikingly, the depolarizing after-potentials of full-blown action potentials evoked by the EPSPs summate with the underlying synaptically evoked depolarization thus enlarging and prolonging underlying the EPSP. This summation suggest that the depolarizing after-potential and the EPSP are generated at different loci or the EPSP and depolarizing after-potential are mediated by different ionic fluxes.

Spontaneously occurring EPSPs were rarely encountered. However,

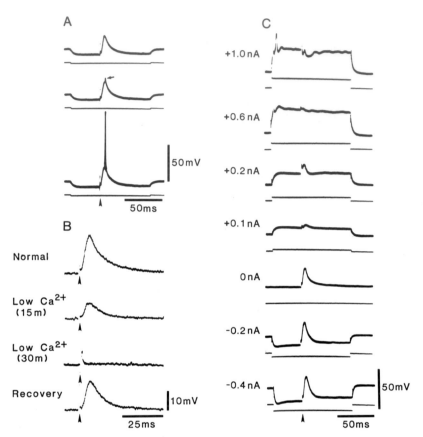

Fig. 6.5. Excitatory synaptic potentials in granule cells evoked by stimulation of the perforant path. A, During the injection of hyperpolarizing currents, the perforant path is stimulated at threshold intensity for the initiation of action potentials. As shown in the second and third traces all-or-none SA-spikes (arrow) or full-blown action potentials can be evoked in this cell. B shows how the perforant-path evoked potentials can be attenuated and eventually suppressed by lowering the concentration of Ca^{2+} to 0.2mM and increasing Mg^{2+} to 8mM in the perfusing medium for 30 minutes. In C, the intracellular injection of depolarizing or hyperpolarizing current through the recording electrode is shown to decrease and increase the evoked potential. The values on the left show the current passed through the recording electrode. Our high-resistance recording electrodes were unstable during the passage of large amounts of current.

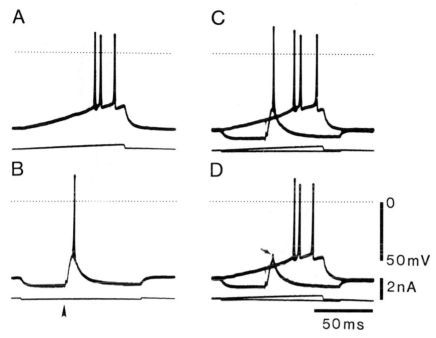

Fig. 6.6. Action potentials in granule cells evoked by stimulating the perforant path and by the intracellular injection of depolarizing current. A, Action potentials evoked by the injection of depolarizing ramp of current (lower sweep) through the recording electrode. In B, action potentials evoked by stimulation of the perforant path at threshold during the passage of hyperpolarizing current through the recording electrode. In C, records similar to those shown in A and B are superimposed and allow a comparison of the peak amplitude and the firing threshold of action potentials evoked by each method. In D, failure of full-blown action potential caused by slightly increasing the hyperpolarizing current reveals the underlying SA-spike (arrow). Proper bridge balance is indicated by the proximity of the peak amplitudes of the first action potentials evoked by the ramp to that evoked by stimulation of the perforant path. In a "good" cell the number of spikes evoked by the ramp remains the same throughout the experiment. · · · ·, indicates zero potential recorded by withdrawing the recording electrode from the cell.

when they did occur they were similar in shape to those evoked by stimulation of the perforant path and initiate spikes in a similar manner. Hyperpolarizing synaptic potentials are not observed in granule cells maintained *in vitro*. This is in agreement with the reports of Dudek *et al.* (1976), who attributed their absence to loss of recurrent collaterals

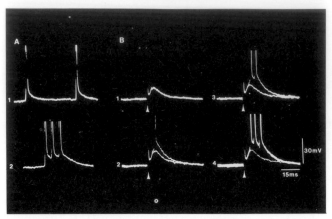

Fig. 6.7. Summation of depolarizing after-potentials and synaptic potentials evoked by stimulation of the perforant path. A, Spontaneously occurring action potentials which are followed by depolarizing after-potentials. The lower trace shows a small burst of action potentials possibly initiated by an underlying randomly occurring synaptic potential. In B, the perforant path is stimulated near threshold for the initiation of full-blown action potentials (15V, 50μs). In trace 1, the stimulus is below threshold and only the synaptic potential is evoked whereas in the remaining traces synaptic potentials which give rise to single or multiple action potentials have been superimposed on the subthreshold synaptic potential. Note the large shift in the apparent amplitude and duration of the synaptic potentials caused by the after-potentials that follow the spikes.

during slicing. However, stimulus-linked inhibition of firing can be evoked by both antidromic and orthodromic stimulation. This inhibition was between 20–50ms in duration and is only one-fifth of that observed *in vivo* (Assaf and Miller, 1978a). The inhibitory potentials IPSP) underlying these changes in excitability can usually be revealed by passing depolarizing current through the electrode. Scholfield (1978b) has recently shown the IPSPs in guinea pig olfactory cortex maintained *in vitro* to be depolarizing. In Section IV A (p. 170), GABA is shown to simultaneously depolarize and inhibit granule cells.

C. So-called "dendritic spikes"

In approximately 33% of granule cells, small amplitude all-or-none spikes (SA-spikes) with amplitudes between 4–15mV were observed. Although SA-spikes can occur spontaneously, they are most often observed during the injection of depolarizing current through the electrodes (see Fig. 6.10). Frequently, SA-spikes precede large full-blown

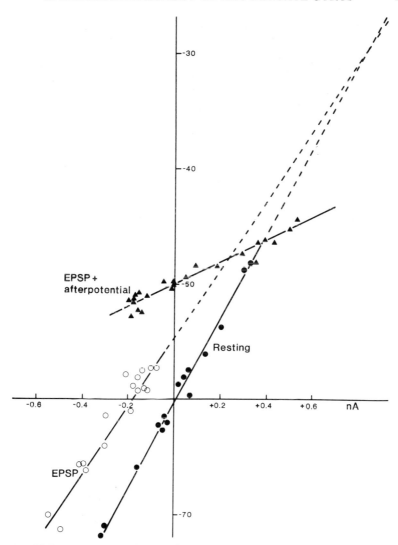

Fig. 6.8. Voltage–current plots to show the greater decrease in input resistance which occurs when depolarizing after-potentials and EPSPs coincide. Hyperpolarizing and depolarizing pulses of current were passed through the electrode during stimulation of the perforant path near threshold so that the EPSP was or was not associated with an action potential. Current–voltage measurements show the resting input resistance to be 51.6MΩ. Similar measurements made during the EPSP show the input resistance to be 43MΩ. During the coincidence of the after-potential and the EPSP the input resistance was 17.5MΩ. The after-potential was readily reversed at − 46mV whereas the EPSP could not be reversed. Extrapolation suggests that the reversal level of the EPSP might be − 30mV.

action potentials by 1–3ms and form a notch on the rising phase of the action potential similar to the prepotentials in hippocampal pyramidal cells (Spencer and Kandel, 1962; Schwartzkroin, 1977). As reported by these authors, hyperpolarizing current passed through the recording electrode suppresses the full-blown action potentials thereby exposing the SA-spike. During depolarization by the injection of intracellular ramps of current, the SA-spikes were evoked at a lower threshold than full-blown action potentials (Fig. 6.9).

As shown in Fig. 6.10, SA-spikes identical to those initiated by direct current injection are also observed on the depolarizing after-potentials which follow full-blown action potentials initiated by antidromic or orthodromic activation. The SA-spikes superimposed on depolarizing after-potentials can give rise to bursts of full-blown action potentials (Figs. 6.9 and 6.10). Occasionally, full-blown action potentials were preceded by SA-spikes evoked by perforant-path stimulation (Fig. 6.11). In four granule cells, trains of SA-spikes were consistently observed during large depolarizations evoked by the iontophoretic application of glutamate or GABA (see Section IV, p. 170, 177). Although trains of SA-spikes occurred during the application of GABA, full-blown action potentials were completely inhibited (see Fig. 6.13). In contrast, SA-spikes which occurred during the iontophoretic application of glutamate often initiate full-blown action potentials (see Fig. 6.20).

The anatomical origin and physiological implications of SA-spikes will be discussed in Section V.

IV. Chemical Transmission in the Dentate Gyrus

On the basis of biochemical and histochemical studies reviewed elsewhere (Assaf and Miller, 1978b), it appears that the dentate receives cholinergic afferents from the septal area, glutaminergic input from the entorhinal cortex, and noradrenergic and serotonergic inputs from the locus coeruleus and median raphe, respectively. Although it has proved difficult to study the intrinsic transmitters of the dentate gyrus, there is a band of neurons located immediately below the granule cell layer that contain glutamate aminodecarboxylase (Storm-Mathisen, 1972) and are thus postulated to be inhibitory neurons using GABA as their transmitter. More recently, intrinsic neurons that contain somatostatin-like immunoreactivity have been observed in the dentate gyrus by Dr. S. Hunt in our Unit.

Although a number of different substances may be contained and

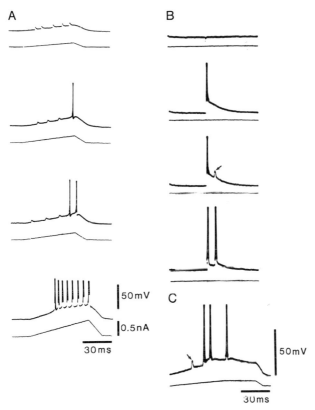

Fig. 6.9. Small amplitude spikes recorded intracellularly from granule cells. A shows trains of SA-spikes to be initiated by intracellular injections of depolarizing ramps of current through the recording electrode. Full-blown action potentials are initiated only at higher thresholds which are reached by increasing the slope of the ramps. In B, antidromic action potentials and associated after-potentials evoked by stimulation in the CA3 region near threshold for antidromic invasion (15V, 50 s). In the first trace, failure of the action potential causes the after-potential to disappear and reveals the stimulus artifact. The after-potential gives rise to either SA-spikes (arrow) or a full-blown action potential. In C, an SA-spike is evoked in the same cell by the intracellular injection of depolarizing current through the recording electrode.

released from afferent nerves to the dentate gyrus, their postsynaptic actions in this region have not been studied with intracellular recordings and their status as synaptic transmitters remains uncertain. Below we describe the effects of applying some of these substances by iontophoresis into the vicinity of granule cells.

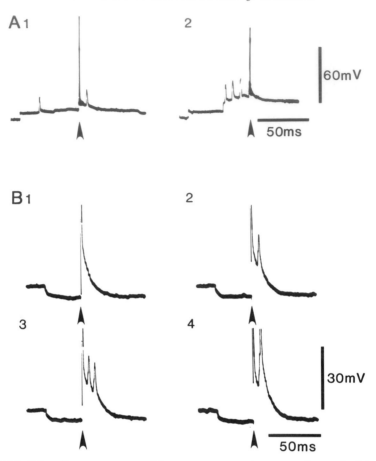

Fig. 6.10. SA-spikes can occur either spontaneously or can be evoked by the intracellular injection of current or depolarizing after-potentials. A shows the spontaneously occurring SA-spike to be identical to those evoked by the depolarizing after-potential of an antidromically evoked action potential (A1) and the intracellular injection of current (A2). B, Higher gain records show that "all or none" SA-spikes evoked by the after-potential can initiate full-blown action potentials (B4). In B, hyperpolarizing current injected through the recording into the electrode enhances the depolarizing after-potential.

A. GABA

Fig. 6.12 summarizes our experimental paradigm and illustrates the typical response of a granule cell to an iontophoretic application of GABA. The top record (A) was made by a continuously moving film to

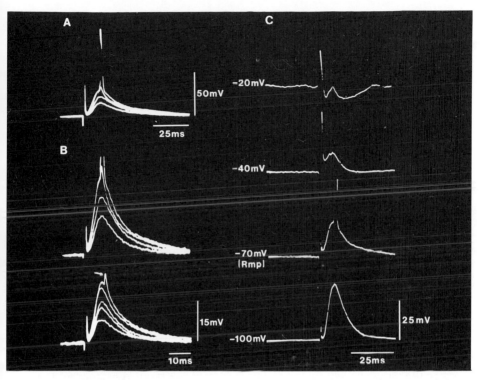

Fig. 6.11. Orthodromically evoked SA-spikes and full-blown action potentials. A shows superimposed intracellular records from a granule cell during stimulation of the perforant path at 5, 10, 15 and 16 V. Threshold for the SA-spike was 15V and between 15 and 16V for the full-blown action potential. B, Higher gain records show that the SA-spike can evoke a full-blown action potential. In the lower record, the SA-spike is again shown to occur in the absence of a full-blown action potential. In C, an attempt was made to reverse the underlying EPSP by hyperpolarizing and depolarizing the membrane to the indicated levels by the intracellular injection of current. Note that the actual peak of this EPSP was not reversed even though the late phase of the EPSP appeared to turn over.

show the way in which depolarizing ramps were used to evoke spike discharge (expanded in B), and were alternated with, every other second, a hyperpolarizing rectangular pulse (C) used to test the input resistance of the cell. GABA ejected close to the cell body inhibited the spike discharge within 0.3–0.5s and produced a large decrease in input resistance (Fig. 6.13). During the application of GABA there is marked depolarization of the membrane. As shown in Fig. 6.14, the depolariza-

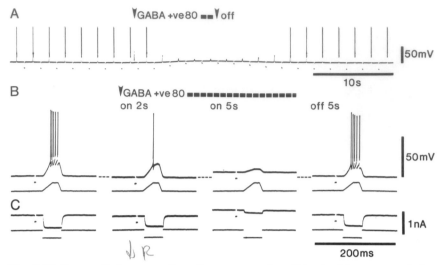

Fig. 6.12. Intracellular recordings from a granule cell in the hippocampal slice to show the response to an iontophoretic application of GABA. A is the voltage record made on moving film to show the way in which an intracellular injection of a depolarizing ramp of current is used to initiate action potentials (enlarged in B) and thus test the cell's excitability. Every alternate second, a hyperpolarizing current pulse is used to measure the input resistance. As shown more clearly in the single shots in B and C, the application of GABA inhibited the spike discharge evoked by the ramp, decreased the input resistance and produced a small depolarization with respect to resting membrane potential (– – – – in B).

tion was dose-dependent and in some experiments could be evoked by the ejection of GABA using currents as low as 4nA.

Since our intention was to eject GABA mainly in the cell body layer of the dentate, the iontophoretic electrode was often positioned below the granule cell layer and thus away from the dendritic tree. However, we cannot eliminate the possibility that GABA diffused into the dendritic tree and thus the depolarizations may be mediated by dendritic receptors. Andersen et al. (1979) have recently reported that CA1, pyramidal cells are hyperpolarized when GABA is applied near the soma and depolarized when applied to the dendrites.

Reversal potentials for GABA were determined using eqn. (1) as described by Ginsborg (1967) and others (Krnjević et al. 1971; Brown and Scholfield, 1979).

$$\Delta V = (E_{gaba} - V_m) \cdot (1 - R_m^*/R_m) \qquad (1)$$

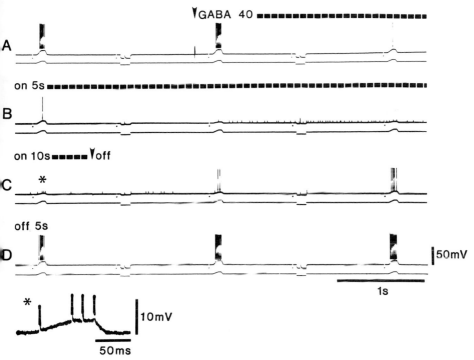

Fig. 6.13. The initiation of small amplitude spikes and the inhibition of full-blown action potentials by the iontophoretic application of GABA. GABA applied near the impaled cell body during the interval indicated by the dashed line caused a marked decrease in input resistance. The voltage response to both the depolarizing current ramp and the hyperpolarizing pulse were greatly reduced. The after-potential of an antidromically evoked action potential is seen superimposed on the response to the injection of hyperpolarizing current. In this particular cell, the inhibition of full-blown action potentials was accompanied by the appearance of SA-spikes. SA-spikes indicated by the asterisks in C are shown below at much higher gain.

Where, ΔV is membrane polarization induced by GABA, E_{gaba} the reversal potential for GABA evoked responses, V_m is the resting membrane potential, R_m the control input resistance and R_m^* is the membrane resistance during the application of GABA. Thus, by simply rearranging eqn. (1), E_{gaba} could be computed even though a constant hyperpolarizing pulse was used.

$$E_{gaba} = V_m + \Delta V (R_m / R_m - R_m^*) \qquad (2)$$

The above equation predicts that the change in potential (ΔV) evoked

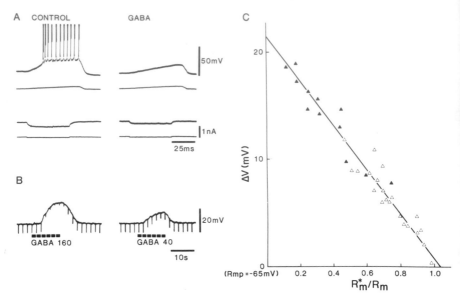

Fig. 6.14. The reversal potential for a GABA-evoked depolarization of a granule cell. A shows the spikes evoked by a ramp of depolarizing current and the voltage response to rectangular pulses of hyperpolarizing current injected through the recording electrode to be greatly attenuated by the iontophoretic application of GABA with 40nA. In B, records from a pen recorder show the depolarization and decrease in input resistance evoked by the application of GABA onto the same cell using ejecting currents of different amplitudes. In C, the GABA-evoked depolarizations (ΔV) are correlated with the change in membrane resistance which occurred during GABA application (R_m^*/R_m). The linear regression drawn through the points obtained during the application of GABA with ejecting currents of 160nA (▲) or 40nA (△) intersects the voltage axis at -44mV and this is taken to be the reversal potential for GABA (Ginsborg, 1967).

by GABA will be linearly related to R_m^*/R_m and is independent of the dose of GABA, and thus remains constant throughout a single application or a number of applications. Indeed, as shown in Fig. 6.14, the depolarization and decrease in R_m evoked by GABA are linearly related and independent of dose. Moreover, the points fall in a straight line with a correlation coefficient of 0.92 which passes very close to the origin.

In addition, the reversal potential for GABA on a few cells was also determined using a more conventional approach in which a family of current pulses was injected through the recording electrode. Current–voltage plots showed the reversal potential for GABA on six cells to range between -39 and -58mV and had a mean value of 17.5 ± 9mV depolarizing with respect to the resting membrane potential.

Brown and Scholfield (1979) have determined similar reversal potentials for bath-applied GABA in neurons of the olfactory cortex maintained *in vitro*. Their results like those of Scholfield (1978b) suggest that the IPSPs are depolarizing and mediated by an increased Cl^- conductance. In a few experiments we have observed that GABA-evoked depolarizations are enhanced by using KCl-filled micropipettes. Moreover, GABA-evoked depolarizations and decreases in resistance are observed in low-Ca^{2+}, high-Mg^{2+} medium suggesting that they are directly mediated by GABA and not secondary to the release of another transmitter.

B. 5-HT

The dentate gyrus is known to receive 5-HT-containing axons which originate in the median raphe nucleus (Moore, 1975). Extracellular recordings in intact rats indicate that stimulation of the raphe inhibits the spontaneous activity of granule cells (Assaf and Miller, 1978a). Following perforant path-evoked activation recurrent inhibition is also observed on the same neurons. However, only the raphe-evoked inhibition is blocked by prior treatment with *p*-chlorophenylalanine which lowered 5-HT content of the dentate gyrus (Assaf and Miller, 1980). In this section, we describe intracellular recordings from 15 granule cells inhibited by the iontophoretic application of 5-HT (Fig. 6.15). The application of 5-HT using positive ejecting currents ranging between 40–160nA caused a clear decrease in the excitability of the cell, a marked decrease in input resistance and depolarization of the membrane potential. The onset latency of the inhibition ranged from 1–10 seconds and was usually 2–3 times longer than that of the GABA-induced inhibition on the same cell. Recovery to the control levels of excitability occurred within 15 seconds of the termination of the ejection (Fig. 6.16). However, GABA was always more potent than 5-HT. The response to GABA always developed faster and much lower GABA ejection currents were required to produce comparable responses. Although these differences may be related to the ease with which GABA leaves the pipette during iontophoresis, similar differences were observed when GABA and 5-HT were bath applied during intracellular recording from the same granule cell (S. Y. Assaf and J.-M. Godfraind, unpublished observations).

As illustrated in Fig. 6.17, the reversal potentials for GABA and 5-HT determined on the same cell are virtually identical. The linear regressions for 5-HT and GABA points were nearly the same for two cells and

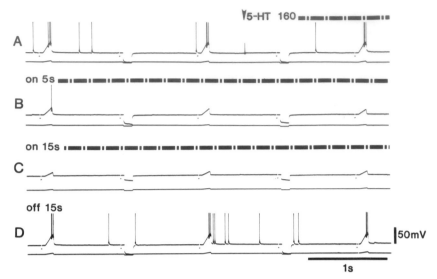

Fig. 6.15. Intracellular recordings from a granule cell in the hippocampal slice to show the response to iontophoretically applied 5-HT. In A, spontaneously occurring action potentials and action potentials evoked by injecting current have the same peak amplitude and are inhibited during iontophoretic application of 5-HT. The inhibition is accompanied by a decrease in the input resistance and a depolarization of the membrane potential.

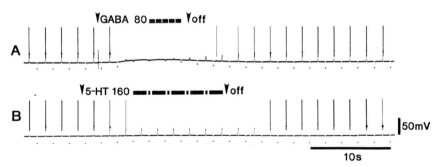

Fig. 6.16. The response of a granule cell to iontophoretic application of GABA and 5-HT. Moving film records from the same granule cell during application of GABA and 5-HT from adjacent barrels of the iontophoretic electrode. Both GABA and 5-HT produced inhibition of action potentials evoked by intracellular injection of current, a decrease in input resistance and depolarization of the membrane. As in all cells examined, lower ejecting currents of GABA produced responses having faster latencies than those observed during 5-HT application.

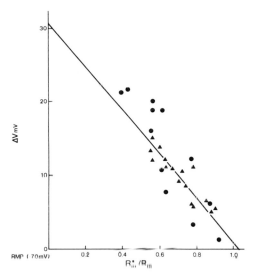

Fig. 6.17. Reversal potential for GABA and 5-HT determined in the same granule cell. The depolarizations (ΔV) evoked during the iontophoretic application of GABA (●) and 5-HT (▲) are related to the changes in membrane resistance produced by the two substances. In this cell, nearly identical regression lines can be drawn through GABA and 5-HT test points. The linear regression intersects the voltage axis at -41mV, the reversal potential for GABA and 5-HT.

differed slightly but not statistically in the remaining four cases. However, there is a trend for GABA points to be skewed to the right and to intersect the voltage axis at a higher reversal potential. The mean reversal potential for 5-HT and GABA respectively, were 15.5 ± 6mV and 17.5 ± 9mV depolarizing with respect to resting membrane potential. The similarities of GABA and 5-HT reversal potentials has led us to suggest that 5-HT responses might be mediated indirectly by the release of GABA (Assaf et al., 1980). Alternatively, 5-HT and GABA may interact with separate postsynaptic receptors but the same ionic channels. We are currently pursuing these possibilities.

C. Glutamate

The actions of iontophoretically applied glutamate to the cell bodies of granule cells are not as clear-cut as the actions of GABA and 5-HT described above. As shown in Figs. 6.16 and 6.18, glutamate applied to the cell body causes an increase in the excitability of the cell and a marked depolarization of the membrane potential. This enhancement of excitability and the concomitant depolarization was a constant feature of glutamate application with currents of between 8 and 320nA. However, the changes in resistance were more complex. In the cell illustrated in Figs. 6.18 and 6.19, glutamate application using low ejecting currents produced an increase in input resistance, whereas higher ejecting currents resulted in a clear decrease in input resistance. In a few cells, biphasic changes in input resistance were observed during

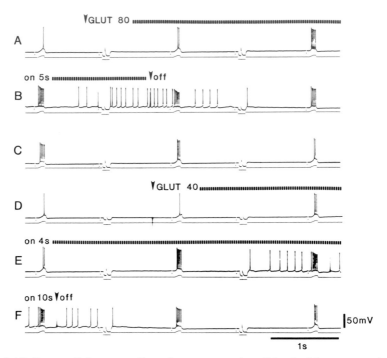

Fig. 6.18. Intracellular recordings from a granule cell in the hippocampal slice during the iontophoretic application of glutamate. Glutamate (GLUT) was applied near the soma of the impaled cell using − ve ejecting currents of 40 and 80nA in this case. In control conditions, intracellular injection of current evoked a single action potential on the depolarizing ramp and antidromic stimulation evoked an action potential on the hyperpolarizing pulse. Unfortunately, the antidromic action potential could not be electronically brightened for photography but its after-potential is clear. Glutamate depolarized the cell and increased the number of action potentials occurring on both the baseline and the ramp. Note that maximal depolarization evoked by the ejection of glutamate with 80nA was associated with a decrease in input resistance and during the depolarization evoked by a 40nA ejection of glutamate there was an apparent increase in a input resistance.

the same glutamate ejection. An increase in input resistance during the initial phase is followed by a clear-cut decrease in resistance during the maximal level of the depolarization (Fig. 6.19). Complex resistance changes in response to iontophoretically applied glutamate on spinal motoneurons were also observed by Bernardi *et al.* (1972), Zieglgansberger and Puil (1973) and Engberg *et al.* (1979).

We failed to determine E_{glut} using the method described for E_{gaba}. As shown in Fig. 6.19, the relationship between ΔV and R_m^*/R_m is not linear during the application of glutamate although the relationship for GABA tested on the same cell is. The non-linearity is particularly striking during the initial phase of depolarization and the late phase of repolarization, i.e. when R_m^*/R_m was greater than 1.0. There are several possible

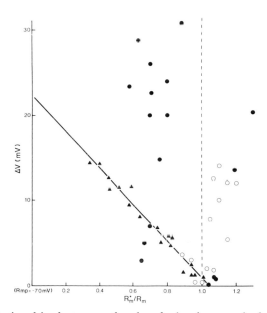

Fig. 6.19. Relationship between the depolarizations and changes in input resistance evoked by iontophoretic applications of glutamate. The depolarizations evoked by iontophoretic application of GABA (▲) and glutamate with 40 (o) or 160nA (●) near the soma are compared, as are the changes in resistance produced by the two substances. As for most cells examined, a linear regression can be drawn through the points plotted for GABA. However, the points for glutamate are widely dispersed and could be on the left or right of the dotted line and correspond either to a decrease or to an increase in input resistance, respectively. Note that during the low dose of glutamate an increase in input resistance predominates.

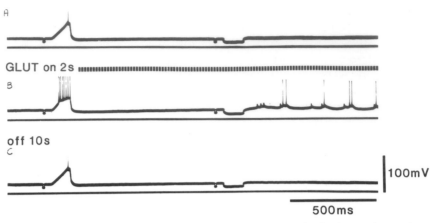

Fig. 6.20. Small amplitude spikes recorded intracellularly during an iontophoretic application of glutamate. Three voltage recordings made before (A), during (B) and after (C) the iontophoretic application of glutamate (GLUT) near the soma. Glutamate ejected with a current of 40nA produced a marked depolarization which initiated SA-spikes some of which gave rise to full-blown action potentials.

explanations for the difficulties we encountered. Receptors for glutamate may be distributed throughout the dendritic tree and therefore some distance from the recording and current injecting electrode. Zieglgansberger and Puil (1973) have suggested that a decrease in resistance only occurs when glutamate interacts with the somal membrane. A second source of complexity may be related to our observations which show that maximal depolarization evoked by glutamate may initiate dendritic spikes (Fig. 6.20) which in turn may give rise to apparent decreases in resistance measured at the soma.

V. Discussion

A. Properties of granule cells

A striking feature of our *in vitro* preparation is the ease with which long-lasting impalements can be made from the small sized granule cells. Unfortunately, there are too few successful impalements of granule cells *in vivo* (Lømo, 1971) to relate our findings to a more 'physiological' preparation. However, the properties of granule cells recorded in the slice appear to be qualitatively similar to those of other neurons in the mammalian brain. Most granule cells had membrane potentials of about -70mV and fast-rising action potentials of large amplitude that overshoot the zero potential. As in other isolated preparations, granule cells only rarely showed spontaneous synaptic activity, but they were more active than hippocampal CA1 cells recorded in the same slice using the same electrodes (Dingledine et al., 1980).

The synaptic potentials evoked by stimulating the perforant path were large and effectively initiated full-blown action potentials. In contrast, hyperpolarizing inhibitory potentials were uncommon. This could be a result of slicing the dentate gyrus and thus sectioning most of the recurrent collaterals to inhibitory interneurons (see Godfraind and Kelly, Chapter 10). Inhibition of dentate granule cells following anti-dromic or orthodromic stimulation is a prominent feature of in vivo preparations (Assaf and Miller, 1978a).

There are some similarities between hippocampal neurons as de-scribed by Kandel and Spencer (1961) and Schwartzkroin (1975) and granule cells in that action potential were followed by pronounced depolarizing after-potentials. Although the prominence of depolarizing after-potentials in our experiments may in part be due to the high concentration of potassium (6.25mM) in the incubating solution, they occur in pyramidal cells in vivo (Kandel and Spencer, 1961).

A second feature of granule cells is the occurrence of small amplitude all-or-none SA-spikes. There are several explanations which may account for SA-spikes. They may be (1) presynaptic potentials, (2) mini EPSPs, (3) axon spikes, (4) the result of electrotonic coupling between adjacent neurons or (5) dendritic spikes. The first two possibilities are highly unlikely since SA-spikes can be initiated by the direct injection of current into the impaled cell and have been observed in low Ca^{2+} media shown to block EPSPs. The possibility that they are axon spikes is also unlikely since SA-spikes occur after the soma is antidromically invaded. Moreover, during orthodromic activation they can occur in the absence of somatic action potentials. Of course this could occur if there were axonal "hot spots" with lower thresholds than the trigger zones on the soma. Since the axons of granule cells are unmyelinated, the hypotheti-cal "hot spots" cannot be nodes of Ranvier as postulated for cat motoneurons (Eccles, 1957).

The possibility that SA-spikes reflect electrotonic coupling between tightly packed granule cells cannot be eliminated at this time. Initiation of SA-spikes by the intracellular injection of current and the stimulation of both the perforant path and mossy fibre pathways could be explained on the basis of electrotonic coupling. However, the observation that trains of SA-spikes are initiated by GABA which inhibited full-blown action potentials is only compatible with electrotonic coupling if one postulates that adjacent neurons are excited or disinhibited by ionto-phoretically applied GABA. However, we have never seen GABA-evoked extracellular spikes corresponding to SA-spikes on withdrawing the recording electrode immediately outside the impaled neurons. Electron microscopic studies of the dentate do not show tight junctions between granule cells (Laatsch and Cowan, 1966). However, in our attempt to demonstrate such coupling we are currently injecting dyes into granule cells that have SA-spikes.

At the present time we favour the suggestion that SA-spikes are regenerative action potentials propagated by dendritic "hot spots". Similar spikes in CA1 pyramidal neurons have been described as dendritic spikes (Spencer and Kandel, 1961; Schwartzkroin, 1975; Wong et al. 1979). Moreover, a recent model of pyramidal cells, which includes actively propagated dendritic spikes, predicts the properties of the SA-spikes outlined above (Traub and Llinas 1979). In particular, if SA-spikes originate in dendrites they might be expected (1) to appear on the EPSP before the threshold for action potentials is reached, (2) to occur after antidromically initiated action potentials and (3) to be initiated by the application of GABA near the soma to inhibit full-blown action potentials. The latter can occur if, as we have suggested elsewhere (Assaf et al., 1980), the depolarization of the soma evoked by GABA can invade the dendrites and initiate SA-spikes. However, the concomitant decrease in somal resistance would prevent SA-spikes from initiating full-blown action potentials.

As discussed by Llinas and Nicholson (1971), such spikes might enhance the integrative capabilities of the neuron (see also Traub and Llinas, 1979). In particular, dendritic spikes in granule cells may amplify and enhance the integration of spatially distinct inputs (Rall and Rinzel, 1973). Such spikes may also serve to repolarize the dendritic tree thus increasing the signal to noise ratio of subsequent synaptic inputs. Somatic currents may invade dendrites (Jefferys, 1979) to initiate dendritic spikes and thereby sustain the bursting behaviour of granule cells.

B. Synaptic transmission in the dentate gyrus

Granule cells maintained in vitro are highly responsive to iontophoretic applications of GABA, 5-HT and glutamate. The first two substances exert similar effects: a reduction of input resistance and depolarization accompanied by inhibition of spontaneous action potentials and a decrease in excitability to intracellular injection of depolarizing current. These effects are probably related and due to increased conductance of the same ions since reversal potentials for GABA and 5-HT were nearly the same. The reversal level values calculated in several granule cell as the product of $\Delta V \cdot R_m / (R_m^* - R_m)$ is unlikely to be due to another process such as anomalous rectification, since the increase in resistance is not secondary to a depolarization. This point is further explained by Krnjević et al. (1971) and Dingledine et al. (1980).

The observation that inhibition and depolarization are concurrent may appear unconventional. However, as Ginsborg (1967) pointed out, even a small increase in conductance will effectively reduce excitability. At the present time, we have not been able to collect enough data on the nature of IPSPs in granule cells to test the possibility that they are depolarizing and mediated by GABA. However, our preliminary experiments indicate that IPSPs are more than likely to be depolarizing in granule cells maintained in vitro. Several studies suggest that depolarizing inhibitory potentials are Cl^--mediated events in mammalian neurons (Adams and Brown, 1975; Scholfield, 1978a, b; Brown and Scholfield, 1979).

The responses of granule cells to iontophoretic applications of glutamate consisted of a marked excitation and depolarization. However, there did not appear to be a correlation between glutamate-evoked depolarizations and changes in resistance as demonstrated for GABA and 5-HT. Thus, we, like others before us (Zieglgansberger and Puil, 1973; Engberg et al.. 1979), failed to show consistent reversal potentials for glutamate. Some of the sources of such difficulties were discussed earlier in Section IV (p. 168). It is our belief that the slice technique is a promising tool for sorting out some of these difficulties. In particular, electrode positioning under visual control and manipulation of the extracellular ionic environment will allow us to pursue some problems that cannot be tackled in vivo.

Acknowledgements

The authors acknowledge the assistance provided by Dr. L. L. Iversen and Dr. J.-M. Godfraind and the technical expertise provided by George Marshall. S.Y.A. is a Fellow of the Canadian MRC.

VI. Conclusions

1. Over 200 intracellular recordings, lasting between 15 minutes and 5

hours, have been made from the 8–10μm granules cells from *in vitro* slices from the dentate gyrus of the rat.

2. The membrane potentials of the granule cells were − 70mV; the action potentials were 70–120mV. The membrane resistance was 57.5MΩ equal to 1.0×10^{14} ohm cm^{-2}.

3. Stimulation of the perforant path evoked large depolarizations (EPSP) which could set up action potentials in the granule cells. However, hyperpolarizing inhibitory potentials (IPSP) were uncommon.

4. Small amplitude (SA) all-of-none spikes, 4–15mV, were observed in 33% of the granule cells. It is suggested that the SA-spikes are regenerative action potentials propagated by electrically active dendritic regions.

5. GABA and 5-HT iontophoresed on to the granule cells show similar effects, causing a reduction in the input resistance and bringing about a membrane depolarization. These are accompanied by an inhibition of spontaneous action potentials and a decrease in excitability of the cell to intracellular injection of depolarizing current.

6. GABA was more potent than 5-HT. The reversal potential for 5-HT and GABA was 15.5 ± 6mV and 17.5 ± 9mV respectively.

7. Iontophoretic application of glutamate caused marked depolarization and excitation of the granule cells. However, there was no correlation between the glutamate-evoked depolarization and the change in membrane resistance.

8. These studies indicate that the hippocampal slice *in vitro* has considerable stability, enabling us to remain inside small neurons for up to several hours, and by means of stimulating afferent pathways and iontophoresis of drugs, allows the study of the nature of the synaptic pathways and transmitters on to granule cells.

References

Adams, P. R. and Brown, D. A. (1975). Actions of γ-aminobutyric acid on sympathetic ganglion cells. *J. Physiol. (Lond.)* **250,** 85–120.
Andersen, P., Bliss, T. V. P. and Skrede, K. K. (1971). Laminar organization of hippocampal excitatory pathways. *Exp. Brain Res.* **13,** 222–238.
Andersen, P., Bie, B., Ganes, T. and Laursen, A. M. (1978). Two mechanisms for effects of GABA on hippocampal pyramidal cells. *In* "Iontophoresis and Transmitter Mechanisms in the Mammalian Central Nervous System",

(R. W. Ryall and J. S. Kelly, eds.), pp. 179–181. North Holland/Elsevier, Amsterdam.

Assaf, S. Y. and Miller, J. J. (1978a). Neuronal transmission in the dentate gyrus: role of inhibitory mechanisms. *Brain Res.* **151**, 587–592.

Assaf, S.Y. and Miller, J. J. (1978b) The role of a raphe serotonin system in the control of septal unit activity and hippocampal desynchronization. *Neuroscience* **3**, 539–550.

Assaf, S. Y. and Kelly, J. S. (1980) On the nature of depolarizing after-potentials in granule cells of the rat dentate gyrus maintained *in vitro*. *J. Physiol. (Lond.)* **296**, 68P.

Assaf, S. Y. and Miller, J. J. (1980). Modulation of neuronal transmission between the entorhinal cortex and the rat dentate gyrus by a serotonin-containing pathway. In press.

Assaf, S. Y., Crunelli, V. and Kelly, J. S. (1980). Similarities between the actions of GABA and serotonin on granule cells in the rat hippocampal slice. *Satellite Symposia to XXVIII International Congress of Physiological Sciences*, Strasbourg, France.

Bernardi, G., Zieglgansberger, W., Herz, A. and Puil, E. A. (1972). Intracellular studies on the action of L-glutamic acid on spinal neurones of the cat. *Brain Res.* **29**, 523–525.

Bloom (1974). To spritz or not to spritz the doubtful value of aimless iontophoresis. *Life Sci.* **14**, 1819–1834.

Brown, D. A. and Scholfield, C.N. (1979). Depolarization of neurones in the isolated olfactory cortex of the guinea pig by γ-aminobutyric acid. *Br. J. Pharmac.* **65**, 334–345.

Burke, R.E. and Bruggencaten, G. (1971). Electrotonic characteristics of alpha motoneurones of varying size. *J. Physiol. (Lond.)* **212**, 1–20.

Coombs, J.S., Eccles, J.C. and Faty, P. (1955). The electrical properties of the motoneurone membrane. *J. Physiol. (Lond.)* **130**, 291–325.

Dingledine, R., Dodd, J. and Kelly, J. S. (1980). The *in vitro* brain slice as a useful neurophysiological preparation. *J. Neuroscience Methods*, **2**, 323–362.

Dudek, F. E., Deadwyler, S. A., Cotman, C. W. and Lynch, G. (1976) Intracellular responses of rat hippocampus: perforant path synapse. *J. Neurophysiol.* **39**, 384–393.

Eccles, J. C. (1957). "The Physiology of Nerve Cells." Johns Hopkins University Press, Baltimore.

Engberg, I., Flatman, J. A. and Lambert, J. D. C. (1979). The actions of excitatory amino acids on motoneurones in the feline spinal cord. *J. Physiol. (Lond.)* **288**, 227–261.

Engel, E., Barcilon, V. and Eisenberg, R. S. (1972). Interpretation of current-voltage relationships recorded in a spherical cell with a single microelectrode. *Biophys.* **12**, 384–403.

Frank, K. and Fourtes, M. G. F. (1956). Stimulation of spinal motoneurones with intracellular electrodes. *J. Physiol. (Lond.)* **134**, 451–470.

Ginsborg, B. L. (1967). Ion movements in junctional transmission *Pharmac. Rev.* **19,** 289–316.

Jefferys, J. G. R. (1979). Initiation and spread of action potentials in granule cells maintained *in vitro* in slices of guinea-pig hippocampus. *J. Physiol. (Lond.)* **289,** 375–388.

Kandel, E. R. and Spencer, W. A. (1961). Electrophysiology of hippocampal neurones: II. After-potentials and repetition firing. *J. Neurophysiol.* **24,** 243–259.

Kelly, J. S., Simmonds, M. A. and Straughan, D. W. (1975). Microelectrode techniques. *In* "Methods in Brain Research" (P. B., Bradley, ed.), vol. 1, pp. 333–377. Wiley, New York.

Krnjević, K., Pumain, R. and Renaud, L. (1971). The mechanisms of excitation by acetylcholine in the cerebral cortex. *J. Physiol. (Lond.)* **215,** 247–268.

Krnjević, K., Lamour, Y., MacDonald, J. F. and Nistri, A. (1979a) Effects of some divalent cations on motoneurones in cats. *Can. J. Physiol. Pharmacol.* **57,** 944–956.

Krnjević, K., Puil, E. and Werman, R. (1979b). EGTA and motoneuronal after-potentials. *J. Physiol. (Lond.)* **275,** 199–224.

Laatsch, H. R. and Cowan, W. M. (1966). Electron microscopic studies of the dentate gyrus of the rat. I. Normal structure with special reference to synaptic organisation. *J. Comp. Neurol.* **128,** 359–396.

Lømo (1971). Patterns of activation in a monosynaptic cortical pathway: the perforant path input to the dentate area of the hippocampal formation. *Exp. Brain Res.* **12,** 18–45.

Llinas, R. and Nicholson, C. (1971). Electrophysiological properties of dendrites and somata in alligator Purkinje cells. *J. Neurophysiol.* **34,** 532–551.

Martin, A. R. and Pilar, G. (1963) Dual mode of synaptic transmission in the avian ciliary ganglion. *J. Physiol. (Lond.)* **168,** 443–463.

Moore, R. Y. (1975). Monoamine neurons innervating the hippocampal formation and septum: organisation and response to injury. *In* "The Hippocampus", Vol. 1: Structure and development, (R. L. Isaacson and K. H. Pribram, eds.), pp. 215–237. Plenum Press, New York.

Nelson, P. G. and Frank, K. (1967). Anomalous rectification in cat spinal motoneurones and effect of polarizing currents on excitatory postsynaptic potential *J. Neurophysiol.* **30,** 1097–1113.

Rall, W. and Rinzel, J. (1973). Branch input resistance and steady attentuation of input to one branch of a dendritic neuron model. *Biophysiol. Journal.* **13,** 648–688.

Scholfield, C. N. (1978a). Electrical properties of neurons in the olfactory cortex slice *in vitro. J. Physiol. (Lond.)* **275,** 535–546.

Scholfield, C. N. (1978b). A depolarizing inhibitory potential in neurones of the olfactory cortex *in vitro. J. Physiol. (Lond.)* **275,** 547–557.

Schwartzkroin, P. A. (1975). Characteristics of CA1 neurones recorded intracellularly in the hippocampal slice preparation. *Brain Res.* **85,** 423–436.

Schwartzkroin, P. A. (1977). Further characteristics of hippocampal CA1 cells *in vitro. Brain Res.* **128,** 53–68.

Skrede, K. K. and Westgaard, R. H. (1971). The transverse hippocampal slice: a well-defined cortical structure maintained *in vitro. Brain Res.* **35,** 589–593.

Spencer, W. A. and Kandel, E. R. (1961). Electrophysiology of hippocampal neurones. IV. Fast prepotentials. *J. Neurophysiol.* **24,** 272–285.

Storm-Mathisen, J. (1972). Glutamate decarboxylase in the rat hippocampal region after lesions of the afferent fibres systems. Evidence that the enzyme is localized in intrinsic neurones. *Brain Res.* **40,** 215–235.

Traub, R. D. and Llinas, R. (1979). Hippocampal pyramidal cells: significance of dendritic ionic conductance for neuronal function and epileptogenesis. *J. Neurophysiol.* **42** (2), 476–496.

Wong, R. K. S. and Prince, D. A. (1978). Participation of calcium spikes during intrinsic burst firing in hippocampal neurons. *Brain Res.* **159,** 385–390.

Wong, R. K. S. and Prince, D. A. and Basbaum, A. I. (1979). Intradendritic recordings from hippocampal neurones. *Proc. Nat. Acad. Sci. U.S.A.* **76** (2), 986–990.

Yamamoto, C. and McIlwain, H. (1966). Electrical activities in thin sections from mammalian brain in chemically-defined media *in vitro. J. Neurochem.* **13,** 1333–1343.

Zieglgansberger, W. and Puil, E.A. (1973). Actions of glutamic acid on spinal neurones. *Exp. Brain. Res.* **17,** 35–49.

7

Electron Microscopic Studies of Brain Slices: the Effects of High-Frequency Stimulation on Dendritic Ultrastructure

KEVIN LEE*, MICHAEL OLIVER,
FRANK SCHOTTLER AND GARY LYNCH

*Max Planck Institute for Psychiatry, Experimental Neuropathology,
Kraepelinstrasse 2, 8 Munich 40, West Germany and Department of
Psychobiology, University of California, Irvine, California 92717, U.S.A.

I. Introduction

The use of *in vitro* brain slices has increased dramatically as it has become evident that these preparations retain many of the physiological properties of the brain *in situ*. For at least three reasons it would seem that

189

anatomical studies of slices are now warranted. First, questions about the relative "normalcy" of slices become more pressing as physiological, pharmacological and biochemical discoveries are made with the *in vitro* method. Information about the gross and selective disturbances produced by the preparation and maintenance of brain slices may be required to interpret the results of a variety of recording experiments— thus, for example, if certain classes of synapses or neurons are differentially affected by the slice procedure then we might well anticipate the occurrence of particular physiological disturbances. Second, detailed and quantitative analyses of slices could prove useful in attempts to improve the method. Physiological measures have already been used for this purpose (e.g., Teyler, 1976), but anatomy offers well-defined measures that can be easily compared to the normal brain in a quantitative fashion. Finally, there is a possibility that slices may be valuable for the investigation of certain types of anatomical problems. Many of the properties of slices which have made them so attractive to physiologists also present new opportunities to neuroanatomists. For example, it has often been suggested that drugs and ionic changes produce effects on the ultrastructure of brain tissue and if slices prove to be "useable" for electron microscopic analyses they might well be of service in testing such ideas.

In the present report, we will describe our ongoing efforts to use the *in vitro* slice in the investigation of a specific neuroanatomical problem. Hippocampal synapses both *in vivo* (Bliss and Lømo, 1973) and in the slice (Schwartzkroin and Wester, 1975; Alger and Teyler, 1976; Andersen *et al.*, 1977; Dunwiddie and Lynch, 1978) exhibit a stable and extremely long-lasting form of synaptic facilitation following brief bursts of high-frequency stimulation. The peculiar persistence of this effect—it can last for weeks (Bliss and Gardner-Medwin, 1973; Douglas and Goddard, 1975)—raises the possibility that it is due to some type of structural change and two studies *in vivo* have presented data that provide some support for this idea. Van Harreveld and Fifkova (1975) reported that lengthy periods of repetitive stimulation induce a pronounced swelling in dendritic spines; however, their studies did not use concommitant physiological recording and it is not clear if the reported effects were due to potentiation or some other effect of stimulation (e.g., seizures or spreading depression). Recently, we found that the induction of long-term potentiation (LTP) by brief bursts of high-frequency stimulation was accompanied by an increase in the numbers of synapses on the shafts of dendrites as well as a reduction in within animal variability of three separate measures of spines (Lee *et al.*, 1980). These latter results indicate that conditions that produce LTP also induce structural change, but it remains to be demonstrated that the two effects are causally related. Slices could prove useful in approaching correlative anatomical–physiological questions of this type. For example, it should be possible to test whether treatments that interfere with the induction of LTP (Dunwiddie *et al.*, 1978; Dunwiddie and Lynch, 1979) concurrently reduce the ultrastructural changes produced by repetitive stimulation.

The research described below was prompted by these considerations and was intended first to explore the possible utility of slices for electron microscopic studies and second as an attempt to replicate the experiments in which the effects of repetitive stimulation were tested using anesthetized rats. A preliminary note describing one aspect of these experiments has been published (Lee *et al.*, 1979).

II. Methods

The following procedures used in the preparation, maintenance and fixation of slices have evolved in this laboratory over the past several years. It must be emphasized that this effort has not been a systematic one but instead represents the gradual accretion of changes, each of which seemed at the time to improve the physiology of the preparation. It is worth noting at the outset that the single most important parameter that we have identified is the practice and skill of the experimenter in removing and slicing the hipppocampus.

A. Preparation of slices

Young adult rats were decapitated and the brain was rapidly removed and immersed in 4°C oxygenated medium. After placing the brain on its dorsal surface the hippocampus was dissected free using blunt, plastic tools, and carefully flattened on the stage of a tissue chopper (Sorvall or McIlwain) with its longitudinal axis at right angles to the cutting blade. A fresh razor was used for each preparation and this is probably of some importance. (We have found when using brain regions other than hippocampus that the height of the tissue block cannot be too much greater than 3–4 mm if optimal results are to be obtained.) Successive cuts were made 500 μm apart and, after each cut, the slice was removed by a single rotation of a sable brush saturated with cold media. The slice was then placed in a dish containing cold oxygenated medium. This procedure was continued until 10–15 slices had been collected after which the slices were transferred to the recording chamber. The temperature of the medium was set at 32–34°C. The entire procedure requires something between 5 and 10 minutes—we have not noticed any qualitative improvements in the slices by accelerating the various steps with the probable exception of removing the brain and placing it in cold medium. In static bath experiments, the level of the medium relative to height of the slice is quite important. Optimal physiology is obtained when the level of the medium is even with, but does not cover, the surface of the slice. It is also of some importance that a dense atmosphere of warmed bubbled O_2 be maintained above the medium (see Fig. 7.1). When perfusion experiments are to be conducted, the level of the medium is lowered below the net and the medium is dripped from a pipette on one slice after the other until the entire collection is covered by 2–3 mm. This procedure causes the slices to adhere to the supporting nylon net. After this, continuously oxygenated medium is perfused at a rate of 3 cc (or one chamber volume) per minute. This procedure is usually begun 30 minutes after the slices have been prepared. The slice chamber itself is quite simple and its design and pertinent dimensions are illustrated in Fig. 7.1.

B. Neurophysiology

After the slices had been in the chamber for 1 hour, a recording pipette was used to sample for "unit" activity in the pyramidal cell body layer of field CA1 and if by this criterion the slices appeared "healthy" the next phase of the experiment was initiated. Two bipolar stimulating electrodes were inserted in the stratum radiatum, one at the border of fields CA1 and CA3, the other in the CA1 field approximately 3 mm medial to the first electrode (Fig. 7.2). The recording micropipette was then placed into the stratum radiatum approximately 100 μm below the pyramidal cell body layer at a point equidistant from the two stimulating electrodes, and lowered into the slice until the maximum synaptic response to the two electrodes was obtained. At 5–10 minutes after the electrodes were positioned, baseline responses to each stimulation electrode were collected for 5 minutes. Stimulation voltages were set so as to produce in the dendritic zone a control monophasic negative response of about 2 mV maximum (baseline to peak); this response was monophasic positive in the pyramidal cell layer and was below the range required to produce a population spike in that layer. Baseline stimulation, consisting of four pulses delivered at 0.2 s^{-1}, was carried out for each electrode once per minute. If the responses proved stable over the 5 minute test period, then the slice was given either a series of three high-frequency

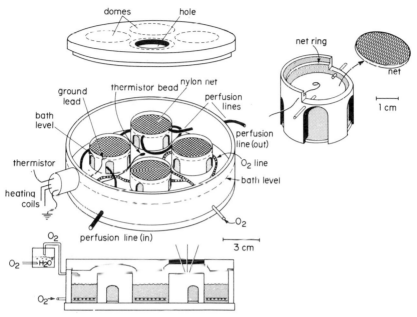

Fig. 7.1. Schematic view of the *in vitro* slice chamber apparatus. The chamber contains four wells each of which has attached perfusion lines. The wells are illuminated from below and the slices visualized with a dissecting microscope. The chamber is filled with water to about the level indicated and is heated to 34°C by submerged heating coils (not illustrated). O_2 is bubbled through the water as well as directly into the overlying atmosphere as shown in the scheme. Nylon is stretched across both surfaces of a thin plastic ring which snaps into a groove on the inner surface of the well (upper right). The upper net supports the slices while the lower net prevents air bubbles, which occasionally enter the perfusion lines, from reaching the slices. The medium level in the wells reaches but does not cover the surface of the slices in " static" experiments, whereas in perfusion studies it reaches the lips of the well approximately 2 mm above the slice. The chamber is covered with a cap containing a hole which can be rotated over the well to be used. It is wise to keep the size of the opening to a minimum in static experiments. The medium is composed of 124 mM NaCl, 4.9 mM KCl, 1.2 mM KH_2PO_4, 1.3 mM $MgSO_4$, 2.5 mM $CaCl_2$, 25.6 mM $NaHCO_3$, and 10.0 mM glucose. In perfusion experiments the medium is continuously oxygenated before it enters the chamber.

bursts (200 s^{-1} for 0.5 s) or a 1 minute train of low frequency (1 s^{-1}) pulses. In order to stimulate the greatest possible number of axons, and hence synapses, the duration of the stimulation pulses was increased by a factor of 10 for the high-frequency bursts. This procedure was carried out for each stimulation electrode in succession. Following this, control stimulation was resumed for 10 minutes and if the slice exhibited stable responses for this period it was removed from the chamber and immediately floated on to cold electron microscopic fixative for 15 seconds and then totally submerged. Before removing the slice from the recording chamber, the still-evident stimulating electrode sites were marked using a broken micropipette filled with Fast Green dye. The procedure was then repeated for the second member of the pair which received the form of stimulation (high or low frequency) that was not used on the first slice. The order of high- or low-frequency application was varied across pairs of slices. The material was then coded and given to the electron microscopist for analysis.

C. Electron microscopy

Fig. 7.2 provides details about the composition of the fixative as well as the procedures followed in preparing appropriate tissue blocks.

Approximately 30 minutes after the slice had been submerged, the zone of the regio superior located between the two recording electrodes was carefully dissected from the remainder of the slice (Fig. 7.2, top right). It must be emphasized that at this point in the procedure the material had been coded by the physiologist, the person performing the dissection did not know the identity of the slice. These blocks remained in fixative overnight and were then treated with osmium tetroxide followed by uranyl acetate and finally dehydrated and embedded in epon araldite. Sections parallel to the orientation of the dendrites of the CA1 cells and equidistant from the stimulating electrode sites (i.e., the cut edges of the block) were collected—these stretched from the cell layer to the fissure and included all of the 500 μm thickness of the slice. These were then stained briefly with lead citrate and uranyl acetate. The core of the section (essentially equidistant from the superficial and deep surfaces of the slice) at a distance 100 μm below the stratum pyramidale was then photographed at 6600 $\times$ and printed at 3·5 times. Micrographs containing severe pathology (degenerating dendrites, swollen glia cells) were not used and numerical analyses were confined to those prints exhibiting normal appearing tissue. Approximately 2600 μm^2 were photographed for each slice. Counts were made of the following: (1) synapses on dendritic spines; (2) synapses located on the shafts of dendrites; (3) presynaptic elements contacting two or more postsynaptic targets (multiple synaptic boutons; MSBs); (4) postsynaptic densities broken by a small space. This last type of contact represents perforated synapses—that is, a disc-shaped postsynaptic specialization

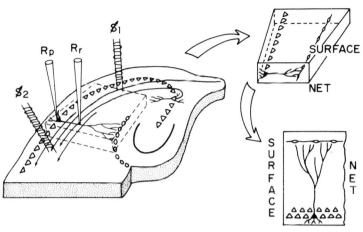

Fig. 7.2. An illustration of the steps followed in the process of preparing slices for electron microscopy. To the left, a slice is shown as it appears in the chamber with stimulating (s) and recording electrodes in place (Rp, stratum pyramidale recording electrode; Rr, stratum radiatum recording electrode). The dotted lines in the slice illustrate the orientation of tissue blocks embedded for electron microscopy. These blocks were cut from the slices (while still immersed in fixative) with the aid of a dissecting microscope. Sections were taken from the face of the blocks and extended from the stratum pyramidale to the obliterated hippocampal fissure (bottom right). The composition of the fixation solution was 0.6M phosphate buffer, 2% glutaraldehyde, 2% paraformaldehyde, 0.18% D-glucose pH 7.35 at 5°C.

containing an irregular hole (Peters *et al.*, 1970). It should be noted that certain of these categories are not exclusive, thus MSBs and perforated synapses were also counted as spine contacts.

Measurements were also made of the following spine characteristics: (1) the length of the postsynaptic densities (PSDs); (2) the total area of the spine as measured using a planimeter; (3) the width of the spine neck. All measurements were done in a blind fashion. Each slice was coded subsequent to the electrophysiological phase of the experiment and wasn't decoded until the collection of measurements was completed.

III. Results

While all experiments were conducted in pairs, in about 50% of the cases one or the other member of the pair could not be used for quantitative analyses because of inadequate fixation, unstable physiology, failure to obtain LTP after repetitive stimulation, and other reasons.

A. Physiology

Since the physiological characteristics of slices prepared and maintained according to the above procedures have been described in several publications and have been recently reviewed (Lynch and Schubert, 1980), our description here will be limited to variables that we suspect are diagnostic of the quality of the slices. Fig. 7.3 illustrates two properties exhibited by hippocampal synapses *in vivo* which we have found to be commonly disturbed in the slices. As shown, the threshold for the population spike is considerably elevated above that for the dendritic field potential; in the case illustrated, the monophasic negative response of the dendrites (and its dipole reflections in the cell body layer) reaches 3–4 mV before the spike appears. Second, paired-pulse stimulation produces a facilitation of the slope of the dendritic response and again its positive counterpart in the stratum pyramidale, but an inhibition of the population spikes, presumably due to the activation of recurrent interneurons. Although formal studies have not been conducted, it is our strong impression that failure to obtain these two aspects of hippocampal physiology indicates that the slice will not behave optimally in terms of thresholds for drug effects and in the maintenance of stable potentials.

The long-term potentiation effect is illustrated in Fig. 7.4. Note that very brief trains of relatively low-voltage stimulation are required to produce the facilitation and that once induced, LTP is nondecremental.

B. Electron microscopy

The quality of tissue preservation in the slices was somewhat variable both between slices and within a given slice. These differences were noticeable with both light and electron microscopic analyses. The superficial and deep aspects of the slice (i.e., those edges exposed to the atmosphere of the chamber and in contact with the supportive net, respectively), demonstrated a traumatic response to the tissue preparation. In these edge regions, distorted cellular elements such as degenerating dendrites, swollen glial cells, clumped synaptic vesicles and vacuolar spaces were seen (Fig. 7.5). Toward the core of the slice, however, the quality of the material improved dramatically (Fig. 7.6). Fig. 7.7 compares the structural preservation seen in the core region of a hippocampal slice maintained *in vitro* for 2.5 hours with that from

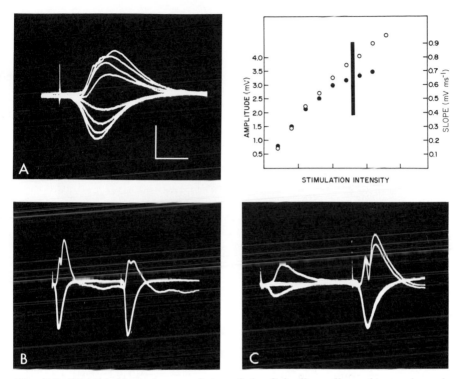

Fig. 7.3. Physiological characteristics of the Schaffer collateral/commissural inputs to CA1 in the hippocampal slice. In A and the adjacent graph (which shows the slope (o) and amplitude (●) of the dendritic response as a function of stimulation intensity), it can be seen that the threshold for the population spike (———) is well above that for the dendritically recorded evoked potential. The results of a typical paired-pulse experiment are presented in B. In this, two stimulation pulses are delivered 35 ms apart to the same pathway and responses are recorded simultaneously in the dendritic field and cell bodies. Note that the negative going population spike response, which is thought to be composed of the synchronous action potentials of several cells, is reduced whereas the slope and amplitude of the dendritic potential is increased. This inhibition is presumably due to recurrent inhibition. A second type of inhibition (presumably feed-forward) is illustrated in C. In this experiment, two separate pathways that innervate the same dendritic field were used (see Fig. 7.2). Low-voltage stimulation of one pathway causes a profound inhibition of the population spike elicited by subsequent higher voltage stimulation of the second pathway (see text and Lynch *et al.*, 1980, for a further discussion). The calibration bars represent 2 mV and 5 ms for A, 2 mV and 20 ms for B and 2 mV and 10 ms for C.

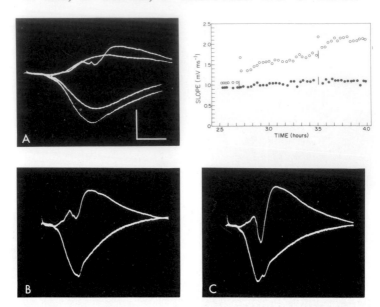

Fig. 7.4. Long-term potentiation in the hippocampal slice. This figure shows the results of a single slice experiment in which two trains of high-frequency stimulation were delivered. In A, the dendritic and cell body responses to single-pulse stimulation are shown before and 10 minutes after a single burst of $100\ s^{-1}$ for 1 s stimulation. In B, the responses are shown 40 minutes after the high-frequency stimulation and as is evident, they are still enhanced. In C, the responses are shown 20 minutes after a second high-frequency burst. The potentials are further facilitated with the population spike exhibiting a marked increase in amplitude. The accompanying graph displays the slope of the initial $850\ \mu s$ of the dendritic responses to the pathway receiving high frequency stimulation (o) and that of a control pathway (●) terminating in the same dendritic region which received only single-pulse stimulation. The short vertical lines indicate the times at which high-frequency stimulation was administered. Note the nondecremental and selective nature of the long-term potentiation effect. The time scale is in hours from the time at which the slices were placed in the chamber. Calibration bars represent 2 mV and 2 ms for A, and 2 mV and 5 ms for B and C. The paradigm followed in this example differs from that used for the electron microscopic studies. Both pathways received repetitive stimulation and the slices were removed 10 minutes later in the cases used for ultrastructural analyses.

material prepared utilizing intracardial perfusion; swelling of the boutons and spines as well as a loss of synaptic vesicles was evident in the slices.

Quantitative analyses of postsynaptic density length, spine area and

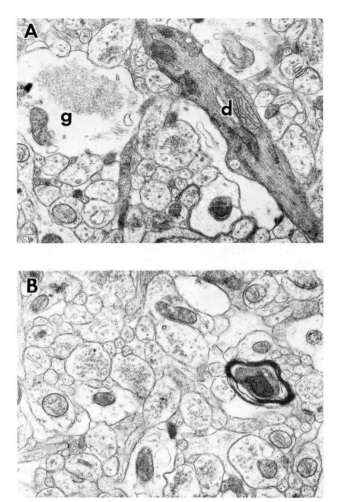

Fig. 7.5. Electron micrographs taken from the peripheral portion of a slice maintained for 3 hours *in vitro*. In A, a degenerating dendrite (d) and a swollen astroglial process (g) are prominent. In B, a degenerating axon, enclosed in its myelin sheath, is seen in a region containing several other typical presynaptic elements.

neck width also indicated that swelling had occurred in the slices. Table 7.1 provides a summary of the percentage differences between spine measures in the control group of the present experiments and those obtained from a set of 11 anesthetized rats prepared for electron microscopy using conventional methods. As is evident, all three spine

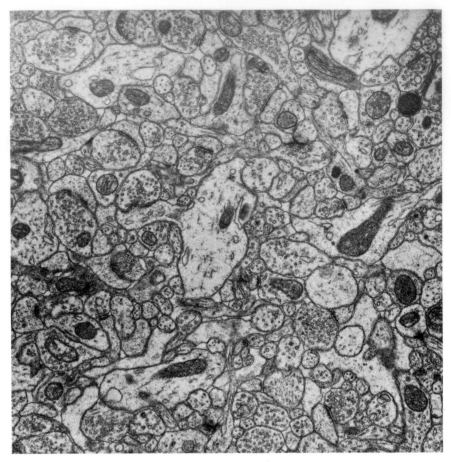

Fig. 7.6. An electron micrograph taken from the stratum radiatum of the core region of a slice. The quality of tissue preservation in this portion of the slices was far superior to that observed in the peripheral regions. Note the prominent dendritic spine protruding from the longitudinal section of a dendritic shaft (center). This spine can be seen making an assymetric synaptic contact with a presynaptic terminal bouton. Further, this contact demonstrates a perforated synapse. For a further discussion of slice ultrastructural quality, see text.

measures were significantly increased in the slices but note also that the effects were not uniform. In particular, the spine necks appear to have been particularly affected by the slice procedure; this comparison suggests that some specific physiological differences between *in vivo* and *in vitro* preparations may be due to disturbances of the normal morpho-

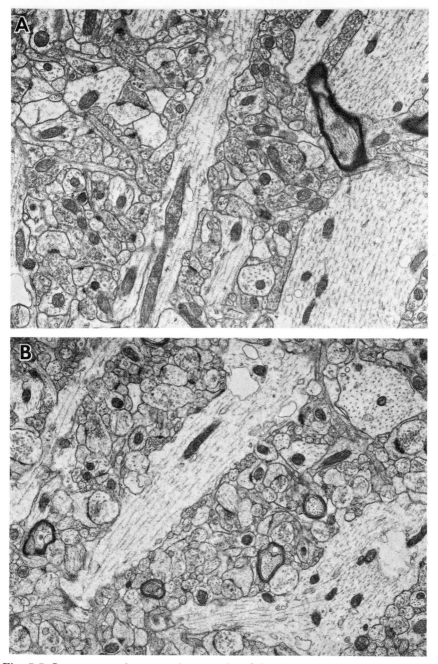

Fig. 7.7. Low-power electron micrographs of the stratum radiatum from CA1 in the rat hippocampus. In A, the material is taken from an animal perfused intracardially and prepared according to conventional electron microscopic techniques. In B, the micrograph was taken from the core region of a hippocampal slice prepared via immersion fixation. Qualitatively, tissue quality appears similar in both micrographs.

Table 7.1. A comparison of *in vivo* and *in vitro* synaptic measures. The synaptic counts are all expressed per 100 μm^2. The dendritic spine area values are given in μm^2 while both the length of spine PSDs and spine neck widths are expressed in μm. The *in vivo* measurements of PSD length, spine neck width and spine area were taken from the control group ($n = 11$) of a previous study of the long-term potentiation phenomenon (Lee *et al.*, 1980). Stimulating and recording electrodes were inserted into the hippocampi of these animals with tissue samples taken approximately 50–100 μm away from the recording electrode tract. Owing to slight differences in sampling techniques between this and subsequent studies (see Lee *et al.*, 1980), the values shown for synaptic counts are derived from intracardially perfused rats ($n = 13$) which did not have electrodes introduced into their brains but which were sampled in a fashion identical to that used in the slice study.

	Perfused animals ($\bar{x} \pm$ S.E.M.)	Immersed slices ($\bar{x} \pm$ S.E.M.)	% Change
Spine synapses	30.26 ± 0.69	27.98 ± 1.67	-7.5
Shaft synapses	0.71 ± 0.11	0.62 ± 0.10	-12.7
Perforated synapses	1.88 ± 0.12	1.59 ± 0.15	-15.4
Multiple synaptic boutons	1.16 ± 0.12	0.70 ± 0.14	-39.7
Spine area	0.115 ± 0.013	0.134 ± 0.003	$+16.5$
Spine PSDs	0.201 ± 0.018	0.213 ± 0.004	$+6.0$
Spine neck width	0.140 ± 0.023	0.217 ± 0.028	$+55.0$

logy as well as to generalized swelling. It also indicates that slices are more appropriate for the study of some anatomical variables than they are for others.

Comparisons of the spine data across the slices demonstrated that different degrees of swelling had occurred. The number of synapses per unit area was negatively correlated with both the length of the PSDs and the area of spines, an effect that would be anticipated if different amounts of generalized swelling had occurred. Beyond this, the standard deviations for spine area and postsynaptic density length were positively correlated with the means for these measures ($R + 0.79$ AND $+0.69$, respectively) across slices and again this would be expected from a generalized swelling of the slices. In three experiments we were able to compare two slices from the same group (control or potentiated) that had been in the chambers for different periods of time before fixation. In each of these cases, the slice that had the longer survival time after preparation had the larger mean spine area ($+3\%$, $+10\%$ and

+ 13%). Although this sample is small, it does suggest that the spines swell with increasing time in the chamber *in vitro*.

It was also clear from quantitative comparisons that certain slices had been more severely affected than others; in particular five of the six slices (two controls, three potentiated) from one experiment had very long PSDs, large spines and a low density of synapses. It was decided therefore that slices exhibiting mean PSD values more than 20% greater than that for mean PSD length of the *in vivo* spines would not be included in the data analysis. On the basis of this criterion, six slices were removed from the study. From an analysis of the experimental protocols, there was no obvious reason for the occurrence of severe pathology in the one experiment.

Table 7.2 summarizes the incidence of various types of contacts found in the slices of the control ($n = 7$) and potentiated ($n = 8$) groups and as is evident there were no detectable differences in the numbers of spine, multiple or perforated synapses. As we reported previously (Lee *et al.*, 1979) the incidence of shaft synapses was substantially increased. This increase is significant for both the absolute number of shaft synapses as well as the shaft to spine ratio. The effects of repetitive stimulation on the areal measurements of the dendritic spine are presented in Table 7.3. The mean of the within-slice distribution of spine areas was comparable for the two groups but the median for this measure was larger in the potentiated slices. Furthermore, the degree of positive skew as well as the coefficient of variation for the spine areas were significantly lower for the slices that had received the high-frequency stimulation. (Since the

Table 7.2. Synaptic counts for control and potentiated slices. All synaptic counts are expressed per 100 μm^2 and values are means $\pm$ s.e.m. The shaft to spine ratio was derived for each slice and the value given is the group mean.

	Control	Potentiated
Spine synapses	27.98 ± 1.67	26.65 ± 1.24
Multiple synaptic boutons	0.70 ± 0.14	0.59 ± 0.07
Perforated synapses	1.59 ± 0.15	1.54 ± 0.13
Shaft synapses	0.62 ± 0.10	$0.98 \pm 0.06^*$
Shaft to spine ratio	2.21 ± 0.31	$3.69 \pm 0.26^{**}$

* $P < 0.02$ on the Mann Whitney U test (2-tail).
** $P < 0.005$.

Table 7.3. Mean, median, coefficient of variation, and skew of the within-slice distributions of the areal measurements of dendritic spines. Values given are group means ± s.e.m.

	Mean	Median	Coefficient of variation (%)	Skew
Control	0.134 ± 0.003	0.104	64.4	1.050
Potentiated	0.137 ± 0.004	0.114	58.4	0.834
P*	n.s.	< 0.03	< 0.02	< 0.01

* Significance measured by the Mann-Whitney U test (2-tail).

standard deviation was correlated with the mean of the spine sizes across slices, the coefficient of variation is the more appropriate measure for comparing variance between slices.) Thus, comparison of the group data indicate that high-frequency stimulation altered the spines such that the skew and variability of random areal measures were reduced while at the same time their median values were increased. Graphical comparisons for the four sets of matched pairs (see above p. 195) also indicated that potentiation was accompanied by a shift in the distribution of the spine areas (Fig. 7.8). Note that for each of these experiments the distribution for the slice that received high-frequency stimulation (and exhibited LTP) had a higher mode (and median) value than did the control, but that the mean for the distributions and the incidence of very large spine areas were not reliably affected. As anticipated from the group data, the skew of the distribution was reduced in the potentiated slices as was the coefficient of variation in three of the four pairs.

The skewdness of the distribution of within-slice measures of post-synaptic density lengths on the dendritic spines was also reduced in the potentiated group but, in contrast to our earlier study *in vivo*, the coefficient of variation was not significantly different between the two groups. The shaft postsynaptic densities did not exhibit the tendencies found for their counterparts on the dendritic spines. As shown in Table 7.4, the coefficient of variation and skew of the within-slice distributions for this measure were clearly not reduced and, if anything, tended to be greater in the potentiated slices.

IV. Discussion

A. Slice Ultrastructure
Qualitatively, the ultrastructural appearance of the core of the slices

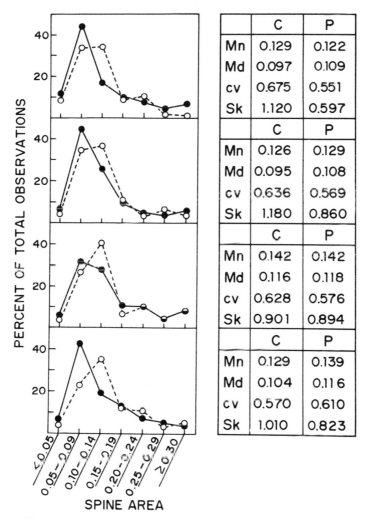

	C	P
Mn	0.129	0.122
Md	0.097	0.109
cv	0.675	0.551
Sk	1.120	0.597

	C	P
Mn	0.126	0.129
Md	0.095	0.108
cv	0.636	0.569
Sk	1.180	0.860

	C	P
Mn	0.142	0.142
Md	0.116	0.118
cv	0.628	0.576
Sk	0.901	0.894

	C	P
Mn	0.129	0.139
Md	0.104	0.116
cv	0.570	0.610
Sk	1.010	0.823

Fig. 7.8. The distributions and statistical values of area measurements are shown for four paired slices from the same animal. All slices were originally paired (one control (C) and one potentiated (P)), but because one member of the pair was lost in several cases due to (1) failure to obtain stable physiology, (2) poor tissue preservation, or (3) excessive swelling (as determined by the criteria described in the text), the data were ultimately analyzed with respect to groups. The abscissae are the spine area sizes (in μm^2) and the ordinates are the frequency (%) of total observations of spine area for a given slice. The comparison of mean values for control (●) and potentiated (○) slices showed no consistent pattern of differences. Median (Md) and mode values (not shown) tended to be greater while the coefficients of variation (cv) (in three of four of the pairs) and degree of skew (Sk) (in four of four pairs) were reduced in the potentiated slices.

Table 7.4. Mean, median, coefficient of variation and skew of the within-slice distributions of postsynaptic density length (in μm)

	Mean ($\pm$ s.e.m.)	Median	Coefficient of variation (%)	Skew
Spines				
Control	0.213 $\pm$ 0.004	0.198	35.5	0.615
Potentiated	0.216 $\pm$ 0.004	0.204	34.0	0.482*
Shafts				
Control	0.287 $\pm$ 0.008	0.282	33.2	0.192
Potentiated	0.304 $\pm$ 0.012	0.284	38.1	0.508

* $P < 0.04$ on the Mann-Whitney U test (2-tail).

proved to be surprisingly normal, the major disturbance being a generalized swelling of neuronal and glial elements as well as the occurrence of degenerating dendrites and terminals. It should be noted that areas of satisfactory neuropil were found surrounded by regions of gross disturbances and that some degree of selection was required in photographing the material. Quantitative analyses of the sizes of the various constituents of the neuropil also pointed to the occurrence of generalized swelling in the slices but provided the additional, and important, information that this had occurred to various degrees in different slices. This variability in swelling poses a serious problem for attempts to study the morphological effects of treatment conditions as within-group variability is sufficiently great to obscure important differences between groups. Information both about the origins of swelling and the reasons why certain slices are less affected than others is badly needed; in particular it would be helpful to know if the variability is due to conditions associated with the preparation and maintenance of the slices or instead reflects a fixation variable (or both). Until a more reliable preparation is achieved, it will be necessary to establish minimal criteria based on anatomical measurements (e.g., a percentage increase beyond *in vivo* values) for inclusion in quantitative comparisons of treatment groups and this was done in the present work. The measurements also produced the disturbing, but not unexpected, information that the slice procedure differentially influenced the size of the various components of the synaptic complex. Thus the length of PSDs and spine areas were for the majority of the slices increased by less than 20% over the value obtained with perfusion–fixation but the spine necks had increased by more than 50%. This could be due to differences between the effects of immersion fixation and those of more conventional procedures but it is perhaps more likely to be caused by the slice procedure itself. If this latter conclusion proves to be correct, then one might expect the slice to exhibit a number of physiological properties that are to some extent altered from those found *in vivo*.

B. The effects of high-frequency stimulation

The pattern of differences found in the comparisons of slices receiving brief bursts of high-frequency stimulation versus those given extended trains of low-frequency stimulation replicated that obtained by Lee *et al.* (1980) using acute, anesthetized rats. The frequency of spine, multiple and perforated synapses was not detectably different between the two groups in the present experiment, nor were these variables influenced by high-frequency stimulation in the previous work. However, the incidence of shaft synapses was substantially increased both in slices, as we reported in a preliminary note (Lee *et al.*, 1979), as well as *in vivo*.

The dendritic spines were also noticeably changed in the potentiated slices. Although the mean of the within-slice distributions of spine areas was not detectably changed by high-frequency stimulation, the median of the population was slightly larger. Furthermore, the coefficient of variation was significantly reduced in the experimental group; the skew of the population was also reduced and to a greater extent than was the coefficient of variation. The spine PSD lengths tended to show the same pattern of effects as did the spine areas but only the reduction in skew achieved statistical significance. In the *in vivo* experiment, we found that both coefficient of variation and skewedness were reduced for the spine PSDs.

The simplest explanation of the observed effects is that a sizeable sample of the dendritic spines were induced to change their shapes by the high-frequency stimulation. If some or all of the innervated spines were altered such that they assumed a rounder or more spherical shape, the variability of random measurements would decrease as would the degree of positive skew of those observations. The incidence of extreme measurements would also be expected to decrease if the PSDs and spines were assuming a rounder and more spherical shape respectively and this appeared to be the case at least for the positive end of the distribution. Fifkova and van Harreveld (1975) have reported that lengthy (30 second) trains of stimulation delivered to the entorhinal cortex induce a swelling of the dendritic spines in the outer molecular layer of the dentate gyrus (the target zone of the entorhinal cortex) but not in the inner molecular layer (which is innervated by non-entorhinal fibers). The pattern of changes that would be anticipated from swelling did not occur in our experiments; thus, swelling would increase mean, median and incidence of extreme observations equivalently (effects that did not occur) and certainly would not produce a reduction in skew and coefficient of variation (two effects that were obtained). It should be noted that we obtained the above described pattern of results using two radically different preparations (perfusion–fixation of anesthetized animals and immersion fixation of slices) and it appears that the swelling that goes on in the chamber over time was detected in one ultrastructural measurement.

The effects of repetitive, high-frequency stimulation on the shaft synapses were dramatically different from those just described for spine contacts. First and foremost, there was a marked increase in the frequency of this category of synapses in the potentiated slices. In addition, the within-slice distribution of shaft postsynaptic densities for the potentiated slices exhibited none of the changes noted for spine PSDs and if anything had greater variability and skew than the controls. These results, as well as those for numbers of spine, multiple and perforated synapses, indicate that the effects of stimulation were selective and could not have involved generalized changes in the neuropil. This agrees well with neurophysiological observations that the effects of potentiating stimulation are restricted to the activated afferents and that other inputs to the target dendritic zones are not affected (Dunwiddie and Lynch, 1978); prolonged trains of low-frequency stimulation (e.g., 15 s^{-1}) do produce more generalized effects (see Lynch et al., 1978; Dunwiddie and Lynch, 1978).

The present results, combined with those of our previous investigations (Lee et al., 1979, 1980), indicate that brief bursts of high-frequency stimulation produce two very different types of effects, (1) an apparent increase in the number of shaft synapses and (2) a change in the shape of the dendritic spines. The spine changes represent a reasonable candidate for the substratum of the LTP effect. A change in the shape of the postsynaptic densities could alter the relationship of spine receptors to the presynaptic element (and possibly increase the area subsumed by the apposition site); although a change in the shape of the spine itself might well reduce the resistance between the site of transmission and the dendritic shaft. On the other hand, it seems less likely that the shaft synapse effect is totally responsible for the increase in postsynaptic field potential and pyramidal cell EPSPs (excitatory postsynaptic potentials) which constitute LTP. First, although the increase represents a substantial increment to the shaft synapses in absolute terms, it is a very small fraction of the total synaptic population of field CA1 (shaft contacts represent only 2–3% of all synapses in the apical dendritic zone). Second, shaft synapses are commonly found concentrated on long

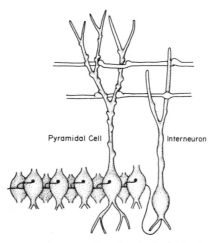

Fig. 7.9. This figure shows the postulated relationship between hippocampal pyramidal cells and one class of interneuron which sends its dendrites into the stratum radiatum in an alignment parallel to that of the pyramidal cell dendrites (see Lorente de no, 1934). This type of interneuron is hypothesized to provide a feed-forward input to the pyramidal cell layer. It is possible that these neurons are the postsynaptic cells on which the shaft synapses are seen to aggregate.

aspinous dendritic processes and it is possible that these are not part of the dendritic tree of the pyramidal cells. Interneurons with cell bodies in the stratum oriens send long straight processes into the stratum radiatum (Lorente de No., 1934) and these could well be the site of the clustered shaft contacts (Fig. 7.9).

It is too early to determine what the ultimate value of *in vitro* slices for neuroanatomical work will prove to be. The present findings demonstrate that qualitative *in vitro* studies are feasible and relatively simple; quantitative experiments of the type described above will require criterion for the rejection of unsuitable material and reliance on internal controls (e.g., expression of values as a percentage of other measures) and comparisons between pairs of slices. In all, however, we feel that these electron microscopic data are encouraging and in light of the obvious advantages of slices for a wide variety of experimental manipulations provide grounds for further efforts to develop the technique.

V. Conclusions

1. The technique is described for preparation and recording from rat hippocampal slices.

2. Two stimulating electrodes were inserted into the slice; one at the border of the CA1 and CA3 field, the other in the CA1 field 3 mm medial to the first electrode. The recording pipette was placed into the stratum radiatum approximately 100 μm above the pyramidal cell body layer at a point equidistant from the two stimulating electrodes.

3. Pairs of slices were used; one member of each pair received three 0.5 s trains of 200 s^{-1} stimulation, the other a single 1 minute train of 1 s^{-1} stimulation. Responses were assessed for 5 minutes before and 10 minutes after the repetitive trains with four pulses delivered at 0.2 s^{-1} at each electrode once per minute.

4. The hippocampal slices were fixed and examined by electron microscopy. Though there was some swelling in the slices, the core of the slice showed good histological and cytological detail.

5. Counts were taken in a "blind" fashion of the following variables: synapses on dendritic spines; synapses on dendritic shafts; presynaptic elements; perforated synapses.

6. Measurements were made of the length of the postsynaptic densities, total area of the spine and the width of the spine neck.

7. The results indicate that the brief bursts of high-frequency stimulation produce (a) an apparent increase in the number of shaft synapses and (b) a change in the shape of the dendritic spine.

8. The results agree with similar experiments carried out on the *in vivo* hippocampus and are discussed in relation to the synaptic responses.

Acknowledgements

This research was supported by NSF grant BNS 76-17370 and a Career Development Award NS-00043 to G.L.

References

Alger, G. and Teyler, T. (1976). Long-term and short-term plasticity in the CA1, CA3 and dentate regions of the rat hippocampal slice. *Brain Res.* **110,** 463–480.

Andersen, P., Sundberg, S. H., Sveen, O. and Wigstrom, H. (1977). Specific long-lasting potentiation of synaptic transmission in hippocampal slices. *Nature (Lond.)* **266,** 736–737.

Bliss, T. and Gardner-Medwin, A. (1973). Long-lasting potentiation of synaptic transmission in the dentate area of the unaesthetized rabbit following stimulation of the perforant path. *J. Physiol. (Lond.)* **232,** 357–374.

Bliss, T. and Lømo, T. (1973). Long-lasting potentiation of synaptic transmission in the dentate area of the anesthetized rabbit following stimulation of the perforant path. *J. Physiol. (Lond.)* **232**, 331–356.

Douglas, R. and Goddard, G. (1975). Long-term potentiation of the perforant path-granule cell synapse in the rat hippocampas. *Brain Res.* **86**, 205–215.

Dunwiddie, T. and Lynch, G. (1978). Long-term potentiation and depression of synaptic responses in the rat hippocampus: Localization and frequency dependency. *J. Physiol. (Lond.)* **276**, 353–367.

Dunwiddie, T. V. and Lynch, G. (1979). The relationship between extracellular calcium concentrations and the induction of hippocampal long-term potentiation. *Brain Res.* **169**, 103–110.

Dunwiddie, T., Madison, D. and Lynch, G. (1978). Synaptic transmission is required for initiation of long-term potentiation. *Brain Res.* **150**, 413–417.

Fifkova, E. and van Harrveld, A. (1975). Long-lasting morphological changes in dendritic spines of dentate granular cells following stimulation of the entorhinal area. *J. Neurocytology.* **6**, 211–230.

Lee, K., Oliver, M., Schottler, F., Creager, R. and Lynch, G. (1979). Ultrastructural effects of repetitive synaptic stimulation in the hippocampal slice preparation: A preliminary report. *Exp. Neurol.* **65**, 478–480.

Lee, K., Schottler, F., Oliver, M. and Lynch, G. (1980). Brief bursts of high frequency stimulation produce two types of structural change in rat hippocampus. *J. Neurophysiol.* **44**, 247–258.

Lorento de Nó, R. (1934). Studies on the structure of the cerebral cortex. II. Continuation of the study of the ammonic system. *J. Psychol. Neurol. (Leipzig)* **46**, 113–177.

Lynch, G. and Schubert, P. (1980). The use of *in vitro* brain slices for multidisciplinary studies of synaptic function. *Ann. Rev. Neurosci.* **3**, 1–22.

Lynch, G., Gall, C. and Dunwiddie, T. (1978). Neuroplasticity in the hippocampal formation. *In* "Maturation of the Nervous System." Progress in Brain Research, Vol. 48 (M. A. Corner, ed.), p. 113–128. Elsevier/North-Holland Biomedical Press, Amsterdam.

Lynch, G., Jensen, R., McGaugh, J., Davila, K. and Oliver, M. Effects of enkephalin, morphine, and naloxone on the electrical activity of the *in vitro* hippocampal slice preparation. *Exp. Neurol.* in press.

Peters, A., Palay, S. L. and Webster, H. de F. (1970) *In* "The Fine Structure of the Nervous System: The cells and their processes". p. 140. Harper and Row, New York.

Schwartzkroin, P. and Wester, K. (1975). Long-lasting facilitation of synaptic potentials following tetanization in the *in vitro* hippocampal slice. *Brain Res.* **89**, 107–119.

Teyler, T.J. (1976) Plasticity in the hippocampus: a model systems approach. *In* "Advances in Psychobiology: Neural Modes of Behavioral Plasticity", vol. III, (A. Riesen and R. F. Thompson, eds.), pp. 301–326: Wiley, New York.

Van Harreveld, A. and Fifkova, E. (1975). Swelling of dendritic spines in the fascia dentata after stimulation of the perforant path fibers as a mechanism of post-tetanic potentiation. *Exp. Neurol.* **49**, 736–749.

8

Characteristics of CA1 Cells in the Rat Hippocampus *In Vitro*. An Evaluation of a Silicon Chip Extracellular Microelectrode Array (SCEMA 9)

H. V. WHEAL

Department of Neurophysiology, School of Biochemical and Physiological Sciences, Southampton University, Southampton SO9 3TU, Hampshire, U.K.

I. Introduction

There have been many studies carried out on the hippocampus, but our understanding of the intrinsic organization of this structure is based on the histology carried out by Ramon Y Cajal (1893) and Lorente de Nó

(1934). Since that time further anatomical studies (Blackstad *et al.*, 1970) and electrophysiological studies have shown that the hippocampal cortex is arranged in parallel lamellae (Andersen *et al.*, 1971; Lømo, 1971). Contained within the laminar are the components of a trisynaptic excitatory pathway (Andersen *et al.*, 1966). The perforant path fibres from the entorhinal cortex terminate in a clearly delimited zone in the outer two-thirds of the apical dendrites of the dentate granule cells (Hjorth-Simonsen and Jeune, 1972). The mossy fibres, axons of the granule cells, pass through the dentate hilus and synapse with the proximal zone of the apical dendrites from the CA3 pyramidal cells (Blackstad *et al.*, 1970). In addition to contributing to the efferent fibres in the fimbria, axons from the CA3 pyramidal cells send (Schaffer) collaterals to synapse with dendrites of the CA1 pyramidal cells. The stratum oriens and radiatum contain afferent fibres from commisural and septal projections in addition to those from the Schaffer collateral (Swanson and Cowan, 1979; Laurberg, 1979).

The laminar organization and well-defined structure of the hippocampus, together with its accessibility make it a perfect candidate for electrophysiological investigation *in vitro* (Skrede and Westgaard, 1971; Yamamoto, 1972). The CA1 pyramidal cells in the slice can be synaptically and antidromically activated, can display excitatory and inhibitory postsynaptic potentials (EPSPs and IPSPs), depolarizing after potentials, and fast prepotentials (Schwartzkroin, 1977). All these intracellular parameters had previously been demonstrated in the intact animal by Kandel and Spencer (1961a, b, c) and Kandel *et al.* (1961).

Many of the intracellular investigations carried out on the *in vitro* or *in vivo* hippocampus have been from guinea pigs, rabbits or cats. This study describes some of the characteristics of CA1 neurons recorded intracellularly in the hippocampal *in vitro* slice preparation from the rat.

Extracellular recordings of population spikes have been made in the hippocampus in response to stimulation of afferent pathways. The population spike is the extracellular potential produced at the site of the recording electrode by the summation of the individual action potentials of many synchronously discharging cells. The population spike is one component of the total field potential that normally also contains slower components due to synaptic current flow (Andersen *et al.*, 1971). The amplitude of the population spike is controlled by several factors including the geometry of the cells and the number of discharging cells. Lømo (1971) has studied the activation of the dentate granule cells by stimulation of the perforant path.

The number of parallel orientated and synchronously discharging cells in the hippocampus and dentate make it possible to investigate the trisynaptic excitatory pathway by analysis of extracellular field potentials and population spikes in several areas simultaneously. However, to introduce two or three glass recording microelectrodes in addition to at least one stimulating electrode, by the use of individual micromanipulators is physically very difficult. This task would be even more difficult if simultaneous iontophoretic application or perfusion of drugs were required to investigate synaptic pharmacology in such a small piece of tissue.

Extracellular recordings suffer from additional problems related to the poor signal to noise ratio. This is due to the high resistance of glass microelectrodes coupled to buffer amplifiers that are some distance away from the electrode tip.

Over recent years, several electrode systems, utilizing microelectronic fabrication procedures, have been described which have attempted to overcome some of the limitations of conventional glass microelectrodes (Pickard, 1979). Both Bergveld (1972) and Wise *et al.* (1970) have tried to improve electrical performance by mounting buffer amplifiers on the substrate that supports the microelectrode tip. Applications of these recording electrodes have been in several different fields; cat cortex (Wise *et al.*, 1970); cat inferior colliculus (Mercer and White, 1978); peripheral nerve (Loeb *et al.*, 1977). Photofabrication techniques have been used by several workers to generate multiple electrode arrays on both flexible and rigid substrates. This approach allows simultaneous recording from several tissue–electrode interfaces (Thomas *et al.*, 1972). Wise and Angell (1975) have combined improvements to the electrical performance of the electrodes as well as having two recording sites available in one device.

This chapter describes the development and evaluation of an active multiple electrode array which combines the advantages of having nine electrode sites with those of having a buffer amplifier at each electrode. Comparisons are made with extracellular potentials recorded with conventional glass microelectrodes. The array was specifically designed for extracellular recording from slices of rat hippocampus. The electrodes and buffer amplifiers were made on a silicon substrate using the technology of planar integrated circuit processing. The method exploites the fact that metal oxide semiconductor transistors (MOST) may be made comparable in size to hippocampal pyramidal cells. Each electrode is part of the gate of an underlying metal oxide semiconductor transistor which is used in the source follower configuration as a buffer amplifier.

II. Methods

The animals used in this study, male Wistar rats (200 g) were first
anaesthetized with halothane and then decapitated using a guillotine
supplied by Luckham Ltd. The brain was quickly removed and placed
in chilled 4–6°C artificial cerebrospinal fluid (ACSF) containing
124mM NaCl, 3.3mM KCl, 1.2mM KH_2PO_4, 0.5mM $MgSO_4$, 3.0mM
$CaCl_2$, 24mM $NaHCO_3$, 15mM D-glucose, pH 7.4, perfused with a
gaseous mixture of 95% O_2/5% CO_2. After 1–2 minutes, the brain was
removed from the media, placed into a petri dish and cut into two along
the saggital line using a sharpened plastic knife. The hippocampus was
then dissected free from each hemisphere using plastic instruments, and
placed on a double thickness of filter paper moistened with cold oxygenated media on a McIlwain tissue chopper. The tissue chopper was
operated manually with the spring tension set to maximum. The tissue
was cut, using a thin stainless-steel razor blade, transversely to the
longitudinal section, in 300–400μm slices. Each slice was placed in cold
oxygenated medium until they were collectively transferred to the
experimental bath.

A. Use of experimental bath

Two different types of experimental bath were tested in this study. The
first bath consisted of a circular slice chamber about 2cm in diameter
with a media volume of 2ml. The slice chamber contains a fine nylon
gauze to support the tissue and is surrounded by an outer heated-water
bath into which was bubbled O_2/CO_2 mixture. The perfusion of the
media (5ml min^{-1}) was controlled by gravity and a 'Dial a flow'
regulator. The media supply tube to the slice chamber passed through
the water bath to allow temperature equilibration of the media. A
thermistor in the slice chamber controls the voltage supplied to two
plastic power transistors which, bonded to a stainless-steel doughnut,
dissipate heat to the outer water bath (Fig. 8.1).

The second experimental tissue bath used in this study is rectangular
in section and functions similarly to the first bath. The main difference is
that instead of a central circular slice chamber, the chamber is open on
one side as shown in Fig. 8.2. This configuration allows greater accessibility to the slices and is simpler to manufacture. The bath was heated
using a closed system with the heat exchanger in the bath, connected to a
circulating water bath. The bath was a modification of the one described
by Haas et al. (1979).

The hippocampal slices were placed on a double layer of lens tissue for

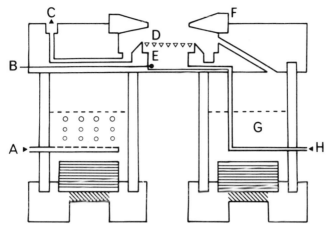

Fig. 8.1. Cross section through the circular experimental bath. A, 95% O_2/5% CO_2 inlet; B, temperature-sensing device; C, media is extracted from the bath using suction; D, nylon mesh covered with lens tissue to support slices; E, silver/silver chloride earth incorporated with thermistor (B); F, removable lid; G, distilled water; H, inlet for ACSF. Horizontal shading indicates the heater device, a stainless-steel doughnut attached to two plastic power transistors (diagonal shading).

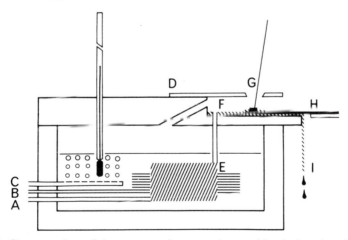

Fig. 8.2. Cross-section of the rectangular experimental bath showing the horizontal side access for the recording device (H). A, Inlet for ACSF; B, circulating hot water; C, 95% O_2/5% CO_2 inlet; D, perspex cover; E, heat exchangers in distilled water; F, inlet for warmed and humidified gas and inlet for warmed ACSF; G, bipolar stimulating electrode penetrating the hippocampal slice; H, active microelectrode array; I, flow of the ACSF over the lens tissue is controlled by the tilting of the bath.

stability and perfused at 1–5 ml min^{-1} with oxygenated medium at 33–37°C. The perfusion rate and level of media in the bath was adjusted until there was a slight concave meniscus between the slices of tissue. Warmed, humidified gas (95% O_2/5% CO_2) passed through the outer water bath provides a suitable atmosphere around the tissue.

B. Stimulating and recording

The conventional glass electrodes used in this investigation were pulled on a Narashige ventrical electrode puller, from 1mm internal diameter prefibred glass, to resistances of 5–10MΩ for extracellular and 50–80 MΩ for intracellular recording. The electrodes were filled with 4M NaCl and 2M potassium acetate respectively. Intracellular signals were led to a negative capacitance amplifier and connected to a bridge circuit for passing current. Hyperpolarizing pulses (100ms, 0.1–1nA) were injected into cells to measure cell input resistance. All responses were recorded on a Racal tape recorder and displayed and photographed on a Tektronix oscilloscope. Extracellular recording was carried out with × 1000 AC amplification filtered below 10Hz and above 10KHz. Pathways were stimulated using a bipolar, twisted, 64μm Ni/Cr wire which was Trimel insulated (Johnson Matthey Metals). A portion (50μm) of the insulation was removed from the tips with the aid of a microscope and scalpel blade. The recording and stimulating electrodes were positioned using Prior micromanipulators with reduction drive.

C. Construction of the microelectrode array

A 3 × 3 array of recording sites with external dimensions of 2 × 2 mm was chosen as this was thought to be a suitable fit for the geometry of the hippocampal slice. Fig. 8.3 shows the relationship between the electrode array and a diagrammatic representation of the slice. The electrode contact areas, 20μm square, were made from gold which was vacuum-deposited on top of normal aluminium gate electrode.

The integrated buffer amplifier design depended on the 40MΩ input impedence presented to it by the gold electrodes. A buffer amplifier with low output impedence was required at the electrode site so that electro-static pick-up, and capacitative loading on the connection to the microelectrodes, would not impair the signal/noise ratio or attenuate the electrical signals detected at the microelectrode. An output impedence of 10KΩ was chosen which would give a maximum pick up noise signal of 50μV.

The electrodes were made, using photolithographic techniques, nine

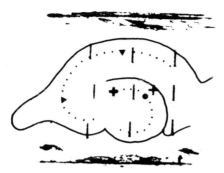

Fig. 8.3. Diagrammatic representation of a hippocampal slice superimposed on a photograph of the array of 9 electrodes on the end of the recording chip. The electrodes are shown as a line (∣). The pyramidal (▼) and granule cell (●) layers are outlined. Each electrode contact area is 20μm square, the remaining apparent electrodes area having been insulated.

arrays at a time on 2-inch diameter circular silicon wafers. The fabrication of the MOST buffer amplifiers and diffused connections to them was based on the standard p-channel MOST process in use at the integrated circuit facility in the Department of Electronics at the University of Southampton. A more detailed description of the device is given by Jobling *et al.* (1980).

The electrodes and buffer amplifiers were at the end of a silicon substrate which was mounted to a larger glass substrate, as shown in Fig. 8.4. Electrical connections between the conductors on the two substrates were made using gold wires thermosonically bonded to metal pads on each substrate and then covered with epoxy. The two-substrate construction was used to minimize the physical size of the device in the tissue preparation bath (see Fig. 8.2). Parts of the device were insulated with polyurethane varnish.

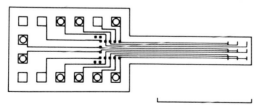

Fig. 8.4. Diagrammatic representation of the silicon substrate (on the right), incorporating the nine electrodes and buffer amplifiers, mounted on to a glass substrate (on the left). Electrical connection between the conductors on the two substrates were made using gold wires thermosonically bonded to metal pads on each substrate and then covered with epoxy. Scale marker is 10mm.

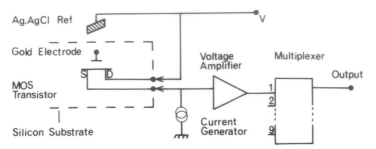

Fig. 8.5. A block diagram of the relationship between one 'on-chip' electrode and 'off-chip' voltage amplifier, current generator and multiplexer.

Off-chip ancillary circuits include a current generator to drive the buffer amplifiers; a compatible amplifier with voltage gain of 450 and a bandwidth of 10Hz–10KHz. A 9 into 1 multiplexer was built to allow the display of all nine channels simultaneously on one beam of an oscilloscope. A block diagram of the chip with ancillaries is shown in Fig. 8.5.

III. Results

A. Intracellular studies on CA1

Intracellular recordings were carried out from 77 cells in the pyramidal cells layer of the CA1 region of the hippocampus. Some 62 of these cells showed characteristic responses associated with CA1 pyramidal cells when they were orthodromically activated by stimulation of the Schaffer/commissural afferents in the stratum radiatum of CA3. About half of these cells were antidromically activated by stimulation of the alvear pathways. On injecting hyperpolarizing current (0.1–1.0nA) for 100 ms, current/voltage plots were made for each of the 62 CA1 pyramidal cells. The range of the input resistance of these cells was 16–47MΩ with a mean of 32.8 ± 10.8MΩ (s.d.). The resting potential of each cell was measured on penetration and on removal of the electrode from the cell. The bridge balance was likewise tested before penetration and checked after the electrode was removed from the cell. The resting potentials of the population of 62 CA1 cells ranged from 40–91mV with a mean at 66.7 ± 12.0mV. The average amplitude of evoked action potentials from these cells was 68.7 ± 12.4mV taken from a range of 50–110mV.

Orthodromic, monosynaptic activation of the CA1 pyramidal cells could be evoked by stimulation of (i) the Schaffer/commissural afferents in the stratum radiatum, (ii) the afferents in the alveus and stratum oriens that arrive via the fimbria, (iii) the alvear pathway from the entorhinal region or (iv) the perforant path or other afferents in the stratum radiatum close to the entorhinal region. All these areas required stimulation with 0.5ms pulses between 10 and 15V to activate the CA1 cells. Latencies of the monosynaptic activation varied between 6 and 9ms, as shown in Fig. 8.6. Polysynaptic activation as shown in Fig. 8.6 was observed when stimulating the mossy fibres in the dentate hilar region. Increased intensity of stimulation frequently yielded a burst of action potentials superimposed on an enlarged excitatory postsynaptic potential (EPSP). The EPSPs were between 20 and 70ms in duration and up to 20mV in amplitude. The spike duration in these experiments was 1.6–3ms. One particularly interesting observation is that on five occasions stimulation of the fimbria/alvear afferents evoked an antidromic spike in the CA1 cell. One of these records is shown in Fig. 8.6.

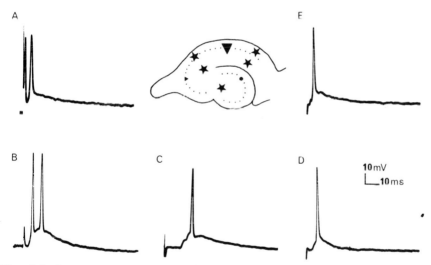

Fig. 8.6. Intracellular recordings from CA1 pyramidal cells (RP −64 to −80mV) (▲) showing EPSPs and action potentials evoked from stimulation (10–14V, 0.5ms) at different sites of the hippocampal slice. The recording sites (★) will be described from left to right. A, Antidromic and orthodromic activation from stimulation of alvear/fimbria afferents. B, Orthodromic activation from stimulating Schaffer/commissural afferents in the stratum radiatum of CA3. C, Long latency (17ms) orthodromic activation from stimulation of the mossy fibres in the dentate hilar region. D, Orthodromic activation from stimulation of the stratum radiatum proximal to the entorhinal region. E, Orthodromic activation from stimulation of the alvear pathway.

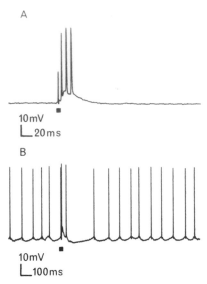

Fig. 8.7. Intracellular recordings from a CA1 pyramidal cell (RP − 82 mV). A, Orthodromic activation from stimulation (15V, 0.5ms) of fimbrial/alvear afferents. EPSP duration of 55ms. B, Inhibition evoked from stimulation of the same pathway. IPSP duration of 200 ms.

In addition to the excitatory phenomena, evoked inhibitory postsynaptic potentials (IPSPs), 5–10mV in amplitude and 200–500ms in duration, could be observed, (Fig. 8.7). On three occasions, rapid firing (> 150Hz) was seen on penetrating cells. The action potentials in these cells were less than 1ms in duration. Intracellular penetrations of CA1 cells were obtained using both types of experimental bath. Penetrations were maintained for up to 4 hours with flow rates < 5ml min⁻¹. The preparation normally remained active for 8–12 hours.

B. Extracellular recordings

Extracellular recordings were carried out using the multiple-electrode array as well as with conventional glass microelectrodes. Extracellular recordings could be made from a single slice for several hours with Ringer flow rates of 5 ml min⁻¹. Fig. 8.8 shows the multiplexed potentials obtained from the nine electrode sites on stimulation of the mossy fibre system. Antidromic potentials, 2ms in latency, were picked up from electrode sites 2 and 5 which coincided with the granule cell layer. The potentials at these electrodes were negative and between 1–1.8mV in amplitude (allowing for × 450 buffer amplifier) and 2ms in duration.

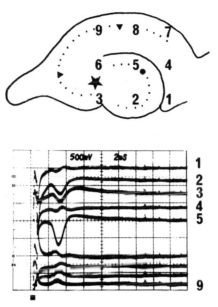

Fig. 8.8. Multiplexed extracellular potentials recorded from all 9 electrode sites after stimulation (12V, 0.5ms) of the mossy fibre system. The position of the electrode sites (★) in relation to the geometry of the slice is shown in the upper diagram. Negative (1.0–1.8mV) antidromic potentials, 2ms in latency can be seen in traces 2 and 5. Electrode sites 2 and 5 coincide with the granule cell layer (●). (Allowing for × 450 buffer amplification, 1 large division is approximately 1 mV.)

The largest potential was recorded on the upper blade of the dentate gyrus at electrode site 5. The small negative wave recorded at site 3 may be due to the antidromic activation of pyramidal cells, or of granule cells on the end of the lower blade of the dentate. It is of particular interest to note that little or no potential was observed from electrode site 6 despite its proximity to the stimulating electrode or to its neighbour electrode 5.

More complex orthodromic responses were recorded from the CA1 pyramidal cell region when the Schaffer/commissural afferents to that region were stimulated (10V, 0.5ms; Fig. 8.9, traces 1, 4 and 7). The potentials recorded from electrodes 1, 4 and 7 were selected for study at a higher amplification and with a larger stimulus (14V 0.5ms; Fig. 8.9).

In order to photograph the multiplexed oscilloscope beam, it is necessary to superimpose 5 or more traces especially when a high gain is being used.

Trace 1 (Fig. 8.9) was recorded from the CA1 pyramidal cell layer, and shows a 2ms duration negative wave with a latency of 4ms. It is thought that this represents a small orthodromically activated population spike. Trace 4 shows a larger and longer (6ms) negative wave recorded from the upper stratum radiatum. Trace 7, recorded from the strata riens CA1 even more distal to the recording electrode, shows the longest latency of 7ms, and a slow positive wave. The slow wave from electrode 7 is a mirror image of the slow negative wave from electrode 6. Both these potentials are thought to be orthodromically activated field potentials and will be discussed later in this chapter.

Extracellular recordings using conventional glass electrode were also carried out at the sites just described (Fig. 8.9). For the same voltage stimulation, the potentials recorded with glass electrodes were half the amplitude of those recorded using the silicon device. In virtually all other respects, the shape of the traces were identical.

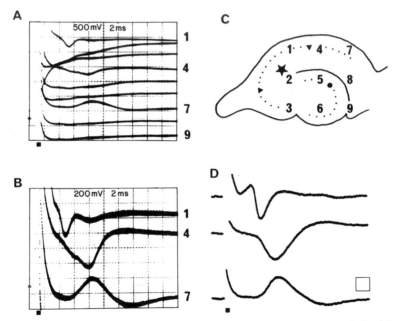

Fig. 8.9. Orthodromically activated extracellular potentials recorded with the active microelectrode array (A, B) in comparison with those from a conventional glass electrode (C). A, Schematic showing the site of stimulation (10–14V, ★ 0.5 ms) and position of the recording electrode sites. B, Recordings from all nine electrodes. B, Recording from sites 1, 4 and 7 which correspond with CA1 pyramidal cell region (▾) shown at higher amplification (negative down). D, Extracellular potentials recorded using a conventional glass electrode positioned at sites 1, 4 and 7 in the CA1 pyramidal cell region. Scale marker: 0.2 mV, 2.0 ms.

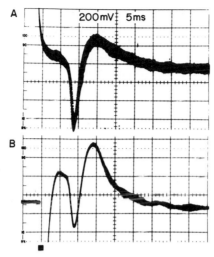

Fig. 8.10. Population spikes and field potentials recorded from the CA1 pyramidal cell layer, evoked by stimulation (■ 14V, 0.5 ms) of Schaffer collateral and commissural afferents in the stratum radiatum. A, Extracellular potential (× 450) picked up from one recording site of the microelectrodes array. B, Extracellular potential (× 1000) recorded using a 5MΩ conventional glass microelectrode.

The response to stimulation of the Schaffer/commisural afferents with the electrode sited on the CA1 pyramidal cells layer is seen in Fig. 8.10. This is a classically shaped orthodromically activated population spike with a late, slow positive potential. This positive potential is enhanced in the records using a glass electrode in the same location. Again, the extracellular potential recorded with the conventional glass electrode is less than half the amplitude of that seen with the electrode array.

IV. Discussion

The characteristics of the rat CA1 hippocampal cells in this investigation are comparable to those of CA1 cells in guinea pig and rabbit (Schwartzkroin, 1975, 1977). The spike amplitudes in all three species are identical. However, the input resistance of CA1 cells from the rat was almost twice that found in guinea pig and rabbit as shown in Table 8.1. *In vitro* studies by Assaf and Kelly (1979) on the granule cells of the rat dentate gyrus show even higher input resistances (80MΩ) than those measured from CA1 cells (33MΩ). The resting membrane potentials recorded in the rat (-66mV) were slightly higher than in other observations, although Kandel et al. (1961) reported resting potentials of -62mV in identified CA2 and CA3 pyramidal cells recorded in the cat hippocampus in vivo. This more negative resting potential may be one explanation for the small (5–10mV) IPSPs recorded in the rat compared with 20mV IPSPs reported by Andersen et al., (1964) in the cat (RP – 50mV). The difference between the inhibition seen in the slice and intact hippocampus have been discussed by Schwartzkroin (1975) who also noticed a lesser degree of inhibition in the in vitro preparation.

Since Raisman's (1966) original anatomical description of the hippocampal projection to the septum there has been considerable discussion on the route of axons from the CA1 pyramidal cells. Raisman (1966) described fibres from the anterior portion of the field CA1 which project through the dorsal fornix, and fibres from the posterior portion of field CA1 and the remaining CA fields which project through the fimbria to the ventral medial, lateral septal nuclei. Andersen et al. (1973) did not find any electrophysiological evidence for this latter projection. There was no antidromic activation of CA1 pyramidal cells on stimulating the fimbria in rabbits. Electrophysiological studies by DeFrance et al. (1973) suggested that the posterior part of the hippocampal region CA1 contributes fibres to the fimbria as well as the fornix.

Stimulation of the fimbria/alvear afferents in the hippocampal slice, usually evokes orthodromic activation of CA1. When using a slightly higher stimulus (14V for 0.5ms instead of 10V), antidromic activation of the CA1 cells was observed in 5 out of the 50 slices tested. In these cases, the site of stimulation was at least 1mm from the activated cell, thus direct stimulation of the basal dendrites seems unlikely. This

Table 8.1. Comparison of CA1 hippocampal cell characteristics measured in vitro.

Animal	Membrane potential (mV)	Input resistance (MΩ)	Spike amplitude (mV)
Rat ($n=62$)	66.6 ± 12.0	32.8 ± 10.8	68.7 ± 12.4
Guinea pig ($n=47$)	54.3 ± 11.7	16.3 ± 5.3	64.9 ± 10.0
Rabbit ($n=62$)	>50	18.1 ± 7.5	68.0 ± 8.2

Values for guinea pig and rabbit were taken from Schwartzkroin (1975, 1977).

observation provides more evidence of a species difference in the efferent pathways from CA1. Recent autoradiographic and HRP histochemical analysis of the rat hippocampus (Meibach and Siegal, 1977; Swanson and Cowan, 1977, 1979; Chronister and De France, 1979) showed that some CA1 cells project their axons out of the hippocampus through the fimbria.

The shape of extracellular potentials recorded from the hippocampus and dentate *in vitro* using the active microelectrode array compare well with those of Andersen *et al.* (1971) and Lømo (1971). Fig. 8.8 shows a monophasic negative-going spike recorded from the granule cell layer in response to antidromic activation via the mossy fibres. The recording of an orthodromically activated field potential and population spike from the CA1 pyramidal cell layer seemed to be missing an early positive component. However, the negative population spike and late positive component show up well. The late positive component has been shown to be correlated with intracellularly recorded IPSPs (Lømo, 1968).

There was good correlation shown between the records from the active microelectrode array and conventional glass microelectrodes for extracellular potentials. This is well illustrated in Fig. 8.9. The ability to record simultaneously from several sites with a rapid display of the potentials from those sites is a major advantage over using conventional microelectrodes. The latter would require repeated individual positioning of the microelectrode with micromanipulators. The use of the array would be suitable for extracellular studies on potentiation (Alger and Teyler, 1976).

In the future it will be possible to accommodate more electrodes into one array enabling a more precise choice of recording site to be made. For example, it would be useful to record from the pyramidal cell layer as well as the apical dendrite field and to monitor population spikes simultaneously with the dendritic EPSP.

Potentials picked up by the arrays were not greatly affected by electrostatic interference. The recordings have not been significantly less noisy than those obtained with conventional glass electrodes, but the signal/noise ratio was much improved, principally as the integrated circuit electrode was twice as sensitive as a $5M\Omega$ glass electrode. One minor inconvenience was that the device is light-sensitive. This was overcome by reducing the background lighting and using a fibreoptic lamp operated by a foot switch to illuminate the preparation.

A major advantage of the device is its reliability. Of a batch of 10 devices, none have shown any deterioration in performance during many hours of experiments. Continuous running of the device in ACSF media for 150 hours did not cause any degeneration.

Of the two types of experimental baths used for maintaining the tissue, the traditional circular bath, as used by several authors in this book, was preferred for intracellular recording. However, the circular bath was found to be unsatisfactory for use with the electrode array. It may be possible in the future to build the array into any experimental bath.

However, the rectangular bath has many advantages, including good accessibility to the tissue slice and simple cheap construction. The rectangular bath is also able to accommodate high (> 5 ml min^{-1}) flow rates which are important when perfusing drugs. Although the electrode array functioned well with flow rates of > 5 ml min^{-1}, its surface is planar. One development that may improve the stability of the tissue on the recording array would be to give the array a three-dimensional aspect. This could best be accomplished by having protruding recording sites which would penetrate about 50μm into the tissue. The contacts would then penetrate through the damaged tissue on the surface of the slice and give an even better signal/noise ratio.

V. Conclusions

1. The rat hippocampus was cut into 300–400μm slices and kept alive for up to 12 hours *in vitro*. Details of the experimental procedure and the bath are provided.

2. Intracellular recordings were made from the CA1 cells using 50–80MΩ glass microelectrodes. The resting potential was 66.2 ± 12mV, action potentials 68.7 ± 12.4mV and input resistance 32 ± 10.8MΩ (mean $\pm$ s.d., $n = 62$).

3. The CA1 pyramidal cells could be monosynaptically driven in the slices by stimulating; (a) Schaffer/commisural afferents in the stratum radiatum; (b) afferents in alveus and stratum oriens from the fimbria; (c) alvear pathway from entorhinal region; (d) perforant path. The latencies varied between 6–9ms. EPSP, IPSP and action potentials could be evoked.

4. Polysynaptic activation of the CA1 cells was obtained on stimulation of the dentate hilar region. The latency of the response was 17 ms.

5. Some CA1 cells could be antidromically activated by stimulation of the alveus and stratum oriens of the CA2/CA3 region of the slice. This provides evidence that, in the rat, some CA1 cells project their axons out through the fimbria.

6. The performance of a 9-electrode MOST system is described. The 9 electrodes were in an array 2 × 2mm on a silicon substrate; each electrode surface area was 20μm.

7. The device incorporates 9 buffer amplifiers, each lead to an off-chip amplifier and then to a nine into one multiplexer. This allowed all 9 channels to be displayed on one beam of an oscilloscope.

8. The isolated hippocampal slice was placed over the microelectrode array and the pathways in the slice were stimulated. The evoked potentials were recorded at the nine sites and compared with the extracellular potentials recorded by a glass microelectrode placed successively at these sites.

9. There was good correlation between the extracellular potentials picked up by the 9-electrode array and that picked up by the glass microelectrode. The active microelectrode array had a better signal/noise ratio due to its inbuilt buffer amplifiers.

10. This multimicroelectrode array should be of value in studying the potentials elicited at various sites in the preparation following afferent stimulation and help in the study of drug and antagonist action at synaptic sites. It could be further developed to contain more electrode sites (16, 25, 36. . . . .100) and allow greater resolution of the potential changes occurring at specific sites in the isolated hippocampal slice.

Acknowledgements

I should like to acknowledge the financial support given by the Wellcome Trust. The recording device was made in collaboration with Dr. D. T. Jobling and Mr. J. G. Smith, Department of Electronics, University of Southampton. My thanks also go to Mr. G. Eastwood and Mr. P. Clampett for their technical assistance and to Miss. J. Taylor for typing the manuscript.

References

Alger, B. E. and Teyler, T. J. (1976). Long-term and short-term plasticity in the CA1, CA3 and dentate regions of the rat hippocampal slice. *Brain Res.* **110**, 463–480.
Andersen, P., Eccles, J. C. and Løyning, Y. (1964). Location of postsynaptic inhibitory synapses on hippocampal pyramids. *J. Neurophysiol.* **27**, 592–607.

Andersen, P., Holmquist, B. and Voorhoeve, P. E. (1966). Excitatory synapses on hippocampal apical dendrites activated by entorhinal stimulation. *Acta Physiol. Scand.* **66,** 461–472.

Andersen, P., Bliss, T. V. P. and Skrede, K. K. (1971). Lamellar organization of hippocampal excitatory pathways. *Exp. Brain Res.* **13,** 222–238.

Andersen, P., Bliss, T. V. P. and Skrede, K. K. (1971). Unit analysis of hippocampal population spikes. *Exp. Brain Res.* **13,** 208–221.

Andersen, P., Bland, B. H. and Dudar, J. D. (1973). Organization of the hippocampal output. *Exp. Brain Res.* **17,** 152–168.

Assaf, S. Y. and Kelly, J. S. (1979). On the nature of depolarizing after-potentials in granule cells of the rat dentate gyrus maintained *in vitro*. *J. Physiol. (Lond.)* **296,** 68P.

Bergveld, P. (1972). Development, operation and application of the ionsensitive field effect transistor as a tool for electrophysiology. *IEEE Trans. Biomed. Engng.* **BME-19,** 342–351.

Blackstad, T. W., Brink, K., Hem, J. and Jeune, B. (1970). Distribution of hippocampal mossy fibres in the rat. An experimental study with silver impregnation methods. *J. Comp. Neurol.* **138,** 433–450.

Cajal, Y Ramon, S. (1893). The structure of Ammon's Horn. Translated by L. M. Kraft, 1968, 78pp., Thomas Springfield, Illinois.

Chronister, R. B. and De France, J. F. (1979). Organization of projection neurones of the hippocampus. *Exp. Neurol.* **66,** 509–523.

De France, J. F., Katai, S. T. and Shimono, T. (1973). Electrophysiological analysis of the hippocampal-septal projections: 1. Response and topographical characteristics. *Exp. Brain Res.* **17,** 447–462.

Haas, H. L., Schaerer, B. and Vosmansky, M. (1979). A simple perfusion chamber for the study of nervous tissue slices in vitro. *J. Neurosci. Meth.* **1,** 323–325.

Hjorth-Simonsen, A. and Jeune, B. (1972). Origin and termination of the hippocampal perforant path in the rat studied by silver impregnation. *J. Comp. Neurol.* **144,** 215–232.

Jobling, D., Wheal, H. V. and Smith, J. G. (1981). An active microelectrode array to record from the mammalian central nervous system *in vitro*. *Med. Biol. Engng. Comput.*, in press.

Kandel, E. R. and Spencer, W. A. (1961a). Electrophysiology of hippocampal neurones. II. After-potentials and repetitive firing. *J. Neurophys.* **24,** 245–259.

Kandel, E. R. and Spencer, W. A. (1961b). Electrophysiology of hippocampal neurones. III. Firing level and time constant. *J. neurophys.* **24,** 260–271.

Kandel, E. R. and Spencer, W. A. (1961c). Electrophysiology of hippocampal neurones. IV. Fast prepotentials. *J. Neurophys.* **24,** 272–285.

Kandel, E. R., Spencer, W. A. and Brinley, F. J. (1961). Electrophysiology of the hippocampal neurones. 1. Sequential invasion and synaptic organization. *J. Neurophysiol.* **24,** 225–242.

Laurberg, S. (1979). Commissural and intrinsic connections of the rat hippocampus. *J. Comp. Neurol.* **184,** 685–708.

Loeb, G. E., Marks, W. B. and Beatty, P. G. (1977). Analysis and microelectronic design of tubular electrode arrays intended for chronic, multiple single-unit recording from captured nerve fibres. *Med. Biol. Engng. Comput.* **15,** 195–201.

Lømo, T. (1968). Nature and distribution of inhibition in a simple cortex (Dentate Area). *Acta. Physiol. Scand.* **74,** 8–9A.

Lømo, T. (1971). Patterns of activation in a monosynaptic cortical pathway: The perforant path input to the dentate area of the hippocampal formation. *Exp. Brain Res.* **12,** 18–45.

Lorente de Nó, R. (1934). Studies on the structure of the cerebral cortex. II. Continuation of the study of the Ammonic system. *J. Psychol. Neurol.* **46,** 113–177.

Meibach, R. G. and Siegel, A. (1977). Efferent connections of the hippocampal formation in the rat. *Brain Res.* **124,** 197–224.

Mercer, H. D. and White, R. L. (1978). Photolithographic fabrication and physiological performance of microelectrode arrays for neural stimulation. *IEEE Trans. Biomed. Engng.* **BME-25,** 494–500.

Pickard, R. S. (1979). A Review of printed circuit microelectrodes and their production. *J. Neurosci. Methods* **1,** 301–318.

Raisman, G. (1966). The connections of the septum. *Brain* **89,** 317–348.

Schwartzkroin, P. A. (1975). Characteristics of CA1 neurones recorded intracellularly in the hippocampal in vitro slice preparation. *Brain Res.* **85,** 423–436.

Schwartzkroin, P. A. (1977). Further characteristics of hippocampal CA1 cells *in vitro*. *Brain Res.* **128,** 53–68.

Skrede, K. R. and Westgaard, R. H. (1971). The transverse hippocampal slice: a well defined cortical structure maintained *in vitro*. *Brain Res.* **35,** 589–593.

Swanson, L. W. and Cowan, W. M. (1977). An autoradiographic study of the organization of the efferent connections of the hippocampal formation in the rat. *J. Comp. Neurol.* **172,** 49–84.

Swanson, L. W. and Cowan, W. M. (1979). The connections of the septal region in the rat. *J. Comp. Neurol.* **186,** 621–656.

Thomas, C. A., Springer, P. A., Loeb, G. E., Berwald-Netter, Y. and Okun, L. M. (1972). A miniature microelectrode array to monitor the bioelectric activity of cultured cells. *Exp. Cell Res.* **74,** 61–66.

Wise, K. D. and Angell, J. B. (1975). A low-capacitance multielectrode probe for use in extracellular neurophysiology. *IEEE Trans. Biomed. Engng.* **BME-22,** 212–219.

Wise, K. D., Angell, J. B. and Starr, A. (1970). An integrated circuit approach to extracellular microelectrodes. *IEEE Trans. Biomed. Engng.* **BME-17,** 238–246.

Yamamoto, C. (1972). Activation of hippocampal neurones by mossy fibre stimulation in thin brain sections *in vitro*. *Exp. Brain Res.* **14,** 423–435.

9

The Amount of Transmitter Released from the Optic Nerve Terminal in Thin Slices from the Lateral Geniculate Nucleus *In Vitro*

C. YAMAMOTO AND S. SAWADA

Department of Physiology, Faculty of Medicine, Kanazawa University, Kanazawa 920, Japan

I. Introduction

When two shocks are applied to the nerve of a curarized nerve–muscle preparation, end-plate potentials (EPPs) elicited by test shocks are reduced in amplitude at short shock intervals (Betz, 1970; Takeuchi, 1958; Thies, 1965). Similarly, the amplitude of successive EPPs declines rapidly during a short train of stimuli at high frequencies (Elmqvist and Quastel, 1965). Depression of the EPP amplitude observed in these experiments has been ascribed to depletion of the transmitter pool immediately available for release, and from the magnitude of depression, release probability of the available pool of quanta has been calculated (Martin, 1977). In the present experiments, we have tried to apply this method to estimate the amount of transmitter released by an impulse from terminals of the optic nerve fibers in the lateral geniculate nucleus (LGN). Although the absolute amount of released transmitter cannot be measured at present, we may estimate the ratio (r) of the released amount to the available pool of the transmitter in the terminals. For this purpose, we have prepared thin slices of the LGN of the guinea pig, incubated them in an artificial medium and elicited excitatory postsynaptic potentials (EPSPs) in the LGN by delivering trains of stimuli to the optic tract. The value of r has been calculated from the rate of fall of the EPSP amplitude during pulse trains (Martin, 1977). In order to block the influence of inhibitory interneurons which exist within the LGN and liberate γ-aminobutyric acid (GABA) as the transmitter (Curtis and Tebecis, 1972), we added bicuculline, a GABA antagonist, to the perfusing medium.

The value of r has also been estimated in the presence of serotonin (5-HT; 5-hydroxytryptamine) in order to clarify the mechanism of action of 5-HT on synaptic transmission in the LGN. It is known that at low concentrations 5-HT suppresses activation of LGN neurons by optic nerve stimulation, but has no effect on excitability of LGN neurons (Curtis and Davis, 1962; Yamamoto, 1974b). Two hypotheses have been suggested to explain this observation. First, 5-HT may reduce the amount of the transmitter released from the presynaptic optic nerve terminals. Second, 5-HT may block access of the transmitter to the postsynaptic receptors on LGN neurons. It remains to be seen which of these hypotheses is correct.

In view of recent, wide interest in the use of slice preparations in neurophysiological studies, detailed descriptions of the methods used are given in order to lessen difficulties for researchers who are experimenting with slice preparations for the first time.

II. Methods

A. Incubation medium

Bicarbonate-buffered medium was used because electrical activities were markedly suppressed in phosphate-buffered media. The standard medium was prepared according to the routine procedures in the Department of Biochemistry at the University of London's Institute of Psychiatry (Prof. H. McIlwain) with some modifications.

1. *Stock solutions*

The following stock solutions a–d were kept in a refrigerator and renewed every 4 weeks.

(a) *5 × Krebs.* In distilled water, 22.7g NaCl, 0.23g KCl, 1.0g $MgSO_4 \cdot 7H_2O$ and 0.53g KH_2PO_4 were dissolved and the volume was made up to 500 ml.

(b) *1.3% NaHCO₃.* In distilled water, 6.5g $NaHCO_3$ was dissolved and the volume was made up to 500 ml. The solution was bubbled with CO_2 for 1 hour. Bubbling was necessary in order to convert contaminating Na_2CO_3 to $NaHCO_3$.

(c) *1 M glucose.* In distilled water, 9g of glucose was dissolved and the volume was made up to 50 ml. Gentle heating facilitated dissolution.

(d) *108 mM CaCl₂.* Since $CaCl_2$ absorbs moisture, a considerable amount of water may be contained in $CaCl_2$ purchased commercially. Therefore, a $CaCl_2$ solution of about 1.1 M was initially prepared and the accurate concentration of $CaCl_2$ in this solution was determined by measuring Cl^- concentration. Then the solution was diluted to make the final concentration of 108 mM.

2. *Standard medium*

In a measuring cylinder (200 ml volume) with a stopper, 20 ml of 5 × Krebs, 80 ml of distilled water, 21 ml of 1.3% $NaHCO_3$ and 1.3 ml of 1 M glucose were placed in succession. The solution was mixed well and 2.7 ml of 108 mM $CaCl_2$ was added. Measuring pipettes were used for adding stock solutions. The mixture was bubbled with 95% O_2/5% CO_2 for 30 min. (Bubbling the solution with pure O_2 was avoided because the pH of the solution increased in the absence of CO_2.) Thereafter, dissipation of the gas from the solution was minimized with the stopper. The composition of the incubation medium thus prepared was (mM): NaCl, 124; KCl, 1; KH_2PO_4, 1.24; $MgSO_4$, 1.3; $CaCl_2$, 2.4; $NaHCO_3$, 26;

glucose, 10. Although the media containing K^+ at 5.88–6.24 mM have been used in neurophysiological and biochemical studies with success (Garthwaite *et al.*, 1979; Kawai, 1970; Yamamoto and McIlwain, 1966), these K^+ concentrations are about twice that in the cerebrospinal fluid (McIlwain and Bachelard, 1971). We have found that in the medium containing 1 mM KCl, slices are as active as or more active than in that containing 5 mM KCl.

B. Instruments for preparation of LGN slices

1. *Microdissection knife (Fig. 9.1a)*

For sectioning brains perpendicularly to the pial surface, a microdissection knife with a sharp tip is versatile and handy. One can be made in the laboratory from a razor blade: the sharp edge of a chisel is placed at an angle across the razor blade, and the razor blade breaks as the chisel is gradually pressed on it. A razor blade of ordinary steel should be used since stainless-steel will not break. A small wooden handle is fixed to the piece of razor blade with glue. The blade is wiped with cotton pads soaked with acetone and with ethanol before use. The knife can be used in about 30 experiments.

2. *Sectioning table (Fig. 9.1b)*

A plastic plate (about 7.5 cm × 7.5 cm) is fixed on the top of an iron cylinder (diam. 5 cm, height 6 cm). The advantage of this table is its stability.

3. *Aluminum hook (Fig. 9.1c)*

This is made of a piece of aluminum wire (about 1 mm in diam., 4 cm long) and used for transferring brain slices. A part of the wire (1 cm long) at one end is flattened and bent at a right angle. The other end is also bent as illustrated.

4. *Filter paper*

A piece of filter paper is placed between the sectioning table and the brain, from which slices are to be prepared, in order to prevent direct contact of the dissection knife with the table. Filter paper which frays minimally when cut is preferred.

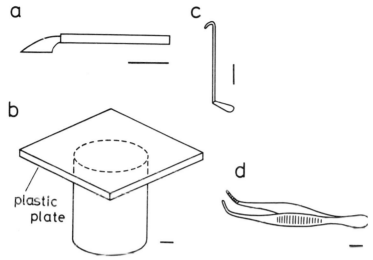

Fig. 9.1. Instruments for preparation of LGN slices. a, microdissection knife; b, sectioning table; c, aluminum hook; d, forceps. Scale = 1 cm.

5. *Iron bar*

An iron bar (35 cm long) with a rectangular cross section (3 cm × 0.5 cm) is used to stun and kill an animal.

6. *Forceps (Fig 9.1d)*

Forceps (about 11 cm long) with curved, blunt tips are preferable.

7. *Stereomicroscope (× 12)*

A micrometer was placed in an eye-piece of a stereomicroscope. A measure was then brought into the focus of the microscope and the number of micrometer scales corresponding to 1 mm in the measure were counted. This information enabled us to section slices at an intended thickness.

C. **Preparation of slices from the LGN**

The most critical factors determining activity of slices seemed to be speed of preparation, thickness of slices and mechanical damage to neurons in the tissue. Therefore, slices were prepared rapidly and transferred into the oxygenated medium within 2–3 minutes of the first blow to stun the animal (see below). Slices were less than 0.35 mm thick to allow free access of O_2 and glucose to the neurons within the tissue. Care was also taken to minimize crushing of the tissue with instruments during preparation of slices.

1. *Removal of the brain*

A guinea pig was placed on a wooden plate and its head was pressed gently from above with the left hand. The animal was stunned by a blow on the back of the neck with the iron bar and then the heart was ruptured by a second blow on the back of the thorax. The second blow prevented bleeding which otherwise occurred on exposure of the brain. The right ear was held with the left hand and the skin covering the right half of the skull and neck was removed in a piece. The skin was similarly removed from the left side. The head was held with the left hand about 10 cm above the wooden plate and the dorsal neck muscles and vertebral column were cut between the occipital bone and the atlas with a pair of straight scissors (18 cm long). The tip of a blade of the scissors was inserted into the foramen occipitale magnum at the midline and the occipital bone was cut along the midline. Damage to the vermis of the cerebellum was usually inevitable during this procedure. The junction between the right and left parietal and frontal bones were divided by the scissors advanced along the sutura sagitalis. Subsequently, the tip of a blade of scissors was inserted beneath the medial edge of the parietal bone and the tip of the other blade was placed on the temporal bone of the same side, and the parietal bone was prized up by upward movement of the blade inserted beneath the bone while the other tip was used as the fulcrum. The parietal bone on the other side and the frontal bones were similarly prized away. Thus the dorsal surface of the cerebrum was widely exposed.

A spatula was inserted under the frontal end of the cerebrum through the junction between the olfactory bulbs and the cerebrum, and the cerebrum was slightly lifted with the spatula. Peering underneath the cerebrum, one could see the optic chiasm with converging optic nerves which were then sectioned with a pair of small scissors. The brain-stem was divided transversely at the level of the colliculi with the spatula and the brain was removed from the skull and placed on its dorsal surface on a piece of filter paper soaked with the standard medium.

2. *Isolation of the LGN with the optic tract*

The left optic tract was cut close to the chiasm (Fig. 9.2a), and then the

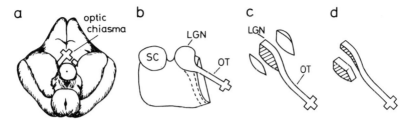

Fig. 9.2. Preparation of a slice from the right LGN. For details, see the text. Abbreviations: OT, optic tract; SC, superior colliculus.

brain was divided along the midline. The right optic tract maintained the connection with the chiasm. The left half of the brain was discarded. The right half of the brain was repositioned on its medial surface and the LGN and the full length of the optic tract were exposed by removing the cerebral cortex covering them. The diencephalon was sectioned with a razor blade along the frontal border of the LGN and the optic tract. The diencephalon frontal to the sectioning was discarded with the remaining cerebral cortex. The chiasm was held with forceps and the optic tract was freed from the underlying diencephalon up to its boundary with the LGN with small scissors (Fig. 9.2b). The LGN was isolated with the incoming optic tract from the rest of the brain and placed on a piece of filter paper which was put on a sectioning table and covered with the standard medium.

3. *Microdissection of LGN slices*

The following procedures were carried out under a stereomicroscope. The frontal and posterior edges of the LGN were cut off with the dissection knife (Fig. 9.2c). The tissue was repositioned on its posterior cut surface, and the LGN was cut almost parallel to its pial surface at a level 0.2–0.3 mm beneath the pial surface (Fig. 9.2d). The ventral part of the LGN was discarded. Since the surface of the LGN curved sharply, the thickness of the section could not be strictly even.

4. *Preincubation*

Immediately after preparation, the tissue was transferred into a small vessel containing the standard medium, and incubated for about 40 minutes at 37°C. The medium was bubbled with 95% O_2/5% CO_2 through a thin polyethylene tubing introduced into the medium. During this preincubation, the tissue recovered a fair proportion of high-energy compounds lost during preparation (McIlwain and Bachelard, 1971).

D. Observation chamber

The tissue was unfolded on the bottom of the tissue compartment (B) of the observation chamber depicted in Fig. 9.3. In a double-walled reservoir (R), the perfusing medium (M) was equilibrated with 95% O_2/5% CO_2 and kept at 37°C with warm water circulating in the external space of the reservoir. The medium was rewarmed in a small compartment (A) and then introduced into the tissue compartment (B) through a thin polyethylene tubing at a rate of about 1.5 ml min^{-1}. The medium flowed past the tissue and was introduced into the drain compartment (D) through a groove (G). The tissue-compartment was 18 mm in length, 13 mm in width and 5 mm in depth. Plastic plates of 1 mm thickness were used for walls and bottoms of the tissue and rewarming compartments. The other parts of the chamber were composed of plastic plates of 3 mm thickness. The tissue and rewarming compartments were warmed with warm water circulating through a water jacket under the compartments. The long tubes connecting the chamber with a perfusion pump for circulating warm water occasionally caused contaminating ripples at 60 Hz (commercial alternating current) in

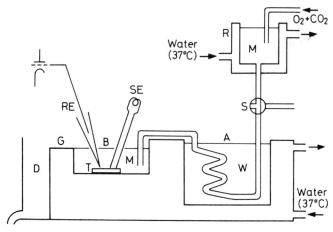

Fig. 9.3. Schematic cross-section of the observation chamber. The various components have not all been drawn to the same scale. A, Rewarming compartment; B, tissue compartment; D, drain compartment; G, groove; M, perfusing medium; R, reservoir; RE, recording electrode; S, 3-way tap; SE, stimulating electrode; W, water. For further details, see the text.

electrical recording. The AC induction could be avoided by the addition of NaCl to circulating water (about 0.5%) and by grounding it at the entrance or exit tubing. A medium containing 5-HT was introduced into the chamber by switching a 3-way tap (S). The tissue was held in position with pieces of silver wire (0.2 mm diam.) placed gently on the edges of the tissue. Since the medium contained glucose, bacteria grew on the inner surface of the polyethylene tubings and the tissue compartment. Every week, therefore, they were cleaned mechanically with pipe cleaners, silver wires and cotton pads. The observation chamber was designed to allow the polyethylene tubing to be removed from the rewarming compartment and straightened for cleaning.

E. Stimulation and recording

The optic tract was stimulated with a pair of silver wires insulated to bare tips and stuck together side by side with a tip separation of 0.5 mm. The stimulus pulses were 0.04 0.2 ms in duration and 1–5 V in strength. When short pulse trains were applied, the interval between trains was kept at 10 s. The field potentials were recorded with glass pipettes of 10–20 μm tip diameter filled with 0.9% NaCl. Single-cell discharges and EPSPs elicited in LGN neurons by optic-tract stimulation were recorded extracellularly with glass micropipettes of about 1.5 μm tip diameter filled with 3 M NaCl. After spikes and EPSPs were recorded from a cell, the cell was destroyed and the field potential was recorded with the same electrode. The amplitude of the S-potentials was measured from the level of the field potential. The reference electrode was a piece of thick silver wire immersed in the medium in the tissue compartment and connected with the ground. Potentials were averaged by an ATAC-350 (Nihon-Kohden) and plotted on a X–Y recorder.

F. Calculation of r and α

The method of estimating r consisted of applying short, rapid trains of stimuli to the optic tract (Martin, 1977) and measuring the reduction of amplitude of successive EPSPs (Fig. 9.4Aa). The amplitude of individual EPSPs were then expressed as a fraction of the first EPSP and plotted against the sum of all previous amplitudes (Fig. 9.4Ab). Provided that the reduction in successive EPSP amplitude was due solely to depletion of the available pool of the transmitter, that r was constant irrespective of transmitter depletion and that replenishment of the transmitter was negligible, all points were expected to distribute on a straight line which intercepted the abscissa at the value of 1/r (Fig. 9.4Ab). In practice,

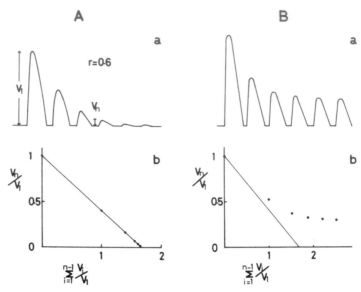

Fig. 9.4. Theoretically expected decline in EPSP amplitude during short-train stimulation of the optic tract. In A, no recovery from depletion of transmitter is assumed to occur between shocks. In B, 20% of net loss of the transmitter pool is assumed to recover between shocks. In b, amplitude of each successive EPSP (V_n), expressed as a fraction of initial EPSP amplitude (V_1), is plotted against sum of all previous amplitude expressed in the same way.

however, recovery from depletion of transmitter during pulse trains was not zero in most LGN neurons and the amplitude of EPSPs approached a steady state (Fig. 9.4Ba). If the rate of recovery was proportional to the net loss of the transmitter pool, the EPSP amplitude was expected to decline during a pulse train as depicted in Fig. 9.4Bb. In this graph, the amplitude of the EPSP (V_2) elicited by the second pulse in a pulse train was expected to be:

$$V_2 = 1 - re^{-\alpha t} \tag{1}$$

where V_2 was expressed as a fraction of the amplitude of the first EPSP, α was rate constant of recovery and t was pulse interval. At the steady state, the amount of transmitter released by a pulse was equal to that replenished during t. Therefore, the EPSP amplitude (V_s), normalized similarly, at the steady level was related to r and α by:

$$1 + (V_s - V_s r - 1)e^{-\alpha t} = V_s \tag{2}$$

Solving eqn. (1) for $e^{-\alpha t}$ and substituting:

$$r = \frac{1 - V_s - V_2 + V_2 V_s}{1 - 2V_s + V_2 V_s} \qquad (3)$$

After the r value was calculated according to eqn. (3), the value of α was calculated according to eqn. (1) using the value of r.

III Results

A. Electrical activities recorded from LGN slices

In response to a single shock to the optic tract, a negative field potential of 2–5 ms duration was recorded from LGN slices (Fig. 9.5A). The potential was suppressed in the medium containing Ca^{2+} in low concentrations (less than 0.8 mM) and Mg^{2+} in high concentrations (more than 7.5 mM). The potential seemed, therefore, to represent postsynaptic activation of LGN neurons in the slice (Yamamoto, 1974b). Although the potential varied to some extent in size and configuration in different slices, it was recorded in almost all slices with no marked sign of deterioration during experiments lasting for several hours. With electrodes of finer tips, it was possible to record extracellularly single-cell

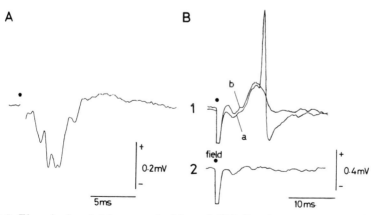

Fig. 9.5. Electrical activities recorded from LGN slices in response to optic tract stimulation. A shows the field potential; five successive potentials were averaged. B1, S-potentials with (a) and without (b) a superimposed spike recorded from a LGN neuron at the critical strength for generation of spike; B2, field potential recorded after death of the neuron. In this and in the following illustrations, dots (●) indicate times at which the optic tract was stimulated.

discharges and to record intracellularly EPSPs (Yamamoto, 1974b; Kelly *et al.*, 1979) followed by inhibitory postsynaptic potentials (IPSPs). In the present experiments, bicuculline was added to the medium at 1 μM for blockade of IPSPs, and Mg^{2+} concentration was increased to 6.5 mM in order to prevent generation of seizure discharges which otherwise took place in the presence of bicuculline (Yamamoto *et al.*, 1980).

Freygang (1958) reported that EPSPs elicited by optic-nerve stimulation were recorded from single LGN neurons extracellularly as small positive waves (S-potentials) *in vivo*. In LGN slices too, the S-potential was observed. This was most evident when the tip of the electrode approached each neuron so close that large biphasic spikes were recorded with a predominant initial positive phase (Fig. 9.5B1). Although spikes appeared in an all-or-nothing manner at the critical stimulus strength, the S-wave increased successively in height as stimulus intensity was gradually increased. This suggested that several optic nerve fibres converged on single LGN neurons to generate EPSPs. The S-potential, which could be recorded easily and maintained for long observation periods, allowed us to study alteration of the EPSP amplitude without difficulties accompanying intracellular recording. In the following experiments, stimulus intensity was adjusted at the level at which a slight increase or decrease in intensity resulted in no appreciable changes in the S-potential amplitude. At such intensities, spikes were often superimposed on the S-potential. The transition between the S-potential and spike was identified on the oscilloscope with an expanded sweep. When the spike obscured the peak of the S-potential, the data were discarded.

B. Reduction of S-potential amplitude during pulse train

In most LGN neurons, the amplitude of the S-potential decreased successively during trains of six stimuli at 29 or 58 Hz and approached steady levels as shown in Fig. 9.6. The decline was faster and the steady level was lower at the higher frequency. As mentioned in the Methods Section (p. 241), values of r and α were calculated from the amplitudes of the second S-potentials and the steady levels. Since r and α were assumed to be constant in individual nerve terminals, these parameters estimated at 29 Hz should agree with those at 58 Hz. The values of r and α calculated in four neurons are listed in Table 9.1 and show that these parameters varied widely from cell to cell but that in individual cells the values obtained at different frequencies agree with each other relatively well in view of the experimental complications. On neuron number 1 in

Table 9.1. Values of r and α (s^{-1}) in four neurons calculated from decline of the S-potential amplitude during pulse trains at 29 and 58 Hz.

	Neuron 1		Neuron 2		Neuron 3		Neuron 4	
	r	α	r	α	r	α	r	α
29 Hz	0.62	4.6	0.74	2.8	0.48	4.4	0.66	6.1
58 Hz	0.64	5.1	0.63	2.0	0.54	2.8	0.81	10

Table 9.1, the values of r and α were, on average, 0.63 and 4.9 s^{-1}, respectively. From these values, decline in the S-potential amplitude during pulse trains at 29 and 58 Hz was calculated and plotted in Fig. 9.6B (——— and – – – –). The S-potential amplitude actually observed in the experiment was similarly plotted (o and ●). Fig. 9.6B shows that the experimental points fell fairly well along the predicted lines.

Fig. 9.7 shows results of two experiments in which pairs of shocks were applied to the optic tract at various intervals, and the difference between the first S-potentials (V_1) and the second (V_2) divided by the amplitude of the first was plotted on a semilogarithmic scale against shock intervals (o and ●). According to eqn. (1), it was expected that if r was constant irrespective of depletion of the transmitter pool, the points obtained in individual neurons would be distributed on a straight line, which intercepted the ordinate at the r value and had a slope of $-\alpha$. The values of r and α were estimated as in Fig. 9.6 with pulse trains at 29 Hz, and theoretically expected lines were drawn utilizing these r and α values (Fig. 9.7; ——— and – – – –), which showed fairly good agreement between observed and theoretical lines.

C. Variation in r and α

In most LGN neurons (group 2), S-potential height declined at moderate rates to reach a plateau during trains of six pulses (Fig. 9.6A). The values of r and α varied widely between these neurons, r ranging between 0.32 and 0.9 and α between 0.05 and 9 s^{-1}. In four other neurons (group 3), reduction of S-potential during pulse trains was so rapid that the second potentials were less than 15% of the first and third S-potentials were undetectable (Fig. 9.8A3 and 9.8B ($\times$)). All of the neurons in group 3 seemed to be innervated by slow-conducting optic fibres because of long latencies of their S-potentials. They had large r values (0.74–0.9) and low α values (almost 0 s^{-1}). The values of r and α

A

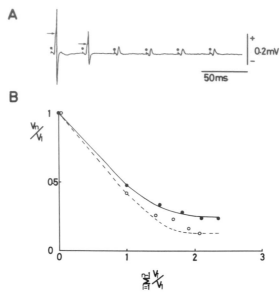

B

Fig. 9.6. Decline of S-potential amplitude during brief pulse train. In A, 25 trains at 29 Hz were applied at a 10 s interval and potentials were averaged. Arrows indicate peaks of first and second S-potentials. In B, amplitude of S-potentials induced by trains at 29 and 58 Hz are plotted as in Fig. 9.4b.——— and – – – – show the theoretically expected decline of S-potential amplitude at 29 (●) and 58 (○) Hz respectively.

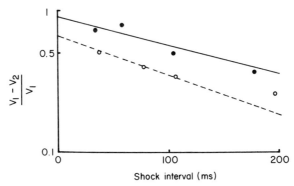

Fig. 9.7. Depression of test S-potentials. ● and ○ indicate results obtained in different neurons 1 and 2. Magnitude of depression is expressed as the difference between the conditioning and the test S-potential amplitudes divided by the amplitude of the conditioning one and plotted on a semilogarithmic scale against shock intervals. The lines (———, – – – –) are drawn according to eqn. (1) utilizing the values of r and α calculated from results of experiments with pulse trains at 29 Hz.

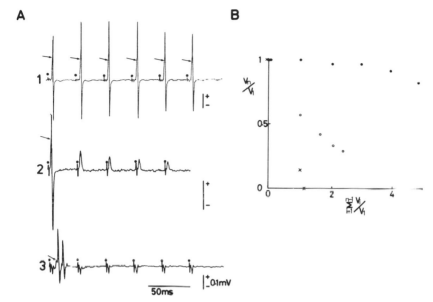

Fig. 9.8. Decline of S-potentials in 3 groups of neurons (A1–A3). A and B were recorded and plotted as in Fig. 9.6A and 9.6B, respectively. B shows plots of S-potential amplitudes from A1 (●), A2 (○) and A3 (×). In A, arrows indicate the peaks of S-potentials. For further details, see the text.

could not be calculated in three other neurons (group 1) in which S-potentials declined slowly and did not reach the steady level during trains of six stimuli (Fig. 9.8A1 and 9.8B (●)).

The values of r and α of all 17 neurons in which estimation of the parameters was possible were plotted in Fig. 9.9. Although no correlation between the two parameters was evident, it may be noted that neurons with low values of α (less than 3 s^{-1}) had relatively high values for r. In 17 neurons, the values of r and α were on the average 0.70 ± 0.17 and 3.3 ± 2.6 s^{-1} (mean ± s.D.), respectively.

D. Effects of 5-HT

As in the standard medium, 5-HT was a potent depressant of the S-potential in the medium containing 1 μM bicuculline and Mg^{2+} in 6.3 mM. As shown in Fig. 9.10, the S-potentials evoked in a neuron by first stimuli in pulse trains at 28 Hz were suppressed to about 23% of control 1 minute after 20 μM 5-HT administration (×). S-potentials elicited by second and subsequent stimuli were less susceptible to 5-HT: the

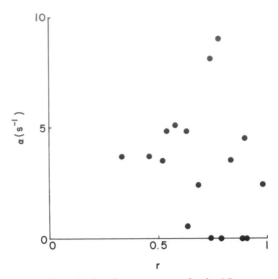

Fig. 9.9. Correlation between r and α in 17 neurons.

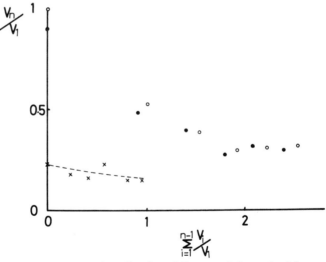

Fig. 9.10. Effect of 5-HT. Amplitudes of S-potentials evoked by trains of six stimuli before (o), 1 minute after administration of 20 μM 5-HT and 6 minutes after removal of 5-HT from the surrounding medium (●) are plotted as in Fig. 9.6B. Each S-potential amplitude during and after administration of 5-HT is expressed as a fraction of initial S-potential amplitude in control. – – – –, Theoretically expected S-potential amplitude under the action of 5-HT if 5-HT solely suppresses release of transmitter without affecting rate of replenishment or access of transmitter to the postsynaptic receptors.

S-potentials elicited by the sixth stimulus were suppressed by only 50%. This phenomenon cannot be explained on the hypothesis that 5-HT blocks access of the transmitter to the postsynaptic receptors on the LGN neurons. On the other hand, the S-potential amplitudes calculated on the assumption that r decreased to 23% and α remained constant in the presence of 5-HT (– – – –) fitted experimental results quite well (×). This result favours the hypothesis that 5-HT suppresses release of the transmitter from terminals of the optic nerve fibres. Identical results were obtained in three other neurons.

IV. Discussion

The method of estimating values of r and α were based on the assumptions that the amount of transmitter released by an impulse was proportional to the transmitter pool, that r was constant irrespective of depletion of the pool, and that the pool was replenished in proportion to the net loss of transmitter with a rate constant of α. In the neuromuscular junction, all of these assumptions have not been fully accepted (Christensen and Martin, 1970). As shown in the present experiments, EPSPs in the LGN decline in amplitude during tetani with the time-course predicted according to the equations derived from these assumptions (Fig. 9.6). In the experiments in which two shocks were applied, test responses recovered as expected from the values of r and α calculated from decline of the EPSP amplitude during tetani (Fig. 9.7). The values of these parameters did not vary widely when measured at different frequencies (Table 9.1). All these results suggest that the underlying assumptions mentioned above are valid in the LGN.

In the present experiments, EPSPs were recorded extracellularly. Because of non-linear summation, amplitude of EPSPs would not increase linearly with an increase in released amount of transmitter. Since the S-potentials were recorded extracellularly, this inaccuracy could not be corrected by conventional means (Martin, 1955) and resulted in underestimations of r values. Amplitudes of S-potentials elicited in these experiments were less than or slightly above inducing action potentials in LGN neurons. Judging from intracellular records already published (Kelly *et al.*, 1979; Yamamoto, 1974b), the depolarization accompanying such weak excitation seemed to be less than 10 mV. In view of large resting potentials of LGN neurons (69 mV) in slices (Kelly *et al.*, 1979), underestimation caused by non-linear summation of the S-potential was probably less than 15%. Therefore, the values of r and α are unlikely to be grossly in error.

250 C. YAMAMOTO AND S. SAWADA

In the neuromuscular junction of the guinea pig (Thies, 1965), the values of r and α are approximately 0.3 and 0.5 s⁻¹, respectively, at 36–37°C in the medium containing 1.8 mM Ca^{2+} and 1 mM Mg^{2+}. The feature of optic nerve terminals in the LGN is, therefore, much higher efficiency in release and recovery of the transmitter. Another feature is a wide variation in r and α. This may reflect the fact that the optic nerve is composed of three groups of axons (Bishop and Claer, 1955; Stone and Fukuda, 1974; Sumitomo et al., 1969) and that individual LGN neurons receive input mainly from one group of optic nerve fibres alone (Cleland et al., 1971).

In comparison with the brains in vivo, slice preparations have several advantages for neurophysiological studies. The present study was an example of the application of slice preparations for analysis of synaptic transmission in the brain. The method used in this study to estimate the amount of released transmitter required that stimulation to the optic tract elicited monosynaptic EPSPs alone. This requirement was fulfilled only with isolated LGN slices maintained in media containing bicuculline and Mg^{2+} in high concentrations, since the LGN in vivo generated polysynaptic EPSPs and IPSPs to optic-tract stimulation and seizure

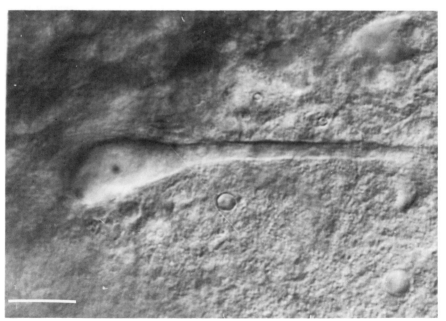

Fig. 9.11. A Purkinje cell in a cerebellar section of 70 μm thickness, observed with a Nomarski microscope. Bar = 20 μm.

discharges in the presence of bicuculline. In designing other studies with slice preparations, the following features of the preparations may be taken into consideration. First, chemical agents can be administered to the tissue at known concentrations. Second, magnitude of synaptic transmission and equilibrium potential for various ions can be controlled at will by changing ionic composition of the perfusing medium. Third, substances liberated by the tissue can be readily collected. Moreover, in slices cut at thickness of 100 μm or less, somata and dendrites of neurons maintaining electrical activities can be clearly visible with the aid of a Nomarski microscope (Fig. 9.11). This has been shown in slices from the cerebellum (Yamamoto and Chujo, 1978) and spinal cord (Takahashi, 1978). (The technique of preparation of such thin sections are given in detail in the Appendix, p. 253.) Therefore, one can record potential activities from and apply chemicals to a desired portion of a neuron under a direct observation. All these features of slice preparations will allow us to carry out experiments that are impossible or quite difficult in the *in vivo* brain.

V. Conclusions

1. The proportion (r) of the amount of transmitter released by an impulse to the available pool in the nerve terminals was estimated *in vitro* in thin sections of the lateral geniculate nucleus of the guinea pig.

2. The value of r varied widely from cell to cell and, on the average, was much larger than that reported in the neuromuscular junction of the guinea pig.

3. The rate of replenishment of the pool was also higher than in the neuromuscular junction.

4. Techniques for preparation of slices were given in detail.

5. Among several advantages of slice preparations discussed, it seemed most promising that neurons in sections thinner than 100 μm could be observed clearly with the aid of a Nomarski microscope.

6. In the Appendix, therefore, techniques for preparation of such thin slices were described.

Acknowledgements

We thank Dr. G. M. Shepherd and Dr. T. Tsumoto for their invaluable suggestions to the manuscript. This study was supported by a grant from the Ministry of Education of Japan.

References

Betz, W. J. (1970). Depression of transmitter release at the neuromuscular junction of the frog. *J. Physiol. (Lond.)* **206**, 629–644.

Bishop, G. H. and Clare, M. H. (1955). Organization and distribution of fibers of the optic tract of the cat. *J. Comp. Neur.* **103**, 269–304.

Christensen, B. N. and Martin, A. R. (1970). Estimates of probability of transmitter release at the mammalian neuromuscular junction. *J. Physiol. (Lond.)* **210**, 933–945.

Cleland, B.G., Dubin, M. W. and Levick, W. R. (1971). Sustained and transient neurons in the cat's retina and lateral geniculate nucleus. *J. Physiol. (Lond.)*, **217**, 473–496.

Curtis, D. R. and Davis, R. (1962). Pharmacological studies upon neurones of the lateral geniculate nucleus of the cat. *Br. J. Pharmacol.* **18**, 217–246.

Curtis, D. R. and Tebecis, A. K. (1972). Bicuculline and thalamic inhibition. *Exp. Brain Res.* **16**, 210–218.

Elmqvist, D. and Quastel, D. M. J. (1965). A quantitative study of end-plate potentials in isolated human muscle. *J. Physiol. (Lond.)* **178**, 505–529.

Freygang, W. H., Jr. (1958). An analysis of extracellular potentials from single neurons in the lateral geniculate nucleus of the cat. *J. Gen. Physiol.* **41**, 543–564

Garthwaite, J., Woodhams, P. L., Collins, M. J. and Balazs, R. (1979). On the preparation of brain slices: morphology and cyclic nucleotides. *Brain Res.* **173**, 373–377.

Kawai, N. (1970). Release of 5-hydroxytryptamine from slices of superior colliculus by optic tract stimulation. *Neuropharmacol.* **9**, 395–397.

Kelly, J. S., Godfraind, J. M. and Maruyama, S. (1979). The presence and nature of inhibition in small slices of the dorsal lateral geniculate nucleus of rat and cat incubated *in vitro*. *Brain Res.* **168**, 388–392.

McIlwain, H. and Bachelard, H. S. (1971). Biochemistry and the Central Nervous System, 4th edn. Churchill Livingston, Edinburgh and London.

Martin, A. R. (1955). A further study of the statistical composition of the end-plate potential. *J. Physiol. (Lond.)* **130**, 114–122.

Martin, A. R. (1977). Junctional transmission II. Presynaptic mechanisms. *In* "Handbook of Physiology", Section 1. Volume 1. (E. R. Kandel, ed.), pp. 329–355. American Physiological Society, Bethesda.

Stone, J. and Fukuda, Y. (1974). Properties of cat retinal ganglion cells: a comparison of W-cells with X- and Y-cells. *J. Neurophysiol.* **37**, 722–748.

Sumitomo, I., Ide, K., Iwama, K. and Arikuni, T. (1969). Conduction velocity of optic nerve fibers innervating lateral geniculate body and superior colliculus in the rat. *Exp. Neurol.* **25**, 378–392.

Takahashi, T. (1978). Intracellular recording from visually identified motoneurones in rat spinal cord slices. *Proc. Roy. Soc. Lond. Ser. B* **202**, 417–421.

Takeuchi, A. (1958). The long-lasting depression in neuromuscular transmission of frog. *Jap. J. Physiol.* **8**, 102–113.

Thies, R. E. (1965). Neuromuscular depression and apparent depletion of transmitter in mammalian muscle. *J. Neurophysiol.* **28**, 427–442.

Yamamoto, C. (1974a). Electrical activity observed *in vitro* in thin sections from guinea pig cerebellum. *Jap. J. Physiol.* **24**, 177–188.

Yamamoto, C. (1974b). Electrical activity recorded from thin sections of the lateral geniculate body, and the effects of 5-hydroxytryptamine. *Exp. Brain Res.* **19**, 271–281.

Yamamoto, C. and Chujo, T. (1978). Visualization of central neurons and recording of action potentials. *Exp. Brain Res.* **31**, 299–301.

Yamamoto, C. and McIlwain, H. (1966). Electrical activities in thin sections from the mammalian brain maintained in chemically-defined media *in vitro*. *J. Neurochem.* **13**, 1333–1343.

Yamamoto, C., Matsumoto, K. and Takagi, M (1980). Potentiation of excitatory postsynaptic potentials during and after repetitive stimulation in thin hippocampal sections. *Exp. Brain Res.* **38**, 469–477.

Appendix

Methods of preparation of cerebellar slices for experiments with a Nomarski microscope

As mentioned in the Discussion Section, somata and dendritic branches of Purkinje cells can be observed clearly in cerebellar slices thinner than 100 μm with the aid of a Nomarski microscope. Such thin slices can be prepared with a Vibratome as follows (Yamamoto, 1974a).

1. *Removal of the cerebellum*

As described in the Methods Section, stun and kill a guinea pig, and remove the skin covering the skull. Since cerebellar sections are prepared from the vermis, the medial portion of the cerebellum must not be damaged. Therefore, after cutting the dorsal neck muscles and the vertebral column, the tip of a blade of scissors is inserted into the foramen magnum about 3 mm lateral to the midline and the occipital bone is cut sagitally. The cleft thus produced in the occipital bone is widened by prizing up each piece of the bone with scissors as described in the Methods Section (p. 238). The parietal bones are also prized away.

Thus, the cerebellum and the posterior part of the cerebrum are exposed.

Insert a spatula vertically through the border between the cerebrum and the cerebellum until the tip hits the basis cranii, then tilt the spatula handle forward. This forces the tip backward so that it slides between the ventral surface of the lower brain-stem and basis cranii. Then, raise the tip of the spatula slowly. The cerebellum with the medulla oblongata will be raised from the skull, mounted on the spatula.

2. *Preparation of tissue blocks from folia*

With a razor blade, isolate the cerebellum from the medulla and divide the cerebellum into anterior and posterior halves. The latter contains the uvula and nodulus from which slices are usually prepared. The posterior half is re-positioned on its cut surface on a piece of filter paper soaked with the standard medium. With the microdissection knife, isolate the vermis from the hemispheres (Fig. 9.12a). To avoid crushing the tissue, make small stabs into the tissue with the knife point along the line of dissection before making the cut. The isolated vermis is placed on its lateral cut surface on a piece of filter paper covered with the standard medium.

With the microdissection knife, isolate the nodulus (Fig. 9.12b) and each folium of the uvula under a stereomicroscope. Then, prepare a tissue block about 2 mm thick from each folium. The tissue blocks are kept in ice-cold medium.

3. *Sectioning with a Vibratome*

Prepare a petri-dish (above 4 cm diam.), agar blocks (4 mm × 4 mm × 3 mm), glue that polymerizes on contact with water (e.g. Aron-Alpha (alpha–cyanoacrylate monomer) Toa-Gosei Kagaku, Tokyo), rectangular strips of filter paper (3 mm × 15 mm), ice-cold standard medium (200 ml), a Pasteur pipette of a large tip diameter (about 5 mm) and ice-cold water. In addition, make a shallow metal dish (about 5 cm × 5 cm × 0.6 cm) with a small metal block underneath to fasten to the chuck of the Vibratome. The dish is used to hold the tissue block in the Vibratome.

Transfer the tissue block of a folium to a petri-dish containing the cold medium and a strip of filter paper. Place the block on its side on the strip of filter paper with the edges in alignment (Fig. 9.12c). Thereafter, spread a strip of glue on the metal dish and fix an agar block on to it. Also spread glue on to the surface of the agar block facing the Vibratome blade. Lift out the filter paper with attached tissue, and blot up the

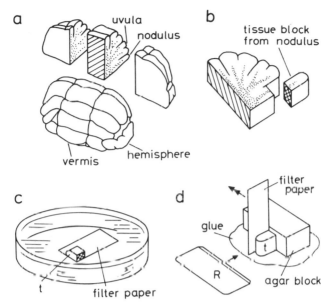

Fig. 9.12. Preparation of thin cerebellar sections with a Vibratome. For details, see the text. t, Tissue block; R, razor blade in Vibratome blade holder. Single arrow in d indicates the direction of advance of razor blade and the double arrow indicates the direction of removal of the strip of filter paper.

excess fluid. Carefully place the tissue in the metal dish against the face of the agar block, with the curved pial surface oriented toward the blade (Fig. 9.12d). The long axis of the folium should stand strictly vertical to the plate of the dish, because the plane of sectioning must be perpendicular to the axis in order to minimize injury to Purkinje cells. Fill the dish with ice-cold medium. The tissue will be stuck to the agar block and plate by the polymerization of the glue. It may be noted that amount of glue is critical: too much glue will spread over the tissue when it polymerizes, and too little will cause uneven adhesion of the tissue. Gently remove the filter-paper carrier, using dissecting instruments if necessary to avoid pulling on the tissue. Place the dish in the Vibratome chuck. Fill the Vibratome chamber with ice cold water. Sections are cut at thickness of 50–100 μm, as desired. Cut with a blade angle of about 18°, using relatively slow cutting speed (2–3 scale units) and moderate vibration amplitude (6–7 scale units). Sections thinner than 80 μm may rupture by surface tension when floated on the surface of the medium. Therefore, they are always kept in the medium and transferred with the Pasteur pipette.

10

Intracellular Recording from Thin Slices of the Lateral Geniculate Nucleus of Rats and Cats

J.-M. GODFRAIND* AND J. S. KELLY†

*Laboratoire de Neurophysiologie, Faculté de Médocine, Université
Catholique de Louvain, B-1200 Bruxelles, Tour Claude Bernard UCL
5449, Avenue Hippocrate 54, Belgium and † Department of Pharmacology,
St. Georges Hospital Medical School, Cranmer Terrace, Tooting, London
SW17 0RE, U.K.

I. Introduction

Although the lateral geniculate nucleus (LGN) has much to recommend it as a source of slices for *in vitro* recording, its popularity has never been great compared with the hippocampus or the olfactory cortex. During the last decade, less than a handful of groups have published accounts of their work performed on this *in vitro* preparation (Kato, 1974; Yamamoto, 1974; Kuhnt and Schaumberg, 1976, 1977; Kelly *et al.*, 1979; Ogawa *et al.*, 1980; Schaumberg, 1977).

From the practical point of view, the lateral geniculate nucleus is easily identified by following the course of the optic tract and slices cut along the line of the optic tract fibres would seem to be fairly representative of the nucleus as a whole. Moreover, a great deal is known about the morphology (Grossman *et al.*, 1973; Hayhow *et al.*, 1962; Laemle, 1975; Szentagothai, 1973) and the anatomical and physiological connections of the LGN cells (see, e.g., reviews by Szentagothai, 1973; Singer, 1977; Burke and Cole, 1978).

From a scientific point of view, the principal cells and interneurons of this particular nucleus have the rare distinction of responding monosynaptically with a relatively pure excitation to stimulation of the optic tract. This observation is based on extracellular (Bishop, 1964; Dubin and Cleland, 1977) as well as intracellular studies in the rabbit (Furster *et al.*, 1965), rat (Burke and Sefton, 1966c), guinea-pig (Yamamoto, 1974; Kuhnt and Schaumberg, 1976 and 1977) and cat (Suzuki and Kato, 1966; McIlwain and Creutzfeldt, 1967; Creutzfeldt, 1968; Maekawa and Rosina, 1969; Singer and Creutzfeldt, 1970; Fertziger and Purpura, 1971; Kato *et al.*, 1971; Coenen and Vendrik, 1972; Singer and Dräger, 1972; Singer *et al.*, 1972; Ono and Noel, 1973; Singer, 1973a, b; Singer and Bedworth, 1973; Singer and Lux, 1973; Eysel and Grüsser, 1975; Eysel, 1976; Ogawa *et al.*, 1978).

Unfortunately, the exact nature of the transmitter involved in the production of monosynaptic EPSP's is still unknown. Some years ago, Cobbin *et al.* (1965) extracted a mixture of biologically active compounds from optic nerves: one fraction corresponded to a completely unidentified chemical, whereas another had acetylcholine (ACh)-like properties. Although ACh "modulates" visual information within the LGN, it does not seem to be the transmitter released by the optic nerve terminals (Phillis and Tebecis, 1967; Satinsky, 1967; Bigl and Schober, 1977; Deffenu *et al.*, 1967; Godfraind, 1978; Berardi *et al.*, 1979). More recently, some workers have suggested that visual excitation is mediated by a glutamate-like substance (Johnson and Aprison, 1971; Morgan *et al*; 1972). This suggestion is based predominantly on the finding that a lesion of the optic nerve is followed by a reduction in the uptake of

[³H]glutamate and [³H]aspartate (Bondy and Purdy, 1977; Karlsen, 1978). Recently, visual responses in the LGN have been shown to be depressed by the iontophoretic application of reputed glutamate-antagonists (Kemp and Sillito, 1979).

Many LGN cells are inhibited by afferent volleys applied to the optic nerve, not monosynaptically, but through inhibitory interneurons. Indeed, electrically evoked hyperpolarizing inhibitory potentials (IPSPs) are delayed by about 1 ms with reference to the monosynaptically evoked EPSPs (Fuster et al., 1965; Suzuki and Kato, 1966; Burke and Sefton, 1966c; Creutzfeldt, 1968; Singer and Creutzfeldt, 1970; Suzuki & Takahashi, 1970; Kato et al., 1971; Eysel, 1976; Sumitomo et al., 1976b; Ogawa et al., 1980). These are predominantly postsynaptic in nature for they can be reversed by the injection of Cl⁻ ions (Burke and Sefton, 1966c) or current (Ono and Noel, 1973). A number of groups have suggested that GABA (γ-aminobutyric acid) may mediate some of the postsynaptic inhibitory phenomena in the LGN (Curtis and Tebecis, 1972; Morgan et al., 1975, Cameron and Sillito, 1977). Other putative transmitters, such as dopamine, noradrenaline, 5-hydroxy-tryptamine (5-HT) and acetylcholine (ACh) may also have an inhibitory role (Curtis and Davies, 1962; Satinsky, 1967; Phillis and Tebecis, 1967; Phillis et al., 1967; Tebecis and Di Maria, 1972; Aghajanian, 1976; Torda, 1978). Among these, special attention has been given to 5-HT for this compound has a powerful blocking effect on synaptically evoked responses in the LGN (Curtis and Davies, 1962; Tebecis and Di Maria, 1972). The action of 5-HT has been examined in LGN slices by Yama-moto (1974). It is also interesting to note here that 5-HT may be released by optic-tract stimulation in slices prepared from the superior collicus. Presumably, a similar release could occur from the LGN (Kawai, 1970; Tebecis, 1973).

Several authors have suggested that all the inhibitory phenomena observed in the LGN cannot be explained solely by IPSPs and that so-called presynaptic mechanisms must also be responsible for a reduction in excitability (Burke and Sefton, 1966b; McIlwain and Creutz-feldt, 1967; Singer et al., 1972; Hayashi, 1972; Coenen and Vendrik, 1972). Indeed, Coenen and Vendrik (1972) have claimed that such a mechanism could account for changes in amplitude of EPSPs. However, others have been unimpressed by the phenomena, either on morphological grounds (Szentagothai, 1968) or physiological grounds (Suzuki and Kato, 1966; Singer, 1973b). Clearly, a slice prepared from the LGN may prove to be a convenient and simple model for the study of excitatory and inhibitory events and the transmitters involved.

We would like to draw attention to the ease with which high-quality recordings can be made from the lateral geniculate of the rat and the cat *in vitro*. Although the project was curtailed by the limited amount of time we have been able to devote to it, it is already clear that slicing the nucleus may well alter the connectivity of the principal cells in a way that may be advantageous to the investigator. For instance, in the rat, slicing may well eliminate the recurrent inhibition thought to be mediated through inhibitory neurons located some distance away in the perigeniculate region of the thalamus (Sumitomo *et al.*, 1976b). This could allow the inhibition evoked by the intrageniculate interneurons to be studied more precisely. Although in this particular instance the elimination of the recurrent pathway did not lead to a dramatic change in the physiology of the individual cells, the ease with which excitation could be evoked by stimulation of the optic nerve in almost every principal cell encountered, raises the possibility that slicing destroys the specificity of the individual cells of the nucleus for particular parts of the retina seen *in vivo*. In other words, slicing may *increase*, as well as decrease, the functional connectivity of the neurons of the slice to levels rarely seen *in vivo*. Only recently have authors drawn attention to the possibility that *in vivo* many synapses may remain silent at a sublimal level of excitability, unless provoked into activity by quite abnormal stimuli, such as denervation (Mendell *et al.*, 1978) or the use of antagonists (Sillito, 1975, 1977; Cameron and Sillito, 1977).

II. Methods

A. Preparation of the lateral geniculate nucleus slice

The technique for the *in vitro* preparation and maintenance of the small pieces of the LGN was similar to that used previously in this laboratory for the hippocampal slice (Dingledine *et al.*, 1977, 1980).

Blocks of tissue containing the LGN were prepared from decapitated rats or cats, anaesthetized with ether, by methods similar to that described by Yamamoto (1974) for the guinea pig. With the brain inverted, the approximate position of the LGN was first located by finding the cut end of the optic nerve and the optic chiasma. The optic tract was then exposed by pushing aside the temporal lobe of the cortex with the tip of a blunt spatula. At this stage, the parietal cortex, and the more laterally situated cerebral cortex were removed by a simple almost parasagital cut placed just lateral to the optic nerve, as it curved upwards over the bulging body of the thalamus. The brain was then righted and the cerebral cortices were pushed apart and the corpus callosum was carefully divided with sharp scissors. The ipsilateral cerebral cortex was then rolled upwards and forwards, away from the cerebellum, to reveal the underlying thalamus. The cortical flap and much of the forebrain was then removed by a single, coronal cut, made right across the brain, just rostral to the thalamus. The 2 mm slice of brain containing the LGN was then separated from the rest of the brain by two parallel cuts in the plane of the optic tract, as it curves over the bulging thalamus and forward over the dorsal surface of the LGN. The cuts lie at approximately 45° to the midline. The first cut passes rostral and lateral to the thalamus and, with practice, skims the LGN and optic tract and severs the large fibres of the internal capsule as they sweep upwards and forwards. The cut was extended over the midline until it met the previous coronal cut. The second cut was made 2 mm caudal to the first. It lay well behind the LGN and passed through the middle of the superior colliculus. This cut crossed the midline and thus the slice contained a large amount of tissue from the opposite half of the brain. The LGN occupied only a small crescent-shaped area at one corner of the upper surface of the slice. The large slice was glued to the metal block of the Vibratome with the corner containing the LGN directed towards the vibrating blade and the rostral surface uppermost. When using such a slice, almost no glue need be placed under the LGN itself. During cutting, the entry of the optic-tract fibres into the LGN was easy to see with a low-powered stereomicroscope. The crescent-shaped region of the LGN was readily distinguished from the rest of the thalamus and the blade was stopped as soon as it had travelled about 2 mm into the slice, and the region of the nucleus was trimmed away with fine scissors. The rest of the cut was not completed and the next slice of the LGN was cut almost immediately. However, the whole process was made easier if the area surrounding the hole left by the removal of the LGN was trimmed back to reveal the boundaries of the LGN in the next slice.

The bath of the Vibratome contained a bicarbonate-buffered saline solution kept at about 35°C, and bubbled with a gas mixture of 95% O_2/ 5% CO_2. The solution was composed as follows (mM final concentration): NaCl, 134; KCl, 5; KH_2PO_4 1.25; $MGSO_4 \cdot 7H_2O$, 2; $CaCl_2 \cdot 2H_2O$, 2; $NaHCO_3$, 16; glucose, 10 (Schwartzkroin, 1975). Before use, the fluid was equilibrated with the gas mixture of O_2/CO_2 during about 20 min.

Usually three to four slices, 250 μm thick, were cut in the plane of the optic tract. The cutting blade (Gilette, valet strip) was set at an angle of 7° and advanced at the lowest possible speed with the greatest possible amplitude of vibration.

After sectioning, each slice was removed from the Vibratome bath in a wide-bore Pasteur pipette, and rapidly transferred to the recording chamber.

B. Perfusion in the recording chamber

The recording chamber used in this laboratory is similar to the one developed in the laboratory of Professor P. Andersen in Oslo, and is a variant of that described by Schwartzkroin (1975) (see also Chapters 2 and 3 of this volume). It consists of a small cylindrical reservoir, covered in part by a nylon mesh. Side vents prevent the appearance of bubbles under the slice. A single layer of Kodak lens paper covers the nylon mesh and directs the perfusion fluid under the slice at approximately 1.5 ml min^{-1} and then into a circular efflux reservoir. The slices lie on the lens paper and are held in the interface between the perfusion medium and an atmosphere of oxygenated humidified gas above the slice. The flow rate of slice perfusion fluid is adjusted so that the slices are just covered by a thin film of fluid.

Less than 25 minutes were required for removal of the brain and the transfer of slices to the recording chamber. The slices were then allowed to recover for about 90 minutes before recording began.

C. Recording

Single omega dot capillaries (Clark Electromedical, cat. No. GC 120F-10) were pulled on a vertical puller (Narishige, model PE2), and filled by capillarity with potassium acetate (1 M) or potassium chloride (1 M). Tip resistances were measured *in situ* with a WPI electrometer (WPI, model 707) and ranged between 60 and 500MΩ. The electrometer was modified so that a simple push button increased the feedback capacitance and "ringing" caused the tip of the pipette to impale the cell under study.

Once a cell was impaled, positive and negative rectangular pulses, or ramps of current of various amplitudes and 200 ms duration were applied intracellularly through the recording pipette. This enabled us to test the excitability and to establish current–voltage relations, for instance, at rest or during synaptic excitatory or inhibitory events.

Voltage recordings and intracellularly injected currents were stored on a FM recorder (Racal Store 4) and played back later. The resistance of the cell membrane was calculated from the respective voltage values and current intensities, or it was estimated by measuring the capacitance charging of the membrane.

D. Stimulation

A monopolar electrode was made with platinum wire insulated with glass almost to the tip (tip size 30 μm) and gently positioned under microscopic control on the optic tract fibres. Pulses of 0.1 to 0.5 ms in duration and 1 to 100 V in intensity were applied to the preparation. When a given cell had been impaled, the threshold stimulus was carefully determined. Paired stimuli were also tested at various intervals ranging from 5 to 400 ms and at various intensities (near threshold, at threshold and ten times threshold). The response evoked by each stimulus of the pair was compared in order to detect any facilitatory, or inhibitory influence of the conditioning stimulus.

III. Results

A. Cell characteristics

In the course of about 6 weeks, intracellular records were obtained from 17 rats (51 stable intracellular and 62 extracellular records); 12 intracellular and 42 extracellular records were also obtained from five kittens. The average resting potential in the rat was 55 ± 2 mV and in the cat 44 ± 2 mV. The mean spike amplitude was 42 ± 1.5 mV and 38 ± 4 mV in the rat and the cat respectively.

Many of the cells classified as extracellular were in fact quasi-intracellular, in that the spikes were accompanied by a DC-shift in excess of 20 mV, and the spikes were monophasic and rarely less than 15 mV in amplitude (McIlwain and Creutzfeldt, 1967).

The membrane resistance of most cells was relatively high and comparable with results obtained on the CA1 region of the hippocampus (Dodd et al., 1981) and in most slices lay between 20 and 40 MΩ. A few cells had resistances in excess of 100 MΩ.

Glial cells were also encountered (15 cells in six rats and 8 in two cats). They were characterized by the great stability of their resting potential and our inability to fire them with large depolarizing ramps of current. In our hands, intracellular recording from glial cells in the hippocampus is a much rarer event (Dingledine et al., 1980). Our experiences in the LGN, however, appear to be similar to those of Schwartzkroin and Prince (1976) in the hippocampus. Constanti and Galvan (1978) found enough glial cells in slices of the guinea pig olfactory cortex to study the actions of amino acids on inexcitable cells.

Although by carefully observing the surface of the slice under high-power microscopy, small groups of cells could be seen surrounded by myelinated fibres, cells were penetrated at all depths in the slice. The majority of cells were penetrated at a depth of about 75 μm.

B. Excitatory synaptic events

Practically all neurons could be driven by electrical stimulation of the optic-tract fibres. Six of the seven cells that could not be discharged by optic-fibre stimulation were encountered in the same slice. Presumably, poor preparation of this slice could account for the failure of optic-fibre stimulation to excite these cells.

In general, the threshold voltage required for adequate stimulation averaged 45 ± 6 V. In both the cat and the rat there appear to be no detectable differences in the threshold voltage between cells recorded with intracellular and extracellular electrodes. As shown in Figs. 10.1, 10.2 and 10.4, good excitations were readily obtained with voltages either below, or near, 10 V and not uncommonly with 3 V (see Fig. 10.7).

Since in the present *in vitro* conditions, the distance between the stimulating and the recording sites was less than 2 mm, the latency to the first response was always short. In the rat, the latency to the first spike was almost identical for the intra- and extracellular records and was 1.1 ± 0.3 ms and 0.9 ± 0.22 ms respectively. In the cat, the response latency for the intracellular records was 0.7 ± 0.2 ms and the extracellular 1.78 ± 0.48 ms. We have no explanation for this difference.

In the rat, optic-tract stimulation evoked only an EPSP, which was only exceptionally followed by an IPSP. This is in contrast with the situation *in vivo* where IPSPs are nearly always evoked by afferent-fibre stimulation (Fuster *et al.*, 1965; Burke and Sefton, 1966c; Suzuki and Kato, 1966; McIlwain and Creutzfeldt, 1967; Maekawa and Rosina, 1969; Singer and Creutzfeldt, 1970; Fertziger and Purpura, 1971; Kato *et al.*, 1971; Coenen and Vendrik, 1972; Ono and Noel, 1973; Eysel, 1976). On 43 occasions (35 in the rat and 8 in the cat), the stability of the penetration and ability of the pipette to pass polarizing current, permitted a careful analysis of the EPSP, using a complete family of depolarizing and hyperpolarizing current pulses through the recording electrodes (Figs. 10.1 to 10.5). The current pulses were 150–200 ms in duration and were applied some 50 ms before optic-tract stimulation, allowing the current voltage analysis to be carried out on the resting membrane, on the EPSP at its peak and at various intervals of time after the stimulus. In most instances, the amplitude of the EPSP was readily altered by the passage of current (Figs. 10.1, 10.2 and 10.3). In Figs. 10.1 and 10.2 the EPSP is shown to grow in size during the passage of hyperpolarizing current and to decrease during depolarizing current. As illustrated in Fig. 10.3 voltage–current curves showed the amplitude of the EPSP to be related to the amount of current passing through the electrode and

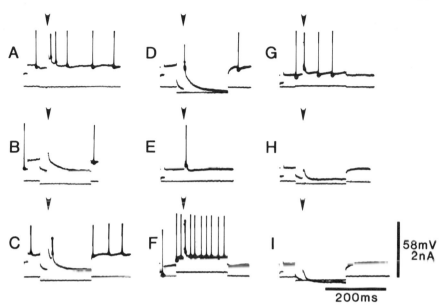

Fig. 10.1. Intracellular records from a cell in the rat dorsal lateral geniculate nucleus to show the excitation and the EPSP evoked by single-shock stimulation of the optic tract. In A–D the stimulus is 10 × threshold (100 V and 0.2 ms). The amplitude of the EPSP is enhanced by passing pulses of hyperpolarizing current through the electrode. Record E shows the typical improvement in resting potential to a steady level of about 70 mV which usually occurred within 2–10 minutes after the initial penetration of the cell. In records E–I, the stimulus is just supra-threshold for a single action potential (10 V and 0.2 ms) and in F and G, depolarizing pulses of current failed to reveal the presence of a hyperpolarizing IPSP or a stimulus-linked pause in the firing evoked by the depolarizing current. In H and I, the short latency EPSP is enhanced by the passage of depolarizing current through the electrode. The cell was recorded at a depth of 70 μm with a potassium acetate electrode, whose resistance was 210 MΩ. The spike height was 43 mV and the resting potential was 69 mV. The cell input resistance was 38 MΩ.

the peak to be associated with a clear change in membrane resistance. However, only on eight occasions was sufficient depolarizing current passed through the electrode to bring about a clear-cut reversal.

C. Inhibitory postsynaptic events

On only five occasions (three in the cat, two in the rat) was a clear-cut hyperpolarizing potential evoked by optic tract stimulation. As shown

266   J.-M. GODFRAIND AND J. S. KELLY

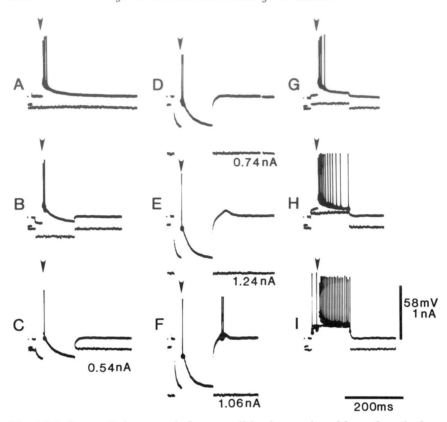

Fig. 10.2. Intracellular records from a cell in the rat dorsal lateral geniculate nucleus to show the non-linear action of current on the EPSP evoked by supramaximal stimulation of the optic tract. Record A was made after the resting potential stabilized to value of about 90mV. In B–F, hyperpolarizing pulses of current enhanced the EPSP in a graded manner, however, in E–I, depolarizing current has little effect on shape or amplitude. Again, the excitation evoked by the depolarizing pulses of current failed to reveal a hyperpolarizing IPSP, or a stimulus-linked pause in the evoked firing. The cell was recorded at a depth of 28 μm with a 200MΩ resistance electrode. The resting potential was 90mV and the spike height 70mV. The latency of the first spike was 0.1ms. Input resistance was 17.4MΩ.

in Fig. 10.4, these hyperpolarizations were associated with a stimulus-linked pause in the spontaneous firing. The passage of hyperpolarizing current through the electrode reduced the amplitude of the hyperpolarizing potential and converted it into a depolarizing response. Fig. 10.5 shows the initial equilibrium potential for this particular IPSP to be

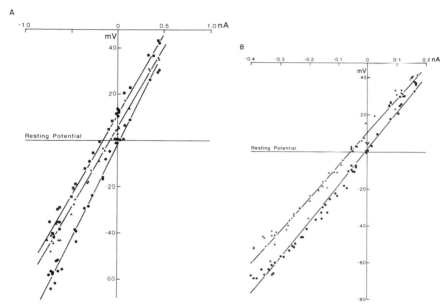

Fig. 10.3. Voltage–current plots drawn to show the influence of hyperpolarizing and depolarizing current pulses on the EPSP. In A, the resting input resistance of the cell shown in Fig. 10.2 is clearly linear (●) throughout the range of current pulses tested. The peak amplitude of the EPSP (■) is dependent on the passage of current through the electrode, and is associated with a reduction in membrane resistance. Measurements made 5 ms after the peak of the EPSP (▲) show the latter part of the EPSP to be associated with a similar change in membrane resistance. The largest depolarizing current pulses were on the point of reversing the polarity of the EPSP. B, in this cell, depolarizing current pulses were not of sufficient intensity to reach the reversal point. In B, the electrode resistance was 370 MΩ and the cell was penetrated at a depth of 38 μm. The cell resistance was 92 MΩ, its resting potential 37 mV and the spike height 56 mV.

hyperpolarizing with respect to the resting potential. After about 30 minutes, however, Cl⁻ ions diffusing into the cell from the 1M KCl pipette, caused the IPSP to become depolarizing with respect to the resting potential. When the series of depolarizing and hyperpolarizing current pulses were repeated, the amplitude of the IPSP was now decreased by depolarizing current and increased by hyperpolarizing current (Fig. 10.4, traces J–K). In Fig. 10.5, the graphs show how the diffusion of Cl⁻ into the cell caused the reversal potential of the IPSP to shift by 44 mV from −64 to −22 mV. The behaviour of this IPSP conforms to the classical description (see Kelly *et al.*, 1969) and is thus

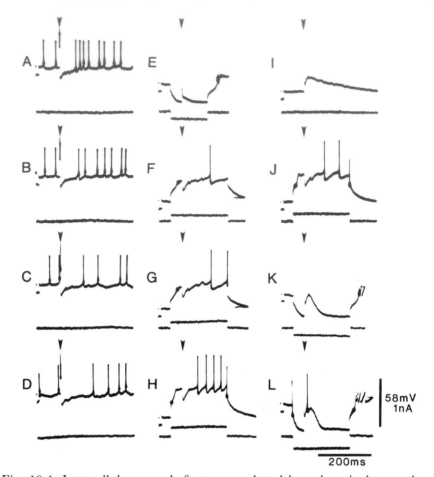

Fig. 10.4. Intracellular records from a cat dorsal lateral geniculate nucleus neuron with an electrode filled with 1M KCl which clearly shows the presence of an IPSP. Records A–D were photographed at 1Hz immediately after intracellular penetration and show the stability of the initial hyperpolarizing IPSP. Record E shows how, at this stage of the experiment, the IPSP could be reversed by the passage of a hyperpolarizing pulse of current through the electrode. Later, the resting potential of the cell improved (note DC shift between records B and F) and the IPSP was only hyperpolarizing during the passage of depolarizing current through the electrode. With time, the depolarizing IPSP grew in size (I) presumably due to the diffusion of Cl⁻ ions from the KCl electrode. However, a hyperpolarizing potential and a stimulus-linked pause in the excitation evoked by a depolarizing pulse could still be revealed by the passage of depolarizing current through the electrode. Hyperpolarizing current enhanced the amplitude of the depolarizing IPSP (K–L). In L, the depolarizing IPSP was of sufficient amplitude to trigger a single spike. The electrode resistance was 400 MΩ and the cell was recorded at a depth of 60 μm. The resting potential was 54 mV and the input resistance 123 MΩ.

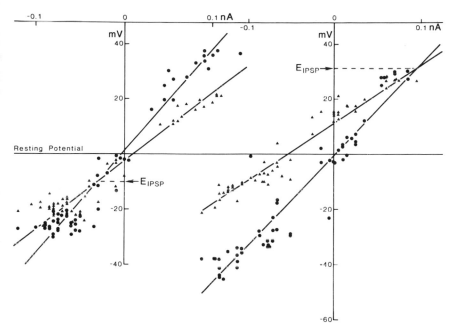

Fig. 10.5. Voltage current plots to show the reversal of an IPSP evoked by stimulation of the optic tract of a slice cut from the LGN of a kitten. The left-hand diagram is drawn from data recorded immediately after the cell was impaled. The resting membrane potential was 54 mV and the input resistance greater than 100 MΩ. Stimulation of the optic tract with a pulse 3 V in amplitude and 0.2 ms in duration evoked a large hyperpolarizing IPSP, which did not appear to be contaminated by an early excitation. The resting input resistance was clearly linear over the range of current pulses tested (●) and the peak of the IPSP (▲) could be reversed in polarity by the passage of hyperpolarizing current through the electrode. The peak of the IPSP was clearly associated by a substantial decrease in membrane resistance. However, the electrode contained 1M KCl and data collected some 30 minutes later showed the IPSP to be depolarizing, even in the absence of the passage of current through the electrode and the reversal level was shown to have moved in a depolarizing direction with respect to the resting potential.

unlike the depolarizing IPSPs observed in the olfactory cortex (Scholfield, 1978) or in the granule cells (see Chapter 6 of this volume) maintained *in vitro*.

D. Other inhibitory phenomena

However, even in the absence of a clear cut IPSP, stimulation of the optic tract in nine cells caused a stimulus-linked pause in the discharge evoked either by injury, or by the passage of depolarizing current through the electrode. In an attempt to detect the presence, and to analyse further the nature of these inhibitory events occurring in the

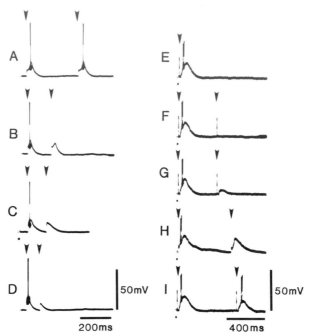

Fig. 10.6. Intracellular records to show the presence of "remote" inhibitory phenomena. Intracellular records (A–D) from a rat cell in which depolarizing pulses failed to evoke a stimulus-linked hyperpolarizing potential, or an inhibitory pause in the excitation evoked by depolarizing current (see Fig. 10.1). Stimulation of the optic tract just above, and just below threshold for a single action potential was accompanied by an inhibition of spontaneously occurring action potentials of at least 300 ms in duration. Two shock studies confirmed the presence of an inhibition lasting approximately 300 ms. In A–D, where the stimulus was just suprathreshold for a single spike, the inhibition was graded and although at longer intervals was just sufficient to inhibit the spike, at shorter intervals it reduces the amplitude of the EPSP.

Records E–I were obtained from the cell from a cat slice shown in Figs. 10.4 and 10.5, after the IPSP inverted and became a depolarization of sufficient magnitude to fire two action potentials. In E–I, a progressive increase in the interval between the two shocks shows the IPSP evoked by the second shock to be completely inhibited by the first shock at intervals as great as 400 ms. At greater intervals, (G–I) the reduction in the second IPSP was graded. This may well be an example of mutual inhibition.

absence of an IPSP, double-shock stimuli were applied to the optic nerve on 52 occasions, and the period of time between the two shocks varied between a few and 400 ms. In 17 cells, no change in the amplitude of the response evoked by the second shock was observed regardless of the stimulus strength or the interval between the paired stimuli. The response evoked by the second shock was inhibited in 11 cells, and potentiated in 16 cells. In the remaining eight cells, mixed effects were observed, consisting of a sequence of events in which potentiation lasted for about 10 ms and was followed by inhibition lasting a further 200 ms.

In Fig. 10.6, intracellular records from two different cells show how

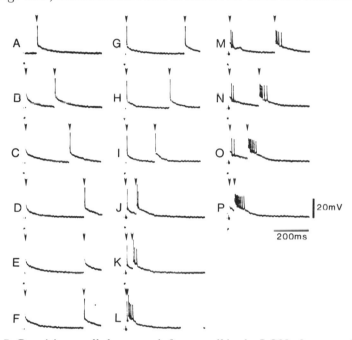

Fig. 10.7. Quasi-intracellular records from a cell in the LGN of a rat to show the facilitation of the response evoked by stimulation of the optic tract by paired stimuli. In records A–F, the stimulus strength of both stimuli was set just below threshold for evoking a single spike by the first stimuli of the pair (40 V at 0.1 ms). At intervals of less than 140 ms the second stimulus fired a spike. In records D–F, the interval between the two stimuli is just critical for the firing of a spike by the second stimulus. At intervals of less than 140 ms the second stimulus fired more than one spike. In records G–L, both stimuli were set at threshold, only at very short intervals was the firing evoked by the second stimulus enhanced. When the stimulation strength was maximal (100 V) (M–P) the second stimulus of the pair fired a greater number of spikes whenever the interval between the two stimuli was 185 ms or less.

272 J.-M. GODFRAIND AND J. S. KELLY

both EPSPs and IPSPs evoked by the test stimulus can be reduced in amplitude as the interval between the two shocks is progressively reduced.

In Fig. 10.7, in which identical test and conditioning shocks were delivered by the same stimulator, the duration of the facilitation was shown to be dependent on stimulus intensity. In the first series of tests

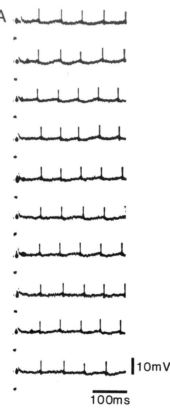

Fig. 10.8. Quasi-intracellular records from a cell in the rat dorsal lateral geniculate nucleus to show rhythmic firing in response to a single shock stimulus to the optic tract. The 10 mV calibration pulse at the beginning of each trace is followed immediately by the shock artefact evoked by a 22 V stimulus to the optic tract. At higher stimulus intensities, a single spike was evoked at a latency of 0.8 ms and was followed by 100 ms inhibition of the initial injury discharge. A two-shock study confirmed the presence and the duration of this inhibition. The synchrony of the spikes from trace to trace was maintained over many hundreds of cycles. The cell was recorded at a depth of 110 μm and the electrode resistance was 110 MΩ.

shown on traces B–F, the stimulus intensity was set just below threshold for a single spike on the first shock and the interval between the two shocks was progressively increased until the spike fired by the second shock disappeared (E). When the stimulus was increased to a level just above threshold, for a spike on the conditioning response, the facilitation of the second response by the first from one spike to two occurred only at very short time intervals, and the duration of the facilitation appears to have been reduced 4-fold. A further increase in the stimulus intensity to a maximum of 90 V caused the duration of the facilitation to be restored. Indeed, on this occasion, a further decrease in the interval between the shocks caused a clear-cut facilitation of the number of spikes evoked by the second stimulus.

E. Rhythmic firing

The spontaneous firing of three cells was observed to be rhythmic (Fig. 10.8), and similar to that observed *in vivo* by Burke and Sefton (1966b) in the rat, and *in vitro* by Kato (1974) in the cat. Of these three neurons, only one was impaled intracellularly, and in this cell there was no sign of inhibitory postsynaptic activity. Rhythmic firing in the olfactory bulb has been attributed to dendro-dendritic reciprocal synapses (see Mori and Tagaki, 1978). The existence of reciprocal synapses in the LGN has also been suggested on anatomical grounds by Famiglietti (1970) and by Lieberman and Webster (1974).

IV. Discussion

Our main aim was to test the viability of a slice prepared from the dorsal lateral geniculate nucleus of rat and cat maintained *in vitro*, and our ability to penetrate cells with fine glass microelectrodes. It is clear from this account that intracellular recording is relatively easy and offers a convenient alternative to intracellular recording *in vivo*.

A. Driven cells

Although the ease with which almost all cells could be excited by stimulation of the optic tract is extremely convenient, in the long run it is rather disconcerting. One possible explanation for this would be that the optic nerve fibres innervate much larger numbers of principal cells than is apparent *in vivo* (Creutzfeldt, 1968; Singer and Creutzfeldt, 1970; Kato *et al.*, 1971; Coenen and Vendrik, 1972; Fertziger and Purpura, 1971). However, *in vivo* the intracellular analysis of the above-mentioned authors favour the assumption that there is a one-to-one relationship between optic tract fibres and the geniculate cells. Earlier, the same conclusion was reached by Bishop (1964) because of the all-or-none properties of the electrically evoked EPSP. On the other hand, additional, but minor, excitatory synaptic inputs from other optic nerve fibres is not altogether excluded (Singer and Creutzfeldt, 1970; Kato et al., 1971). Indeed, Coenen and Vendrik (1972) insist that divergence must occur, since the number of optic tract fibres is estimated to be 125,000 and the number of LGN principal cells 450,000 (Bishop, 1953). This also agrees with another comment made by Singer and Creutzfeldt (1970), who mention that degeneration and Golgi studies suggest that one optic tract fibre is connected to several geniculate cells and that a considerable overlap of the brush-like endings from several fibres exists. Thus, presumably, much of the apparent specification of the LGN cells for one particular cell in the retina is not only due to the morphological connectivity of the afferent fibres, but the way in which inhibitory and facilitatory processes are distributed.

B. Graded EPSPs

In several intracellular studies, it has been noted that a gradual increase in stimulus strength, often, but not always, resulted in a stepwise increment of the amplitude of the EPSP (Maekawa and Rosina, 1969; Kato et al., 1971; Yamamoto, 1974; Eysel, 1976). Eysel (1976) has proposed that the increase in EPSP amplitude which occurs with increasing stimulus strength, is due to convergence, or possibly non-synchronous excitation. If convergence occurs, isolation of the individual components of EPSP would be difficult, since the differences in threshold between the different optic-tract axons involved is probably small (Eysel, 1976). In this study, the amplitude of the EPSP was independent of the stimulus strength and remained the same whether the stimulus was just below threshold, or ten times threshold. However, the EPSP amplitude was not fixed and could be increased, for instance by repetitive stimulation. It is therefore possible, in the present experiments, that the stimulus inadvertently excited all the afferent fibres simultaneously.

C. Identification of I- and P-cells

In all probability, most of our records were from principal cells, for these are much more numerous than interneurons that make up less than 10% of the total cells that can be recorded from in the rat (Burke and Sefton, 1966a; Sumitomo and Iwama, 1977) and in the cat in vivo (Dubin and

Cleland, 1977) and *in vitro* (Ogawa *et al.*, 1980). In the cat, Ono and Noel (1973) claimed to have recorded from an unusually high percentage of interneurons *in vivo*. However, recent histological studies appear to agree with the former estimate (Le Vay and Fester, 1979). In addition, principal cells that are only about 15–20 μm in diameter are by far the largest cell present and thus easier to penetrate than interneurons (Hayhow *et al.*, 1962; Grossman *et al.*, 1973; Guillery, 1966; Laemle, 1975; Szentagothai, 1973). Indeed, even simple extracellular recording from interneurons is difficult (Sumitomo and Iwama, 1977). However, other workers have differentiated between relay cells and interneurons, by the way in which the cells respond to an orthodromic, or antidromic volley. Burke and Sefton (1966a) state that following an orthodromic volley P-cells are characterized by a short latency EPSP, or action potential, followed by an IPSP, while I-cells are characterized by a long-lasting EPSP, on which a burst of spikes is superimposed. Similar criteria were also used by Ono and Noel (1973) in a study in which they showed 45% of the cells to be interneurons. Ogawa *et al.* (1980) carried out similar tests *in vitro*. In the present study, no such differentiation was possible, since very few IPSPs were observed. This is, in all probability, a consequence of using much thinner slices than those used by Ogawa *et al.* (1980). Presumably, slicing interrupts the connectivity of the interneurons. However, the exact location of the interneurons that respond with the characteristics described by Burke and Sefton (1966a) has recently been re-examined more precisely. Histological reconstruction of the electrode tract (Sumitomo *et al.*, 1976a; Ogawa *et al.*, 1978), showed interneurons not to be localized in the LGN itself, but in an adjacent region of the nucleus reticularis thalami (see above). Nevertheless, the intrageniculate interneurons can be recognized *in vivo* both in the cat (Dubin and Cleland, 1977) and the rat (Sumitomo and Iwama, 1977) since they are trans-synaptically excited by electrical stimulation of the visual cortex, whereas the principal cells are excited anti-dromically. Indeed, others have tried to apply the antidromic invasion test *in vitro*, by placing a second set of stimulating electrodes on the edge of the slice (Kuhnt and Schaumberg, 1976, 1977). Unfortunately, the absence of an antidromic spike does not confirm that a cell under study is an interneuron. It has also been suggested that the small interneurons of the LGN may also send their axons to the cortex and that there are no true interneurons within the LGN (Laemle, 1975).

D. Inhibitory events in the slice

In the present study, optic-tract stimulation has occasionally been shown to produce rather long-lasting inhibitory events. This not only confirms the viability of the LGN inhibitory interneurons, but also suggests that the slice preparation can be used to explore the mechanisms involved in the production of inhibitions. In this context, the paucity of IPSPs and the ease with which both inhibitory and facilitatory phenomena can be demonstrated by the double-shock technique, supports the idea that *in vivo* inhibition of principal cells of the LGN is mediated by two different groups of interneurons (Fig. 10.9) (cf Burke and Cole, 1978; Singer, 1977). The first group is thought to lie within the nucleus, and to be responsible for feed-forward inhibition (Dubin and Cleland, 1977; Sumitomo and Iwama, 1977). In contrast, the second group of neurons are not thought to be present within the LGN, but located in the perigeniculate reticular area (Sumitomo *et al.*, 1976a, b; Schmielau, 1979). However, more recent data has suggested that at least two subgroups of cells lie within the perigeniculate reticular area (Sanderson, 1971; Dubin and Cleland, 1977; Ahlsen and Lindström, 1978; Godfraind, 1978; Schmielau, 1979; Sumitomo *et al.*, 1977), and that they are responsible for both feed-forward and feedback inhibition of LGN principal neurons. Since the inhibition of most cells in the slice was not accompanied by a change in membrane resistance, our data supports the hypothesis that most of the IPSPs observed *in vivo* are mediated either by a recurrent pathway via the reticular cells (equivalent to feedback inhibition) (Ahlsen *et al.*, 1978; Dubin and Cleland, 1977; Jones, 1975; Schmielau, 1979; Sumitomo *et al.*, 1976a, 1977; Szentagothai, 1972) or a feed-forward pathway in which the perigeniculate cells are fired monosynaptically by the optic-tract fibres (Sumitomo *et al.*, 1976a, 1977; Schmielau, 1979). This view is consistent with the experiments of Sumitomo *et al.* (1976a, b) who showed *in vivo* electrical stimulation of the perigeniculate–reticular region evoked IPSPs in the LGN cells and that destruction of the same zone virtually abolishes postsynaptic inhibition in the same cells. As mentioned above, slicing would presumably eliminate inhibition by interrupting the actions of interneurons within the LGN, or by excluding the perigeniculate reticular cells from the slice. However, the intrageniculate neurons may not be conventional interneurons (Lieberman, 1973; Hamori *et al.*, 1978; Le Vay, 1971) for instance, it has also been proposed that the dendritic appendages of the principal cells may behave as if they were "small quasi interneurones", and act virtually independently (Fetziger and Purpura, 1971; Maekawa and Rosina, 1969; Szentagothai, 1973; Hamori *et al.*, 1974). This anatomical specialization could well be the origin of the "remote" or the "presynaptic" inhibition revealed by the double-shock technique. Clearly, in the present study, the precise nature of the inhibition cannot be defined more precisely. Even in the spinal cord, the existence of presynaptic inhibition has recently been challenged by a study in which a small increase in membrane conductance was shown to be associated with the inhibitory phenomena (Carlen *et al.*, 1980).

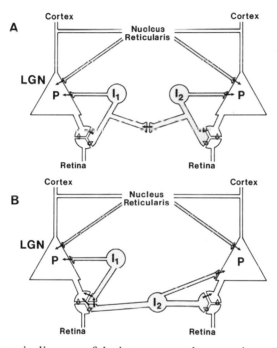

Fig. 10.9. Schematic diagram of the interneuronal connections within a lateral geniculate nucleus slice. Diagram shows the possible connections between the various elements of the slice, optic fibres, principal cells (P) and interneurons (I), which might account for the inhibitory and facilitatory events observed during a two-shock study. Open and closed arrows indicate the presence of an excitatory, or an inhibitory synapse respectively. Slicing 250 μm thick slice is assumed to sever the axon collaterals of the P cells and of the inhibitory interneurons of the nucleus reticularis. Although the actual circuitry in A and B is almost identical, in the B the interneuron I_2 is more intimately involved in the connections of the left-hand glomerulus (gemmule). Inhibition or facilitation of the response to the second stimulus will depend not only on the size of the optic fibre input to the two inhibitory interneurons I_1 and I_2, but the contribution of each interneuron to the individual glomeruli and the connectivity between the two interneurons.

E. Facilitatory events in the slice

As illustrated in 10.9A, the retinal afferent may activate the dendrites of the intrageniculate I-cell, which in turn will inhibit locally the post-synaptic principal cell. Of course, it is also conceivable that the inhibition would be mediated at other dendritic appendages within the same cell, up to distances of 400 μm away from the site of excitation. Clearly, local interactions (Jahr and Nicoll, 1980) between neighbouring dendritic appendages of the same, or different cells (Fig. 10.9B) could result in disinhibition and thus explain the facilitatory event observed in the two-shock study. Similar phenomena were observed by Mori and Takagi (1978) in the olfactory bulb.

V. Conclusions

1. In 250 μm slices prepared from the dorsal lateral geniculate nucleus (LGN) of rat and cat, intracellular and extracellular recording showed virtually all 51 LGN cells to be driven by electrical stimulation of the optic tract. The mean response latency was around 1 ms.

2. In only five cells was the EPSP followed by an IPSP. The amplitude of the EPSPs could be enhanced or attenuated by passing current pulses through the recording electrode. However, true reversal occurred on only eight occasions.

3. IPSPs were observed only rarely. However, when they occurred they became hyperpolarizing and their amplitude and polarity was readily influenced by the passage of current through the electrode. They were readily reversed by the diffusion of Cl^- ions from the KCl recording pipette.

4. Double-shock studies revealed the inhibition and facilitation of the second response by the first.

5. In 11 cases, the response evoked by the second shock was inhibited in the absence of any detectable change in membrane resistance.

6. This inhibitory response may be an example of "remote" or "pre-synaptic" inhibition.

7. In another 16 cells, the response evoked by the second shock was potentiated.

8. This facilitation may be the consequence of mutual inhibition at dendro-dendritic synapses.

Acknowledgements

The majority of the experiments described in this report have not been published elsewhere and were carried out in the MRC Neurochemical Pharmacology Unit in Cambridge, where J.S.K. was a member of staff. J.-M.G. was in receipt of an award from the British Council and the F.N.R.S. Belgium. We are indebted to Dr. S. Maruyama of the University of Niigata, who participated in many of the experiments, and to George Marshall, for his patient assistance and expert attention to detail.

References

Aghajanian, G. K. (1976). LSD and L-bromo-LSD: comparison of effects on serotonin neurones in two serotonergic projection areas, the lateral geniculate and amygdala. *Neuropharmacol* **15**, 521–528.

Ahlsen, G. and Lindström, S. (1978). Axonal branching of functionally identified neurones in the lateral geniculate body of the cat. *NELED Suppl.* **1**, 156.

Ahlsen, G., Lindström, S. and Sybirska, E. (1978). Subcortical axon collaterals of principal cells in the lateral geniculate body of the cat. *Brain Res.* **156**, 106–109.

Berardi, N., Kemp, J.A., Milson, J.A. and Sillito, A.M. (1980). Cholinergic modulation of dorsal lateral geniculate nucleus (dLGN) activity. *J. Physiol.* (Lond.) **301**, 15P.

Bigl, V. and Schober, W. (1977). Cholinergic transmission in subcortical and cortical visual centres of rats: no evidence for the involvement of primary optic system. *Exp. Brain Res.* **27**, 211–219.

Bishop, P. O. (1953). Synaptic Transmission. *Proc. R. Soc. Ser B.* **141**, 362–392.

Bishop, P. O. (1964). The neuronal organization of the visual pathways in the cat. *Prog. Neurobiol.* **6**, 149–184.

Bondy, S. C. and Purdy, S. L. (1977). Putative neurotransmitters of the avian visual pathway. *Brain Res.* **119**, 417–420.

Burke, W. and Cole, A. M. (1978). Extraretinal influences on the lateral geniculate nucleus. *Rev. Physiol. Biochem. Pharmacol.* **80**, 105–166.

Burke, W. and Sefton, A. J. (1966a). Discharge patterns of principal cells and interneurones in lateral geniculate nucleus of rat. *J. Physiol. (Lond.)* **187**, 201–212.

Burke, W. and Sefton, A. J. (1966b). Recovery of responsiveness of cells of lateral geniculate nucleus of rat. *J. Physiol. (Lond.)* **187**, 213–229.

Burke, W. and Sefton, A. J. (1966c). Inhibitory mechanisms in lateral geniculate nucleus of rat. *J. Physiol. (Lond.)* **187**, 231–246.

Cameron, N. E. and Sillito, A. M. (1977). The effect of bicuculline on receptive field organization of cells in the dorsal lateral geniculate nucleus. *J. Physiol. (Lond.)* **271**, 55P.

Carlen, P. L., Werman, R. and Yaari, Y. (1980). Post-synaptic conductance increase associated with presynaptic inhibition in cat lumbar motoneurones. *J. Physiol. (Lond.)* **298**, 539–556.

Cobbin, L. B., Leeder, S. and Pollard, J. (1965). Smooth muscle stimulants in extracts of optic nerves, optic tracts and lateral geniculate bodies of sheep. *Br. J. Pharmacol.* **25**, 295–306.

Coenen, A. M. L. and Vendrik, A. J. H. (1972). Determination of the transfer ratio of cat's geniculate neurones through quasi-intracellular recordings and the relation with the level of consciousness. *Exp. Brain. Res.* **14**, 227–242.

Constanti, A. and Galvan, M. (1978). Amino-acid-evoked depolarization of electrically inexcitable (neuroglial?) cells in the guinea-pig olfactory cortex slice. *Brain Res.* **153**, 183–187.

Curtis, D. R. and Davis, R. (1962). Pharmacological studies upon neurones of the lateral geniculate nucleus of the cat. *Br. J. Pharmacol.* **18**, 217–246.

Curtis, D. R. and Tebecis, A. K. (1972). Bicuculline and thalamic inhibition. *Exp. Brain Res.* **16**, 210–218.

Creutzfeldt, O. D. (1968). Functional synaptic organization in the lateral geniculate nucleus and its implication for information transmission. *In* "Structure and function of inhibitory neuronal mechanisms" (C. Von Euler, S. Skoglund and V. Södeberg, eds.), pp. 117–122. Pergamon Press, Oxford.

Deffenu, G., Bertaccini, G. and Pepeu, G. (1967). Acetylcholine and 5-Hydroxytryptamine levels of the lateral geniculate bodies and superior colliculus of cats after visual deafferentation. *Exp. Neurol.* **17**, 203–209.

Dingledine, R., Dodd, J. and Kelly, J. S. (1977). Intracellular recording from pyramidal neurones in the in vitro transverse hippocampal slice. *J. Physiol. (Lond.)* **269**, 13–15P.

Dingledine, R., Dodd, J. and Kelly, J. S. (1980). The in vitro brain slice as a useful neurophysiological preparation for intracellular recording. *J. Neurosci. Meth.*, **2**, 323–362.

Dodd, J., Dingledine, R. and Kelly, J. S. (1981). The excitatory action of acetylcholine on hippocampal neurones of the guinea pig and rat maintained in vitro. *Brain Res.*, in press.

Dubin, M. W. and Cleland, B. G. (1977). Organization of visual inputs to interneurones of lateral geniculate nucleus of the cat. *J. Neurophysiol.* **40**, 410–427.

Eysel, U. Th. (1976). Quantitative studies of intracellular post-synaptic potentials in the lateral geniculate nucleus of the cat with respect to optic tract stimulus response latencies. *Exp. Brain Res.* **25**, 469–486.

Eysel, U. Th. and Grüsser, O. J. (1975). Intracellular post-synaptic potentials of cat lateral geniculate cells and the effects of degeneration of the optic terminals.

Famiglietti, E. V., Jnr. (1970). Dendro-dendritic synapses in the lateral geniculate nucleus of the cat. *Brain Res.* **20**, 181–191.

Fertziger, A. P. and Purpura, D. P. (1971). Diphasic-PSPs during maintained activity of cat lateral geniculate neurones. *Brain Res.* **33**, 463–467.

Fuster, J. M., Creutzfeldt, O. D. and Straschill, M. (1965). Intracellular recordings of neuronal activity in the visual system. *Z. Vergl Physiol.* **49,** 605–622.

Godfraind, J.-M. (1978). Acetylcholine and somatically evoked inhibition on perigeniculate neurones in the cat. *Br. J. Pharmacol.* **63,** 295–302.

Guillery, R. W. (1966). A study of golgi preparations from the dorsal lateral geniculate nucleus of the adult cat. *J. Comp. Neurol.* **128,** 21 50.

Grossman, A., Lieberman, A. R. and Webster, K. E. (1973). A golgi study of the rat dorsal lateral geniculate nucleus. *J. Comp. Neurol.* **150,** 441–466.

Hamori, J., Pasik, T. and Pasik, P. (1978). Electron-microscopic identification of axonal initial segments belonging to interneurones in the dorsal lateral geniculate nucleus of the monkey. *Neuroscience,* **3,** 403 412.

Hayashi, Y. (1972). Terminal depolarizations of intra-geniculate optic tract fibres produced by moving visual stimulus. *Brain Res.* **42,** 215–219.

Hayhow, W. R., Sefton, A. J. and Webb, C. (1962). Primary optic centres of the rat in relation to the terminal distribution of the crossed and uncrossed optic nerve fibres. *J. Comp. Neurol.* **118,** 295 321.

Jahr, C. E. and Nicoll, R. A. (1980). Dendro-dendritic inhibition: demonstration with intracellular recording. *Science* **207,** 1473–1475.

Johnson, J. L. and Aprison, M. H. (1971). The distribution of glutamate and total free amino acids in thirteen specific regions of the cat central nervous system. *Brain Res.* **26,** 141–148.

Jones, E. G. (1975). Some aspects of the organization of the thalamic reticular complex. *J. Comp. Neurol.* **162,** 285–308.

Karlsen, L. (1978). Neurotransmitters of the mammalian visual system. *In* "Amino-acids as Chemical Transmitters" (F. Fonnum, ed.), pp. 241–256. NATO Advanced Study Institute series. Plenum Press, New York.

Kato, H. (1974). Types of neurones in the sliced LGB of cat. *Brain Res.* **66,** 332–336.

Kato, H., Yamamoto, M. and Nakahama, H. (1971). Intracellular recordings from the lateral geniculate neurones of cats. *Jap. J. Physiol.* **21,** 307–323.

Kawai, N. (1970). Release of 5-HT from slices of superior colliculus by optic tract stimulation. *Neuropharmacology* **9,** 395–397.

Kelly, J. S., Godfraind, J.-M. and Maruyama, S. (1979). The presence and nature of inhibition in small slices of the dorsal lateral geniculate nucleus of rat and cat incubated *in vitro. Brain Res.* **168,** 388–392.

Kelly, J. S., Krnjević, K., Morris, M. E. and Yim, G. K. W. (1969). Anionic permeability of cortical neurones. *Exp. Brain Res.* **7,** 11–31.

Kemp, A. J. and Sillito, A. M. (1979). Action of excitatory amino acid antagonists on the retinal input to the cat lateral geniculate body. *J. Physiol. (Lond.)* **292,** 46P.

Kuhnt, U. and Schaumberg, R. (1976). Spontaneous and evoked activity in slices maintained *in vitro* for the lateral geniculate body and the cerebral cortex of the guinea-pig. *Exp. Brain Res. Suppl.* **1,** 394–396.

Kuhnt, U. and Schaumberg, R. (1977). Extrinsic afferent activation of identified neurones of the lateral geniculate body and the visual cortex in the guinea-pig. An *in vitro* study. *XXVII Int. Union Physiol. Sci.* Abstract.

Laemle, L. K. (1975). Cell populations of the lateral geniculate nucleus of the cat as determined with horseradish peroxidase. *Brain Res.* **100,** 650–656.

Le Vay, S. (1971). On the neurones and synapses of the lateral geniculate nucleus of the monkey, and the effects of eye nucleation. *Z. Zellforsch. Mikrosk. Anat.* **113,** 396–419.

Le Vay, S. and Fester, D. (1979). Proportion of interneurones in cat's lateral geniculate nucleus. *Brain Res.* **164,** 304–308.

Lieberman, A. R. (1973). Neurones with presynaptic perikaya and presynaptic dendrites in the rat lateral geniculate nucleus. *Brain Res.* **59,** 35–59.

Lieberman, A. R. and Webster, K. E. (1974). Aspects of the synaptic organization of intrinsic neurones in the dorsal lateral geniculate nucleus. An ultrastructural study of the normal and of the experimentally deafferented nucleus in the rat. *J. Neurocytol.* **3,** 677–710.

McIlwain, J. T. and Creutzfeldt, O. D. (1967). Microelectrode study of synaptic excitation and inhibition in the lateral geniculate nucleus of the cat. *J. Neurophysiol.* **30,** 1–21.

Maekawa, K. and Rosina, A. (1969). Synaptic transmission in the sensory relay neurones of the thalamus. *In* "Progress in Brain Research" (K. Akert and P. G. Waser, eds.), vol. 31, pp. 259–264. Elsevier, Amsterdam.

Mendell, L. M., Sassoon, E. M. and Wall, P. D. (1978). Properties of synaptic linkage from long ranging afferents onto dorsal horn neurones in normal and deafferented cats. *J. Physiol. (Lond.)* **255,** 299–310.

Morgan, R., Vrbova, G. and Wolstencroft, J. H. (1972). Correlations between the retinal input to lateral geniculate neurones and their relation response to glutamate and aspartate. *J. Physiol. (Lond.)* **224,** 41–42P.

Morgan, R., Sillito, A. M. and Wolstencroft, J. H. (1975). A pharmacological investigation of inhibition in the lateral geniculate nucleus. *J. Physiol. (Lond.)* **246,** 93–94P.

Mori, K. and Takagi, S. (1978). An intracellular study of dendrodendritic inhibitory synapses on mitral cells in the rabbit olfactory bulb. *J. Physiol. (Lond.)* **279,** 569–588.

Ogawa, T., Takimori, T. and Takahaski, Y. (1978). Intracellular recording and staining of cat's lateral geniculate neurones. *Brain Res.* **139,** 35–41.

Ogawa, T., Ito, S. and Kato, H. (1980). P- and I-cells in *in vitro* slices of the lateral geniculate nucleus of the cat. *Tohoku J. Exp. Med.,* **130,** 359–368.

Ono, T. and Noel, W. K. (1973). Characteristics of P- and I-cells of the cat's lateral geniculate body. *Vision Res.* **13,** 639–646.

Phillis, J. W. and Tebecis, A. K. (1967). A study of cholinergic cells in the lateral geniculate nucleus. *J. Physiol. (Lond.)* **192,** 695–713.

Phillis, J. W., Tebecis, A. K. and York, D. H. (1967). The inhibitory action of monoamines on lateral geniculate neurones. *J. Physiol. (Lond.)* **190,** 563–581.

Sanderson, K. J. (1971). The projection of the visual field to the lateral geniculate and medial interlaminar nuclei in the cat. *J. Comp. Neurol.* **143,** 101–118.

Satinsky, D. (1967). Pharmacological responsiveness of lateral geniculate nucleus neurones. *Int. J. Neuropharmacol.* **6,** 387–397.

Schaumberg, R. (1977). Morphologisch-physiologisch Untersuchungen zur organisation des Nucleus Geniculatum Laterale Dorsalis des Meerscheinchen an *in vitro* Praparationen. Doctoral Thesis, Göttingen, 1977.

Schmielau, F. (1979). Integration of visual and nonvisual information in nucleus reticularis thalami of the cat. *In* "Developmental Neurobiology of Vision" (R. D. Freeman, ed.), pp. 205–226. Plenum Press, New York.

Scholfield, C. N. (1978). A depolarizing inhibitory potential in neurones of the olfactory cortex *in vitro*. *J. Physiol. (Lond.)* **275**, 547–557.

Schwartzkroin, P. A. (1975). Characteristics of CA1 neurones recorded intracellularly in the hippocampal *in vitro* slice preparation. *Brain Res.* **85**, 423–436.

Schwartzkroin, P. A. and Prince, P. A. (1976). Microphysiology of human cerebral cortex studied *in vitro*. *Brain Res.* **115**, 497–500.

Sillito, A. M. (1975). The contribution of inhibitory mechanisms to the receptive field properties of neurones in the striate cortex of the cat. *J. Physiol. (Lond.)* **250**, 305–309.

Sillito, A. M. (1977). Inhibitory processes underlying the directional specificity of simple, complex and hypercomplex cells in the cat's visual cortex. *J. Physiol. (Lond.)* **271**, 699–720.

Singer, W. (1973a). The effect of mesencephalic reticular stimulation on intracellular potentials of cat lateral geniculate neurones. *Brain Res.* **61**, 35–54.

Singer, W. (1973b). Brain-stem stimulation and the hypothesis of presynaptic inhibition in cat lateral geniculate nucleus. *Brain Res.* **61**, 55–68.

Singer, W. (1977). Control of thalamic transmission by corticofugal and ascending reticular pathways in the visual system. *Physiol. Rev.* **57**, 386–420.

Singer, W. and Bedworth, N. (1973). Inhibitory interaction between X- and Y-units in the cat lateral geniculate nucleus. *Brain Res.* **43**, 291–307.

Singer, W. and Creutzfeldt, O. D. (1970). Reciprocal lateral inhibition of on- and off-centre neurones in the lateral geniculate body of the cat. *Exp. Brain Res.* **10**, 311–330.

Singer, W. and Dräger, V. (1972). Postsynaptic potentials in relay neurones of cat lateral geniculate nucleus after stimulation of the mesencephalic reticular formation. *Brain Res.* **41**, 214–220.

Singer, W. and Lux, H. D. (1973). Presynaptic depolarization and extracellular potassium in the cat lateral geniculate nucleus. *Brain Res.* **64**, 17–33.

Singer, W., Pöppel, E. and Creutzfeldt, O. D. (1972). Inhibitory interaction in the cat's lateral geniculate nucleus. *Exp. Brain Res.* **14**, 210–226.

Sumitomo, I. and Iwama, K. (1977). Some properties of intrinsic neurones of the dorsal lateral geniculate nucleus of the rat. *Jap. J. Physiol.* **27**, 717–730.

Sumitomo, I., Iwama, K. and Nakamura, M. (1976a). Optic nerve innervation of so-called interneurones of the rat lateral geniculate body. *Tohoku, J. Exp. Med.* **119**, 149–158.

Sumitomo, I., Nakamura, M. and Iwama, K. I. (1976b). Location and function of the so-called interneurones of the rat lateral geniculate body. *Exp. Neurol.* **51**, 110–123.

Sumitomo, I., Sugitani, M. and Iwama, K. (1977). Disinhibition of perigeniculate reticular neurones following chronic ablation of the visual cortex in rats.
Tohoku, J. Exp. Med. **122,** 321–329.

Suzuki, H. and Kato, H. (1966). Binocular interaction at cat's lateral geniculate body. *J. Neurophysiol.* **29,** 909–920.

Suzuki, H. and Takahashi, M. (1970). Organization of lateral geniculate
neurones in binocular inhibition. *Brain Res.* **23,** 261–264.

Szentagothai, J. (1968). Synaptic structure and the concept of presynaptic
inhibition. *In* "Structure and Functions of Inhibitory Neuronal
Mechanisms" (C. Von Euler, S. Skoglund and U. Soderberg, eds.), pp.
15–31. Pergamon Press, Oxford.

Szentagothai, J. (1972). Lateral geniculate body structure and eye movements.
In "Cerebral Control of Eye Movements and Motion Perception. (J. Dich-
gans and E. Bizzi, eds.), Bibl. Ophtal. Vol. 82, pp. 178–188. Karger, Basle.

Szentagothai, J. (1973). Architecture of the lateral geniculate nucleus. *In*
"Handbook of Sensory Physiology" (R. Jung, ed.), vol. VII. 3B, pp.
141–176. Springer, Berlin.

Tebecis, A. K. (1973). Studies on the identity of the optic nerve transmitter.
Brain Res. **63,** 31–42.

Tebecis, A. K. and Di Maria, A. (1972). A re-evaluation of the mode of action
of 5-hydroxytryptamine on lateral geniculate neurones: comparison with
catecholamines and LSD. *Exp. Brain Res.* **14,** 480–493.

Torda, C. (1978). Effects of noradrenaline and serotonin on activity of single
lateral and medial geniculate neurones. *Gen. Pharmacol.* **9,** 455–462.

Yamamoto, C. (1974). Electrical activity recorded from thin sections of the
lateral geniculate body, and the effects of 5-hydroxytryptamine. *Exp. Brain
Res.* **19,** 271–281.

11

An *In Vitro* Preparation of the Diencephalic Interpeduncular Nucleus

D. A. BROWN AND J. V. HALLIWELL

*Department of Pharmacology, The School of Pharmacy,
29/39 Brunswick Square, London WC1N 1AX, U.K.*

I. Introduction

The presence of well-defined anatomical pathways in thin slices of mammalian central nervous tissue facilitates their electrophysiological analysis when maintained *in vitro* under suitable incubation conditions.

This was a key factor in the success of the olfactory cortex slice (McIlwain, 1966; Yamamoto and McIlwain, 1966): this slice incorporates a discrete bundle of afferent fibres, the lateral olfactory tract, which may be activated with the minimum of stimulus spread. The analysis of this preparation has reached a high level of sophistication (see chapters 4 and 5 of this volume). The well-ordered serial excitatory projections between the spatially separated cell groups in the hippocampal formation (Andersen et al., 1971b) are responsible for slices of this structure (Skrede and Westgaard, 1971) becoming the paradigm for this technique (see Chapters 2, 3 and 4 in this volume).

The interpeduncular nucleus (IPN), which is situated on the midline at the ventral surface of the midbrain, receives bilateral inputs, originating largely in the habenular nuclei (Herkenham and Nauta, 1979), which enter the structure from each side on its antero-dorsal aspect after travelling across the brain within the myelinated Fasciculi Retroflexi of Meynert (FRM). On reaching the nucleus, the fibre input 'loops' to and fro, bundles of axons travelling parallel to the transverse axis of the IPN in a manner which, to Ramon y Cajal (1911) and his student Calderon (1927–8) was reminiscent of cerebellar parallel fibres (see Fig. 11.1). This similarity was reinforced by the way the fibres were seen to perforate the parasagittally aligned dendritic arrays of the interpeduncular neurons in Golgi material. Recent electron microscope studies (Lenn, 1976; Hattori et al., 1977) have described serial en passant synapses between FRM fibres, which become unmyelinated within the IPN, and many neurons across the lateral extent of the nucleus. The FRM/IPN system is nearly co-planar in the rat brain and therefore a suitably angled slice of the midbrain includes the anatomical features mentioned above.

These structural details, and the localization within the IPN of an interesting form of synapse, the 'crest', where inputs from each fasciculus are paired, sandwiching a postsynaptic dendritic profile (Lenn, 1976; Murray et al., 1979), are justification enough for subjecting this area to the slice technique, since a study of the FRM fibre properties would be extremely difficult in vivo. However, in addition, much biochemical data have been accumulated to suggest that the FRM-input is exceptionally rich in cholinergic (Kataoka et al., 1973; Kuhar et al., 1975; Hattori et al., 1977; Gottesfeld and Jacobowitz, 1978; McGeer et al., 1979) and peptidergic (Mroz et al., 1976; Cuello et al., 1978) fibres. The IPN contains a high concentration of binding sites for the nicotinic acetylcholine (ACh) receptor marker α-bungarotoxin (Arimatsu et al., 1978) and fewer, but intensely localized, muscarinic antagonist binding sites (Rotter et al., 1979). Recently, release of ACh from the IPN in vivo has been demon-

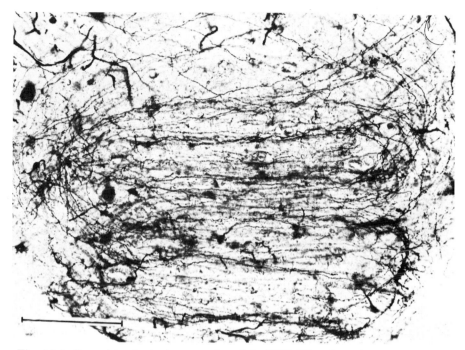

Fig. 11.1. Some histological properties of the IPN. The plate shows an oblique frontal section through the lower end of the FRM and the IPN, cut with the orientation used in preparing the brain slice (see Fig. 11.2). The brain, from an 11-day-old rat, was processed according to the Golgi–Kopsch procedure. Prominent features include the point of entry of one of the FRM (top right) and the obvious criss-crossing of its fibres within the nucleus; occasional IPN neurons have also been impregnated. Scale = 0.2mm.

strated following stimulation of the habenular region (Sastry *et al.*, 1979). The additional presence of opiate binding sites has been indicated (Atweh and Kuhar, 1977) and terminals immunohistochemically positive for LH-RH (luteinizing hormone releasing hormone; luliberin) (Silverman and Krey, 1978) have been detected. The IPN therefore represents a site where many putative neurotransmitters may interact at the cellular level and this further justifies its inclusion in the growing list of *in vitro* slice preparations.

We report here a method of slice preparation developed for the IPN bilateral FRM-system, but applicable to other situations where carefully orientated slices of *internal* brain structures are required, and illustrate results demonstrating the structural and functional integrity of the IPN slice.

II. Methods

A. Slice preparation

To exploit the obvious advantages of this brain slice, namely the well-defined bilateral input pathways (Brown and Halliwell, 1979), a method of cutting was devised which optimized the inclusion of these features within the tissue preparation. When brain slices that include discrete fibre tracts were used, these pathways were usually observable either on the outer brain surface (e.g. Yamamoto and McIlwain, 1966: lateral olfactory tract; Kawai and Yamamoto, 1967: optic tract) or close to the surface (Gardner-Medwin 1972b: cerebellar parallel fibres; Bliss and Richards, 1971: perforant path fibres). The pathways that we incorporate in this slice, the FRM, traverse the mid-brain rostrocaudally from the dorsally positioned habenulae to the ventrally placed IPN (Fig. 11.2); consequently, their position within the brain has to be interpolated from an atlas (König and Klippel, 1963) and from external

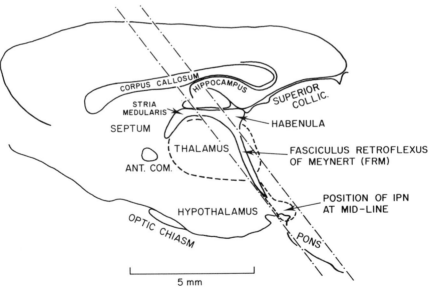

Fig. 11.2. A parasagittal section of the rat brain redrawn from König and Klippel (1963) showing the orientation of the FRM as it leaves the habenula. This section is 740μm lateral to the midline. The position of the IPN (as it appears on the ventral midline) is marked in dashed lines. The brain slice was prepared by cutting across the brain at the angles indicated −·−·−·−, including the FRM/IPN system in one planar slice.

landmarks during the slicing procedure. The angle and position of the first cut is crucial and therefore a standardized method for holding the brain has been employed. This involves the manufacture of a flat-bottomed agar cup, shaped exactly to contain the dorsal half of a freshly removed rat brain. A silicone rubber facsimile, prepared from an acrylic casting of a formalin-fixed whole rat brain, was lowered, dorsal surface lowermost and horizontal, into a shallow pool of molten 2% (w/v) agar placed on a horizontal surface. The facsimile was removed when the agar had set, leaving the required shape of cup. A range of differently sized silicone rubber formers ensured that slices could be ultimately prepared from rats of different ages.

The rats were killed by cervical dislocation and their brains rapidly removed and then placed, ventral surface uppermost, in an agar cup at room temperature. The agar cup and brain was positioned on an inclined plane with a variable pitch, set within a pair of vertical guide bars (Fig. 11.3). The angle of the plane to the horizontal ($\sim 40°$) was calculated and adjusted so that the FRM-IPN system would lie, within the brain, parallel to the guide bars. In this position, the brain was transected just caudal to the IPN, by means of a bow-cutter held against the guides. The rostral segment of the brain was then placed on a cutting block, trimmed and the exposed IPN–FRM removed by subsequent plane sections with the bow-cutter and a guide of $500\mu m$ nominal thickness. Finally, the slices so prepared were transferred to a petri-dish containing pre-oxygenated medium, where the cortical tissue was trimmed off, leaving just upper brainstem material, and thence to a vessel containing oxygenated Krebs solution.

B. Incubation

Sections were incubated in oxygenated Krebs' solution at ambient temperature for 2 or more hours, and then transferred to the recording bath. (As usual with brain slices, the IPN slices suffered some form of 'shock' immediately after preparation, so that at least 2 hours incubation was necessary before electrical activity recommenced.) The incubation bath consisted of a small chamber (0.5ml) with a nylon-mesh base for the slice to rest on (Fig. 11.4); the latter was gently restrained on the mesh by means of nylon cross-wires fixed to a stainless-steel ring that could be raised or lowered by a small dove-tail screw slide built onto the external water jacket. The slice was kept totally submerged in the bath which was constantly perfused from beneath the mesh at up to 5ml min^{-1} (range 2–5ml min^{-1}) with oxygenated Krebs' solution prewarmed to the required temperature in a water-jacketed coil; the usual

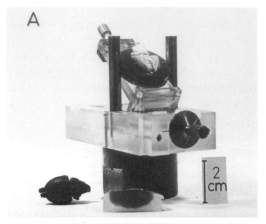

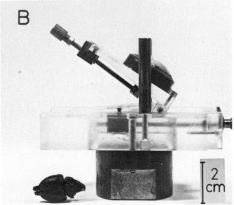

Fig. 11.3. The brain sectioning jig. Three-quarter and elevation views of the cutting table used to angle accurately the rat brain before sectioning are shown. See text for full details. In each plate, the rat brain in its agar cup (coloured to maximize photographic contrast) is shown positioned on the inclined plane, and near the apparatus, the silicone-rubber former used to prepare the agar cup. The item directly in front of the cutting jig in A is a cup filled with contrasting agar and sectioned longitudinally to show the relationship of the brain-shaped recess to the flat bottom of the cup's exterior. The methods of adjusting both the pitch of the plane and its longitudinal relationship with the guide bars are obvious from B.

temperature selected for experiments was 28°C, to ensure adequate oxygenation for the metabolic demands of the tissue (see, Harvey *et al.*, 1974). The solution flowed into the coil under gravity from a small-capacity (~0.2ml) reservoir supplied with Krebs' solution from one

Fig. 11.4. The recording chamber. This figure shows an oblique top-view of the upper section of the incubation apparatus. When operational, this component is fixed to the subjacent water bath. Incubation medium, prewarmed in the latter by means of a coiled tube, enters the inflow channel from below at position A. This point is also the locus of drug injection (see text). The open channel round the central chamber, with which it may be seen to connect, serves to eliminate any bubbles that may form in the medium during warming. The central mesh supports the slice; the latter is restrained by the ring positioned by screw slide B. The second slide C controls the outflow nozzle and thereby the fluid level in the central chamber. Other electrodes, etc., may be supported by the multiple holder D.

channel of a multi-channel peristaltic pump (Watson-Marlow, U.K.); in this way, the pulsatile flow of the pump was eliminated. Two channels of the same pump were used to remove the perfusion medium from the chamber via a tube that could be raised or lowered by a second slide, thereby controlling the fluid level in the chamber; the level was usually kept just above the top surface of the slice. The Krebs' solution had the following composition (in mequiv. l^{-1}): Na^+, 118; K^+, 3 or 6; Ca^{2+}, 2.5;

Mg^{2+}, 1.2; Cl^-, 128 or 125; HCO_3^-, 25; $H_2PO_4^{2-}$, 1.0; D-glucose, 5.5 or 11 (pH 7.4 when bubbled with 95% O_2/5% CO_2). To block synaptic responses in some experiments, Ca^{2+} was omitted and replaced with 10 mM Mg^{2+}. Solutions could be recirculated, run to waste or changed as necessary between four alternatives. The dead time for a total fluid change was about 5 minutes; to allow more rapid and brief application of drugs, the drug solution was injected at controlled rates against the perfusion stream close to the point where it entered the bath, so that the steady-state concentration could be calculated from flow rates. Temperature was monitored with a thermocouple placed in the recording chamber, and was constant to within 0.2°C.

C. Stimulation and recording

Stimuli were applied at the required loci using (i) bipolar stainless-steel electrodes insulated to their tips, (ii) electrolytically sharpened tungsten electrodes (2μm tip diam.) insulated to within 10μm of their tips or (iii) broken-back glass micropipettes (25μm diam.) filled with saline solution. Stimulation through the two latter types of electrode was monopolar, tip-negative; the anode was a conducting wick filled with 2% (w/v) agar in 0.9% NaCl, 2mm in diameter, placed on the slice in a position that minimized the stimulus artefact. Constant current stimuli (1–500μA, 50–200μs) were supplied by an optically coupled isolated stimulator (Neurolog, Digitimer Ltd., U.K.). Evoked activity was recorded using extracellular glass microelectrodes (2–5μm diam.) filled with 10% NaCl; the wire retaining ring served as an indifferent electrode. Potentials were followed by an AC preamplifier (Neurolog) with a time constant of 0.1 or 10 s and displayed on an oscilloscope (Tektronix, U.S.A.). Signals were further processed by means of a Neurolog Signal Averager to provide peristimulus histograms and averaged evoked potentials and a Neurolog Spike Trigger and Pulse Integrator were used to obtain integrated spike discharge rates.

III. Results

In spite of considerable care exercised in the preparation of the slices, slight variations in the angle of cutting produced IPN slices with different lengths of FRM. Two examples at each end of the spectrum are shown in Fig. 11.5. However, the results presented below are typical of the electrical responses observed from slices falling within the range bounded by these extremes. Furthermore, the data in Fig. 11.9 suggest that, even when both fasciculi have been shorn from the nucleus, the latter remains a preparation worthy of electrophysiological investigation.

Slices were electrically silent for the first 2 hours or so after sectioning. Thereafter, the electrical activity described below could be recorded for up to 36 hours after isolation.

A. Fibre input

Stimulating the FRM at the habenular end evoked compound action potential volleys which could be recorded along the length of the macroscopically visible tract. These took the form of di- or triphasic complexes, predominantly negative-going (Hubbard et al., 1969) and showed a systematic latency shift with the distance of the recording site from the point of stimulation. When a near-full thickness of the FRM was included in the slice, the afferent volley close to the IPN displayed two distinct negative peaks (Fig. 11.6c). Plotting the latency shift of these peaks against recording position on the tract revealed a linear relationship for both components but different slope values for each; these suggest modal conduction velocities at 28°C of about 1.3 and 0.4 m s^{-1} for two major fibre populations (Fig. 11.6A). Stimulating at different points in the body of the tract indicated a slightly different disposition of the two fibre groups within the FRM, such that the faster-conducting axons appear to occupy a more medial location than the slow fibres (Fig. 11.6B). Sometimes, a double-peaked response could be observed within the nucleus itself, although a single component was more common. Tracking the recording electrode transversely across the IPN in the plane of the slice normally gave linear relationships between latency and recording position (slope values 0.16–0.29m s^{-1}) consistent with anatomical observations of FRM fibres crossing the nucleus in parallel bundles (Ramon y Cajal, 1911; Calderon, 1927–8; see Fig. 11.1). On occasion, however, both non-linear plots and those with slopes of different sign for the two negative peaks were observed (Fig. 11.7); these results suggest a deviation away from parallelism amongst the FRM fibres within the nucleus and may reflect the looping 'figure-of-eight' behaviour of the habenulo-interpeduncular afferents. Volleys recorded both in the FRM and in the IPN were insensitive to removal of Ca^{2+} and its replacement with high Mg^{2+} and followed stimulus rates up to about 50Hz. Closely spaced, paired identical stimuli to the FRM strongly depressed the volley in response to the second stimulus (Fig. 11.8): recovery from relative refractoriness was complete when the inter stimulus interval was greater than 10ms; the absolute refractory period was

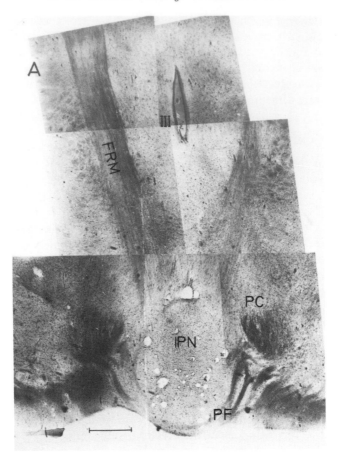

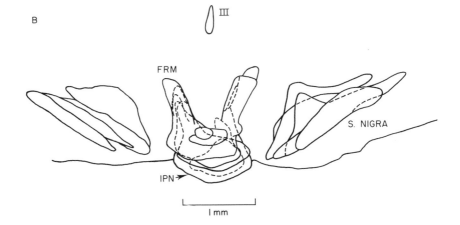

about 1ms. In contrast, volleys in the IPN evoked by paired stimulation of both FRM summed linearly, even with interpulse intervals of 1ms, indicating no presynaptic interaction between left and right fasciculus fibres within the nucleus.

By placing a stimulating electrode in the body of the IPN it was possible to excite narrow beams of fibres (Fig. 11.9). A single-peaked, predominantly negative deflection was observable along the lateral axis of the nucleus; distal to the stimulating point it increased in both latency and width with a concomitant reduction in amplitude. The slope value of the latency/conduction distance curve was about $0.2m\ s^{-1}$, lying roughly halfway between the values for components produced in the IPN by stimulating the FRM. Presumably the longer conduction paths from the FRM allow a greater temporal dispersal of the volleys in differently conducting fibres, whereas the latency shift over the shorter intranuclear distance indicates an average value for the conduction velocity of the fibre spectrum present. The marked reduction in amplitude and increase in the width of the volley at greater distances from the point of stimulation is consistent with desynchronization due to a range of conduction velocities. Further electrophysiological evidence for the 'looping' of fibre pathways in the IPN is presented in Fig. 11.9C trace vi: on the margin of the beam of excited fibres ($200\mu m$ away from the maximally recorded volley position Y' in Fig. 11.9) a much attenuated volley with a prolonged latency could be observed. This accords with the conduction path performing a 'U-turn' and leading the volley back towards the stimulation site.

B. Evoked fields in the IPN

Fig. 11.10 shows examples of longer latency components present in averaged evoked responses recorded in the middle of the IPN. Unlike

Fig. 11.5. The structures present in the slice. In A, is shown a composite photograph of a luxol-stained section of a brainstem slice from which evoked activity had been recorded. The near full extent of one FRM is evident within the orientation of the section. B shows a reconstruction of the IPN and FRM from camera-lucida drawings of Nissl-stained sections taken from another experimental brain slice. The diagram takes the form of a contour map of the area. Eight $60\mu m$, serial sections of the brain slice were needed to reconstruct the IPN tissue present; as for A these had been cut from the slice on a freezing microtome after prior formalin fixation and embedding in egg yolk. Abbreviations: III, third ventricle; FRM, fasciculus retroflexus of Meynert; IPN, interpeduncular nucleus; PC, cerebral peduncle; PF, pontine fibres; S. nigra, substantia nigra. Scale for A = 0.5mm.

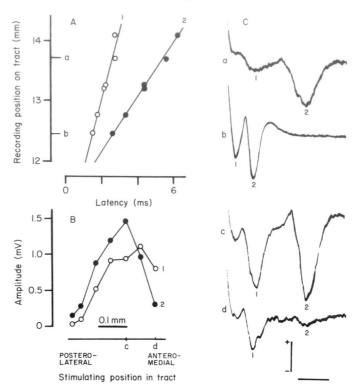

Fig. 11.6. Propagating volleys in the FRM. The graph in A shows the latencies to the peaks of the two components (1 and 2) recorded at different positions along one FRM following a single 150μA, 100μs stimulus. Sample records of the potentials corresponding to positions a and b on the ordinates are illustrated in C. In B, the *stimulating* electrode was moved across the FRM while recording from a fixed point near to the IPN. The graph shows that the amplitude of each of the two components of the compound spike varies differentially with stimulating position. The records c and d in C illustrate responses associated with the corresponding stimulus positions marked on the abscissa of the graph. Calibration: 0.5mV (a, c, d), 1mV(b); 2ms.

the presynaptic responses described above, these field potentials were modified in form or amplitude by increasing the frequency of stimulation (Fig. 11.10A), by a previous conditioning pulse (Fig. 11.10B) or by substituting Mg^{2+} for Ca^{2+}; this is suggestive of a postsynaptic origin for this component of the response. Assuming the residual components of the responses evoked by FRM stimulation at high frequency or in Ca^{2+}-free solution to be substantially presynaptic, the postsynaptic

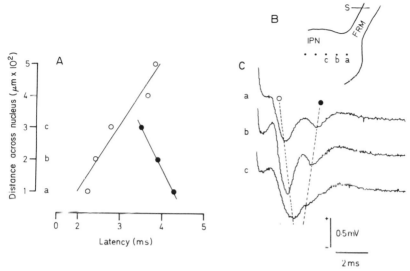

Fig. 11.7. Bi-directional volleys within the IPN. The graph A shows latencies to the peak of two components (●, ○) recorded along a track across the IPN (indicated in B) following stimulation of the FRM, plotted against recording position. The latency distribution suggests that the first component propagates from right to left and the second component in the opposite direction, although the latter was only distinctly observable at position a, b and c. The conduction velocities were 0.21m s^{-1} (right to left) for the first component and 0.27m s^{-1} (left to right) for the second. The averaged responses (eight sweeps) recorded at positions a, b, c are reproduced in C: dashed lines have been fitted and drawn as closely as possible to the peaks of the two components to indicate the latency shifts.

fields schematized in Fig. 11.10D might be deduced: a negative field (N) with a superimposed 'population spike' (P), the latter representing synchronous neuronal discharges (Andersen et al., 1971a). Similar field potentials have been described in the cat IPN in vivo (Lake, 1973). The effects of higher frequency stimulation on the N-component were mixed; sometimes the response was depressed (Fig. 11.10A) and at others it exhibited frequency potentiation (see Fig. 11.14, trace e). In the example illustrated in Fig. 11.14 potentiation of the N-component was associated with the enhanced excitability of interpeduncular neurons close to the recording electrode. No long-term effects of high-frequency stimulation (cf. Andersen et al., 1977) were observed on return to regular, low rates of stimulation.

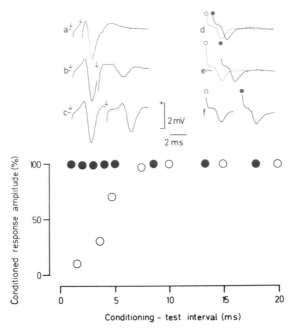

Fig. 11.8. The refractory period of FRM fibres. The graph shows how the amplitude of a second response, when expressed as a percentage of the first, varies with interstimulus interval when paired identical stimuli (arrowed) were delivered to one FRM and the recording site was close to the nucleus (○: traces a, b and c); ● depict the same measure, but when the recording site was in the centre of the IPN and the first stimulus (○) delivered to the left FRM, and the second (●) to the right FRM (traces d, e and f). The traces are averages of eight sweeps; in d and e the trace represents the difference between the averaged response to both stimuli and the averaged response to the first stimulus alone. – – – –, indicates the position and magnitude of the first (conditioning) response.

C. Unit activity

1. *Spontaneous activity*

Most preparations yielded a number of spontaneously active units within the IPN, firing at rates of between 1 and 30Hz for prolonged periods. In some cases, single units could be held for several hours and spontaneous activity was still encountered on the second day of incubation, more than 24 hours after slice preparation. Extracellularly recorded unit spikes varied in amplitude between 100μV and several

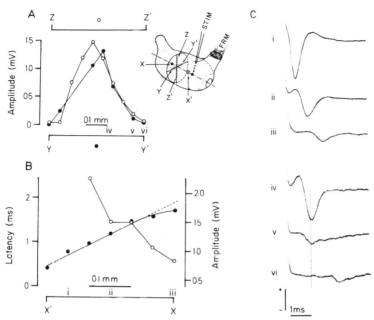

Fig. 11.9. Propagating volleys initiated within the IPN. A focal stimulating electrode was placed in a fixed position at the lateral edge of the IPN and evoked potentials recorded with a roving micro-electrode placed at different points along the three axes shown in the diagram. Graph A shows the peak amplitude of the positive/negative deflexion (see, for example, trace iv) recorded along axes Z–Z' (o) and Y–Y' (●). Graph B shows the change in latency (●) and amplitude (o) of the same potential as the electrode was moved along the horizontal axis X–X'. Traces i–iii show the increase in latency and decrease in amplitude demonstrated in graph B and are recordings made at the points indicated on the abscissa between points X and X'. Traces iv–vi show the limited A–P spread of the fibre volley and are recordings made at the points indicated on the axis Y–Y' in graph A. Note that on the margin of the fibre beam (trace vi) a late volley is apparent, much later than would be expected from a *direct* conduction pathway (cf. dashed line through traces iv–vi) from the stimulation site. Calibration: i–iii, 1.5mV; iv–vi, 0.75mV.

millivolts and took the form of monophasic positive or negative transients or positive–negative complexes; in the latter case, inflexions on the rising phase of the potential (Fig. 11.11) were commonly observed, indicative of a sequential activation of the neuron from some trigger point. Spontaneous activity was still recorded in preparations perfused with a Ca^{2+}-free solution (to block synaptic activity) or with a medium containing a reduced concentration of K^+ (3mM).

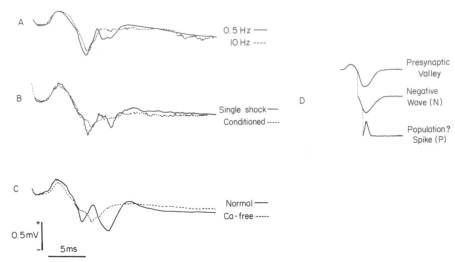

Fig. 11.10. Averaged evoked field potentials recorded in the IPN following FRM stimulation. In records A–C, ——— shows the averaged response to eight stimuli at 0.5Hz; – – – – show the responses recorded during 10Hz stimulation (A), 50ms after a pre-conditioning stimulus (B) and also at 0.5Hz, but after replacing the medium with a Ca^{2+}-free/high-Mg^{2+} solution (C). (In C, the solid line shows the potential recorded after replacing the Ca^{2+} rather than before its omission—cf. solid line in A.) D shows the scheme of summed potentials deduced from the records, as described in the text.

2. *Evoked activity*

The behaviour of many IPN neurons could be influenced by stimulation of the FRM or by activating fibre 'beams' (cf. Fig. 11.9) within the nucleus. *Quiescent* units were recruited with variable latency by such treatment (Fig. 11.12a, b, c); in general these units showed low synaptic efficiency: that is, their probability of firing after a single stimulus never attained unity. However, it was greater at higher rates of stimulation (10–20Hz) than at the lower rates routinely used (0.2–1Hz). The average latency (about 8ms) of units recruited thus coincided with the P-complex of the evoked field potential.

    The discharge rate of *spontaneous* IPN units could often be accelerated by stimulating the FRM (Fig. 11.14B). Higher frequency stimulation was more effective in increasing the firing probability than low frequency, as the peri-stimulus histograms in Fig. 11.13 show; moreover, as with the recruited units, the period of enhanced cell excitability coincided with the P-complex. In the example illustrated in Fig. 11.14, where 1Hz stimulation was ineffective in discharging the cell regularly,

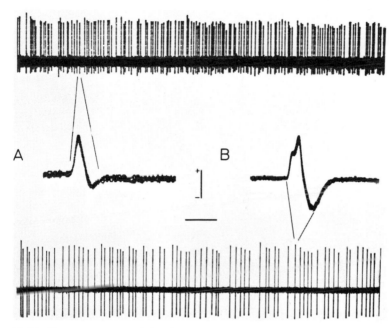

Fig. 11.11. Spontaneous activity from two separate cells in the lateral IPN is shown. The expanded sweep (10 superimposed traces) in each case was captured by playing back an F.M. tape recording of the activity through a delay-line, which was triggered by the occurrence of each spike discharge. Calibration: A, 0.5 mV; B, 1mV; 2ms 2 s.

the effects of 10Hz stimulation of the FRM took effect slowly and dissipated quickly. However, nearly all the unit discharges occurring at high stimulus rates were time-locked to the stimulus with a short but variable latency (Fig. 11.14A). Similar enhanced firing could be achieved by local stimulation within the IPN; in addition, periods of inhibited firing after an initial acceleration were encountered (Fig. 11.12). In other cells, intranuclear stimulation produced pure inhibition; in the latter case, a displacement of the stimulating electrode by 100μm or so rendered the stimulation ineffective. Many units were not clearly affected by stimulation in either the FRM or the nucleus itself.

IV. Discussion

The major conclusion to be drawn so far from our tests is that the FRM–IPN system can be isolated and maintained *in vitro*, and that the

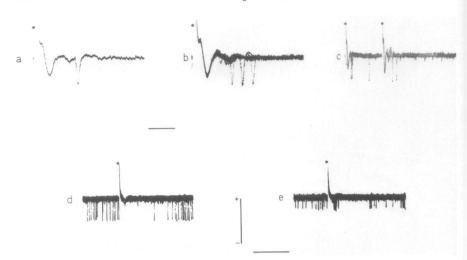

Fig. 11.12. Evoked and modified unit activity in the IPN. Traces a and b show records of single (a) and 10 superimposed sweeps (b) from the midpart of the IPN following FRM stimulation at 0.5Hz. During the 10 successive stimuli, a cell responded with a variable latency discharge after four of these, whereas the presynaptic volley was invariant. Trace c, multiple discharges in the middle of the IPN were evoked in this case by paired stimuli 12 ms apart (the trace shows two superimposed sweeps). Traces d and e show the effect of stimulating interpeduncular (cf. Fig. 11.9) fibres on two spontaneously active cells in the lateral third of the IPN. Each record shows 30 superimposed sweeps. In trace e, stimulation first accelerated and then inhibited the ongoing spontaneous discharge; in trace d, pure inhibition of the largest unit is seen. Calibration: 0.5mV; a, b, 5ms; c, 12 ms; d, e, 30 ms.

preparation so obtained demonstrates electrophysiological properties consistent with known anatomical information and with previous electrophysiological studies *in vivo* (*cf.* Lake, 1973). The latter include (i) spontaneous unit activity, (ii) evoked short-latency postsynaptic fields, and (iii) modification of unit discharge activity by FRM stimulation. (We have also regularly recorded acceleration of unit discharge rates by cholinergic agonists, as previously observed *in vivo* (Lake, 1973; Sastry, 1978) which is consonant with a cholinergic input but may also be susceptible to other interpretations.) To this extent, the present preparation offers a fair prospect for further studies.

The preservation of such contorted fibre pathways as the FRM indicates that the present technique of slicing, which utilizes a stereotaxic approach, might be applicable in the study of other orientated fibre pathways in brain slice preparations. Functional properties of both

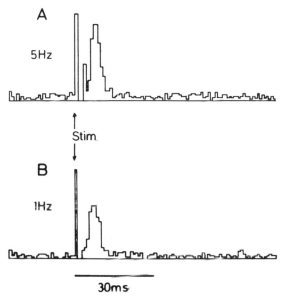

Fig. 11.13. Peri-stimulus histograms from a spontaneously active unit in the lateral IPN, showing acceleration of the discharge rate evoked by FRM stimulation at 5 and 1 Hz. The record in A represents 64 sweeps, that in B 32 sweeps. The probability of firing can be judged against the height of the stimulus artefact (arrowed), which represents unit probability. Note that increasing the stimulus frequency increased the firing probability in the 10ms period immediately after the stimulus. The histogram bin-width was 1ms.

myelinated and unmyelinated axons have been described above; accumulation of such data would be impossible *in vivo* because of the lack of the facility to be able to place stimulating and recording electrodes under direct visual control. Thus, the investigation of the functional characteristics of fibre systems need not be limited to those close to the brain's surface *in vivo* (Gardner-Medwin, 1972a; Merrill *et al.*, 1978) or *in vitro* (e.g. Gardner-Medwin, 1972b). The shorter conduction pathways aligned within the hippocampus have been exploited in a similar manner (Andersen *et al.*, 1978).

Ogata (1979a, b and c) has described another preparation of this region from the guinea pig, in which the brain is cut medially to provide a longitudinal thick (0.5–0.6mm) section of the IPN with one FRM. Cells in the IPN also showed spontaneous activity and were monosynaptically excited by FRM stimuli. This preparation is clearly viable but suffers from the disadvantage that the bilateral criss-crossing input to the centre of the IPN is disrupted. Further, current evidence shows that

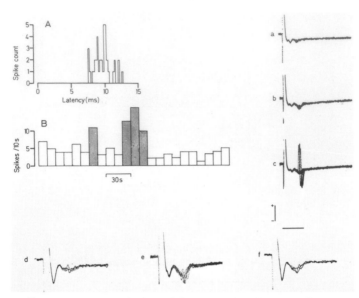

Fig. 11.14. Frequency potentiation of the synaptic effects of FRM stimulation. The single neuron in the lateral IPN depicted here was spontaneously active but unaffected by 1Hz stimulation. The histogram in B is a portion of the continuous spike discharge-rate record. The stippled portions signify periods when the rate of FRM stimulation was changed from 1Hz to 10Hz. Traces a–c show evoked potentials in the vicinity of the unit a, just before the longer period of 10Hz stimulation; b, the first 10 stimuli of the 30s train; c, the second 10 stimuli of the train. Note that first the postsynaptic negative component was potentiated by the faster stimulation rates, and then the unit was recruited with a short latency. The histogram inset in A shows the latency distribution of 39 of the 41 unit discharges during the 30s of 10Hz stimulation. Although, at the higher stimulus rate, the synaptic efficiency was low, the unit discharge became time-locked to the stimulus. Traces d–f show evoked potentials closer to the beam (cf. Fig. 11.9) of excited presynaptic fibres, recorded 150μm away from the unit mentioned above; they were the responses to FRM stimulation just before (d, five superimposed sweeps) during (e, ten superimposed sweeps) and immediately after (f, three superimposed sweeps) a 20s period at 10Hz; the basal rate, as before, was 1Hz. Calibration: a, b, c, 1mV, 15ms; d, e, f, 0.5mV; 7.5 ms. Histogram bin-width in A, 0.2ms.

each of the paired crest synapses on the neural dendrites receive contra-lateral inputs (Lenn, 1976; Murray *et al.*, 1979), suggesting that the bilateral input is a crucial and unique feature in the organization and functioning of the IPN. Although we have not yet explored in detail the response of IPN neurons to bilateral stimulation, our slice preparation does appear to preserve the appropriate looping circuits to allow such an analysis.

Intracellular recording from IPN neurons has also been reported (Ogata, 1979a and b) but the cell somata are not large, and their large dendritic tree and complex organization may limit the scope of intracellular analysis, as may be apparent from the recordings so far published. The principal interests of this preparation obviously lie in (i) its interesting neuroanatomy and (ii) its pharmacology. Fine extracellular recording coupled with focal stimulation may be more appropriate analytical techniques for both areas of study.

V. Conclusions

1. This chapter describes a slice preparation from the rat brain in which afferents from the fasciculi retroflexi of Meynert (FRM) synapse onto cells in the interpeduncular nucleus (IPN). This is a bilateral input and originates largely in the habenular nuclei.

2. The FRM–IPN system is almost coplanar and contains synapses where the inputs from the fasciculi are paired, sandwiching a postsynaptic dendritic profile within the nucleus.

3. To prepare the slice the dorsal half of the rat brain was held in a moulded agar cup on a cutting jig so that the FRM–IPN was aligned with a pair of guide bars. Using these guides, the brain was transected just caudal to the IPN. The rostral segment of the brain was placed on a cutting block and slices of the exposed FRM–IPN were removed with a bow-cutter and a guide approximately 500μm thick.

4. The slices were electrically silent for about the first 2 hours after sectioning, but thereafter electrical activity could be studied for up to 36 hours.

5. Stimulation of the FRM led to compound afferent volleys along the tracts with two modal conduction velocities of 1.3 and 0.4m s^{-1} at 28°C. The faster conducting fibres were more medially positioned within the tract than were the slower conducting fibres.

6. Electrophysiological evidence was obtained for the traversal of the IPN by parallel bundles of FRM fibres; bidirectional volleys initiated at one stimulation site were encountered within the nucleus in accord with the histological demonstration of the looping of the afferent tracts into a figure-of-eight.

7. Evoked postsynaptic potentials were recorded extracellularly in the middle of the IPN, consisting of an N-field with a superimposed population spike.

8. Many spontaneously active units were seen in the IPN with action potential frequencies from 1–30Hz. The activity was still seen in slices perfused with Ca^{2+}-free solution.

9. The discharge rate of the spontaneously active units in the IPN could be accelerated by stimulation of the FRM.

10. This preparation should allow investigation into the importance of the complex looping afferent pathway and into the pharmacology of the synaptic systems of the IPN.

Acknowledgements

This work is supported by the Medical Research Council. We thank Mr. D. McCarthy and Mr. D. King for photographic assistance and Mr. D. J. Tullett for constructing some of the apparatus.

References

Andersen, P., Bliss, T. V. P. and Skrede, K. K. (1971a). Unit analysis of hippocampal population spikes. *Exp. Brain Res.* **13,** 208–221.

Andersen, P., Bliss, T. V. P. and Skrede, K. K. (1971b). Lamellar organisation of hippocampal excitatory pathways. *Exp. Brain Res.* **13,** 222–238.

Andersen, P., Sundberg, S. H., Sveen, O. and Wigstrom, H. (1977). Specific long-lasting potentiation of synaptic transmission in hippocampal slices. *Nature (Lond.)* **266,** 736–737.

Andersen, P., Silfvenius, H., Sundberg, S. H., Sveen, O. and Wigstrom, H. (1978). Functional characteristics of unmyelinated fibres in the hippocampal cortex. *Brain Res.* **144,** 11–18.

Arimatsu, Y., Seto, A. and Amano, T. (1978). Localization of α-bungarotoxin binding sites in mouse brain by light and electron microscopis autoradiography. *Brain Res.* **147,** 165–169.

Atweh, S. F. and Kuhar, M. J. (1977). Autoradiographic localization of opiate receptors in rat brain. II. The brain stem. *Brain Res.* **129,** 1–12.

Bliss, T. V. P. and Richards, C. D. (1971). Some experiments with *in vitro* hippocampal slices. *J. Physiol. (Lond.)* **214,** 7–9P.

Brown, D. A. and Halliwell, J. V. (1979). Neuronal responses from the rat interpeduncular nucleus *in vitro. J. Physiol. (Lond.)* **292,** 9–10P.

Calderon, L. (1927–28). Sur la structure du ganglion interpedonculaire. *Trav. du Lab de Rech. Biol.* **25,** 297–306.

Cuello, A. C., Emson, P. C., Paxinos, G. and Jessell, T. (1978). Substance P containing and cholinergic projections from the habenula. *Brain Res.* **149,** 413–429.

Gardner-Medwin, A. R. (1972a). An extreme supernormal period in cerebellar parallel fibres. *J. Physiol. (Lond.)* **222,** 357–371.

Gardner-Medwin, A. R. (1972b). Supernormality of cerebellar parallel fibres: the effects of changes in potassium concentration *in vitro. Acta physiol. Scand.* **84,** 38–39A.

Gottesfeld, Z. and Jacobowitz, D. M. (1978). Cholinergic projection of the diagonal band to the interpeduncular nucleus of the rat brain. *Brain Res.* **156,** 329–332.

Harvey, J. A., Scholfield, C. N. and Brown, D. A. (1974). Evoked surface-positive potentials in isolated mammalian olfactory cortex. *Brain Res.* **76,** 235–245.

Hattori, T., McGeer, E. G., Singh, V. K. and McGeer, P. L. (1977). Cholinergic synapse of the interpeduncular nucleus. *Exp. Neurol.* **55,** 666–679.

Herkenham, M. and Nauta, W. J. H. (1979). Efferent connections of the habenular nuclei in the rat. *J. Comp. Neur.* **187,** 19–48.

Hubbard, J. I., Llinas, R. and Quastel, D. M. J. (1969). 'Electrophysiological Analysis of Synaptic Transmission'. Edward Arnold; London.

Kataoka, K., Nakamura, Y. and Hassler, R. (1973). Habenulo-interpeduncular tract: a possible cholinergic neuron in rat brain. *Brain Res.* **62,** 264–267.

Kawai, N. and Yamamoto, C. (1967). Effects of γ-aminobutyric acid on the potentials evoked *in vitro* in the superior colliculus. *Experientia* **23,** 822–23.

König, J. F. R. and Klippel, R. A. (1963). 'The Rat Brain and Stereotaxic Atlas of the Forebrain and Lower Parts of the Brain Stem'. Williams and Wilkins, Baltimore.

Kuhar, M. J., DeHaven, R. N., Yamamura, H. I., Rommelspacher, H. and Simon, J. R. (1975). Further evidence of cholinergic habenulo-interpeduncular neurons: pharmacologic and functional characteristics. *Brain Res.* **97,** 265–275.

Lake, N. (1973). Studies of the habenulo-interpeduncular pathway in cats. *Exp. Neurol.* **41,** 113–132.

Lenn, N. J. (1976). Synapses in the interpeduncular nucleus: Electron microscopy of normal and habenular lesioned rats. *J. Comp. Neur.* **166,** 73–100.

McGeer, E. G., Scherer-Singler, U. and Singh, E. A. (1979). Confirmatory data on habenular projections. *Brain Res.* **168,** 353–376.

McIlwain, H. (1966). Stimulation of mammalian cerebral cortex *in vitro* from the lateral olfactory tract: Metabolic changes associated with evoked potentials. *J. Physiol. (Lond.)* **185,** 65–66P.

Merrill, E. G., Wall, P. D. and Yaksh, T. L. (1978). Properties of two unmyelinated fibre tracts of the central nervous system, lateral lissauer tract, and parallel fibres of the cerebellum. *J. Physiol. (Lond.)* **284,** 127–145.

Mroz, E. A., Brownstein, M. J. and Leeman, S. E. (1976). Evidence for substance P in the habenulo-interpeduncular tract. *Brain Res.* **113,** 597–599.

Murray, M., Zimmer, J. and Raisman, G. (1979). Quantitative electron microscopic evidence for re-innervation in the adult rat interpeduncular nucleus after lesions of the fasciculus retroflexus. *J. Comp. Neur.* **187,** 447–468.

Ogata, N. (1979a). Substance P causes direct depolarization of neurones of guinea-pig interpeduncular nucleus *in vitro. Nature (Lond.)* **277,** 480–481.

Ogata, N. (1979b). Electrophysiology of mammalian hypothalamic and interpeduncular neurons *in vitro. Experientia* **35,** 1202–1203.

Ogata, N. (1979c). Effects of substance P on neurons of various brain regions *in vitro. Brain Res.* **176,** 395–400.

Ramon y Cajal, S. (1911). Histologie du Systeme Nerveux de l'Homme et des Vertebres'. Vol. II. Maloine, Paris.

Rotter, A., Birdsall, N. J. M., Field, P. M. and Raisman, G. (1979). Muscarinic receptors in the central nervous system of the rat. II. Distribution of binding [^{3}H]propylbenzilyl choline mustard in the midbrain and hindbrain. *Brain Res. Rev.* **1,** 167–183.

Sastry, B. R. (1978). Effects of substance P, acetylcholine and stimulation of habenula on rat interpeduncular neuronal activity. *Brain Res.* **144,** 404–410.

Sastry, B. R., Zialkowski, S. E., Hansen, L. M., Kavanagh, J. P. and Evoy, E. M. (1979). Acetylcholine release in the interpeduncular nucleus following the stimulation of habenula. *Brain Res.* **164,** 334–337.

Silverman, A. J. and Krey, L. C. (1978). The luteinizing hormone-releasing hormone (LH-RH) neuronal networks of the guinea-pig brain. I. Intra- and extra-hypothalamic projections. *Brain Res.* **157,** 233–246.

Skrede, K. K. and Westgaard, R. H. (1971). The transverse hippocampal slice: a well-defined cortical structure maintained *in vitro. Brain Res.* **35,** 589–593.

Yamamoto, C. and McIlwain, H. (1966). Electrical activities in thin sections from the mammalian brain maintained in chemically-defined media *in vitro. J. Neurochem.* **13,** 1333–1343.

Note Added in Proof

B. R. Sastry (1980, *Life Sci.* **27,** 1403) has recently described two forms of synaptic potential in interpeduncular neurons following FRM stimulation: a short-latency "fast" EPSP after low-frequency stimulation, and a long-latency "slow" EPSP after high-frequency stimulation. The latter might be responsible for the recruitment of excitation following repetitive stimulation described in the present paper.

12

Characteristics of Neuronal Activity in Striatal and Limbic Forebrain Regions Maintained *In Vitro*

J. J. MILLER

Department of Physiology, Faculty of Medicine, University of British Columbia, Vancouver, British Columbia, V6T 1W5, Canada

I. Introduction

The unravelling of complex synaptic networks and the description of their pharmacological properties are major objectives in the study of cellular mechanisms underlying central nervous system function. The

in vitro slice preparation provides a unique method by which to approach
these objectives in that it allows for investigation of relatively simplified
neuronal models together with an increased flexibility of experimental
manipulations which previously could not be considered using more
traditional *in vivo* preparations. However, the preparation of *in vitro* slices
involves the separation of central nervous system tissues from their
normal connections with other neurons and transferral to another
physicochemical environment. This raises the important question of
whether there are alterations in neuronal activity of these tissues which
may be reflected by an imbalance in the normal excitatory or inhibitory
properties of their synaptic networks. The extent to which such abnormalities in activity are present or the degree to which the *in vitro* tissue
bears any similarities to an intact preparation is of critical importance in
establishing the usefulness of the *in vitro* technique.

The functional integrity of cerebral explants has been substantiated
by a variety of experiments on diverse regions of the brain. One structure, the hippocampus, has received the most attention to date. Since
the original experiments of Bliss and Richards (1971), Yamamoto
(1972) and Skrede and Westgaard (1971), there have been a number of
detailed electrophysiological investigations of the synaptic processes in
this region (Alger and Teyler, 1976; Schwartzkroin, 1975, 1977;
Schwartzkroin and Wester, 1975; see also references in this text). These
studies served first to demonstrate the viability of this particular tissue
and secondly the essential similarity of the synaptic responses to those
recorded from the intact anesthetized preparation. It is also clear from
these investigations that considerable knowledge has been gained about
the local circuitry of this region which has in turn facilitated the
examination of the pharmacological properties of these networks. As a
first step, therefore, in the analysis of any new structure using the *in vitro*
technique, it is critical to establish that there is a preservation of
functional synaptic processes in the tissue and that these are comparable
to the intact preparation. The most unequivocal demonstration of
whether a particular neuronal assembly is operating normally is by
electrophysiological analysis and the present report is concerned with
such an analysis of two regions maintained *in vitro*, the corpus striatum
and septal area.

Unlike the well-defined morphological organization of some structures, such as the hippocampus, the striatal and septal regions do not
have a simplified anatomical arrangement in which distinct cellular or
dendritic layers may be visualized for ease of recording, nor do they have
distinct fiber bundles for stimulation. Both structures possess intricate
networks of local circuit neurons which together with long projecting

output cells have profuse collateral arborizations (Carpenter, 1976; Kemp and Powell, 1971; Pasik *et al.*, 1976; Hassler *et al.*, 1977; Swanson and Cowan, 1976; Swanson, 1978). It is, in fact, the relative degree of homogeneity of the neuronal aggregates in these regions that has caused them to remain as functional enigmas within the central nervous system. Using conventional preparations, electrophysiological studies have examined some of the diverse afferent systems that innervate target elements in these regions (Assaf and Miller, 1978; McLennan and Miller, 1974, 1976; Richardson *et al.*, 1977; Kocsis *et al.*, 1977; Purpura, 1976; Miller *et al.*, 1975). However, little is known about the interaction of these target neurons with the intrinsic and/or output cells. This study was undertaken in an attempt to characterize some of the intrinsic and extrinsic synaptic responses present in the striatum and septum maintained *in vitro*, thereby providing some indication of the viability and potential use of these regions for investigations of the pharmacological properties of local circuit neurons.

II. Methods

A. Corpus striatum

Male Wistar rats weighing between 150–200 g were sacrificed by decapitation and the brains quickly removed from the skull and divided along the midline. One hemisphere was placed in oxygenated medium at 25–27°C while the other was positioned on moistened filter paper for dissection. The medial surface of the striatum was exposed by peeling away the brainstem and midline structures including the thalamus and septal area and cutting through the fibers of the internal capsule. The tissue was blocked ventrally at the level of the anterior commissure and posteriorly using the tail of the striatum as a marker. The overlying dorsal and lateral regions of the cortex remain intact. Using a Sorvall tissue chopper, slices 350–400 μm thick were cut nearly perpendicular to the longitudinal axis of the striatum. Attempts were made to keep the plane of section parallel to the fiber radiations emanating from the confluence of the internal capsule by periodic adjustments of the moveable stage of the tissue chopper (Fig. 12.1A). The slices were rapidly transferred to the recording chamber where they were suspended on a nylon net bathed in oxygenated (95% O_2/5% CO_2) artificial cerebrospinal fluid which was maintained at 36–37°C and consisted of 124 mM NaCl, 5mM KCl, 1.25 mM KH_2PO_4, 2mM $CaCl_2$, 24 mM $NaHCO_3$, 2mM $MgSO_4$ and 10 mM D-glucose. The pH of the medium was 7.4 and its osmolarity ranged between 295–305 mosmol. In some experiments, the calcium concentration was lowered or reduced to zero and in others the chloride was replaced with equivalent amounts of proprionate. A humidified gas mixture (95% O_2/5% CO_2) was constantly directed over the slices.

The recording chamber is similar to the design originally described by Schwartzkroin (1975), except that temperature is controlled by a power transistor embedded in an aluminum heater ring positioned in the outer water bath. The feedback monitoring by thermoprobes in the inner and outer baths, together with the heater coil, allows for rapid changes in temperature. The slices are constantly perfused by gravity feed of the medium at rates of 2–3 ml min^{-1} controlled by a variable-flow valve (Dial-A-Flo). Since they are positioned on a net which is lower than the rim of the central recording chamber, the medium that enters through the bottom of the chamber moves up over the slices where it is removed by capillary action using a tissue-paper wick. Difficulties with floating slices were overcome by adjustments of the wick so that lemniscal flow just covers them. The slices were incubated for at least 1 hour prior to recording.

Stimulating and recording electrodes were positioned in the tissue under direct visual control using a dissecting microscope suspended above the chamber. Twisted bipolar nichrome wire (62 μm) or mono-polar tungsten (10 μm) electrodes were used for stimulation, which consisted of single square-wave pulses of 0.1 ms duration and 1–50 V intensity delivered through an isolation unit. Glass micropipettes filled with 4 M NaCl and having tip diameters of 1–2 μm were used for recording extracellular unit activity and field responses. The electrodes were lowered in 1 μm steps through the tissue using a micromanipulator (Inchworm Controller). Electrical activity was led through a high-pass filter, amplified and displayed on an oscilloscope. Action potentials and evoked field responses were also passed through an amplitude discrimi-nating window the output of which was fed to a PDP 11/10 computer for averaging and histogram analysis.

A semi-schematic diagram of the striatal slice indicating some of the distinguishing features is shown in Fig. 12.1B. The majority of slices examined were taken from the middle one-third of the striatum along the plane illustrated in Fig. 12.1A since it is this region that the fascicles of the corticofugal projections form continuous bundles extending through the slice. At the medial edge is located the confluence of the internal capsule characterized by a darkened core of these fibers. From this region, fiber radiations course through the tissue like spokes of a wheel. In close proximity to the capsular area is a slightly darkened zone

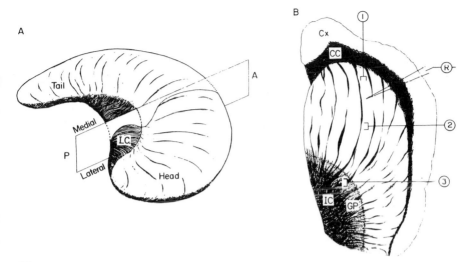

Fig. 12.1. A, Diagrammatic representation of the corpus striatum dissected free from surrounding tissues with the medial surface uppermost. B, Section of the striatum cut in plane indicated in A showing fiber radiations coursing through the tissue from the overlying dorsal and lateral surfaces of the cortex (Cx) to the confluence of the internal capsule (IC). Typical recording Ⓡ and stimulating electrode placements: ①, towards the cortical surface; ②, towards the internal capsule; ③, in the globus pallidus (GP).

which corresponds to the globus pallidus. Further lateral, the fibers penetrate through the cellular regions which appear as strips of light tissue. The dorsal and lateral border of the striatum is bounded by the dark fibers of the callosum and the overlying cortex. Various orientations of stimulating and recording electrodes in the slice were used. However, the most frequent and successful one was a medial–lateral orientation such that the two electrodes were located along the same plane and usually within the same cellular strip (Fig. 12.1B).

B. Septum

Similar procedures to those described previously were used to prepare slices of the septal area. The excised brain tissue was blocked posteriorly at the level of the descending columns of the fornix and on each side using the lateral ventricles as markers. The overlying cortex and callosum were also removed and the exposed septum was rapidly transferred to the cutting stage of a vibratome (Oxford) for sectioning into coronal slices of 300–400 μm. Initially, a tissue chopper was used. However, because of the thickness of the septum along the cutting plane, significant compression of the tissue resulted in poor preparations. About four or five slices are obtained from one septum. The majority of cells recorded were restricted to the medial septal–diagonal band region (MS–DBB) which is illustrated in Fig. 12.2. Stimulation sites were positioned according to the known anatomical organization of fiber trajectories and nuclei within the septum (Swanson, 1978): S_1-major ascending afferents coursing via the median forebrain bundle from various brainstem regions; S_2-fornix fibers and efferents of cells of origin of the MS–DBB region; S_3-dorsolateral septal nuclei including fimbrial fiber afferents.

III. Results

A. Corpus striatum

1. Spontaneous and locally evoked responses

Extracellular recordings were obtained from 243 neurons located in the light bands of tissue interspersed between the radially oriented fascicles coursing through the striatum (Fig. 12.1B). As shown in Fig. 12.3C, 22% of these neurons fired spontaneously in an irregular or random pattern at rates between 1 and 14 Hz ($\bar{X} = 6.7$ Hz). This activity was recorded in all regions of the striatum and at depths ranging from 50–250 μm below the surface of the slice. No spontaneous activity was recorded at temperatures below 33°C.

The majority of neurons recorded in the striatal tissue are 'silent' (78%) and require local stimulation in order to be identified. Single-pulse stimulation at relatively low intensity (2–10 V) evoked spikes with a mean latency of 4.1 ms (Fig. 12.3A). This spike activation faithfully followed trains of repetitive stimuli at rates between 50 and 80 Hz. Although the latency variability of individual cells was minimal (< 1.5 ms), the range for the whole population varied between 2.4 and 5.7 ms and was largely due to the distance between stimulating and recording electrodes. Responses could be obtained 3–5 mm from the stimulation site as long as the electrodes were positioned within the same "strip" of radially oriented tissue (Fig. 12.1B). Little if any activity was recorded from placements in immediately adjacent parallel "strips" indicating that the spread of stimulus current in the slices was minimal.

Stimulation of one site usually resulted in the activation of a number of different cells as the recording electrode penetrated through the tissue.

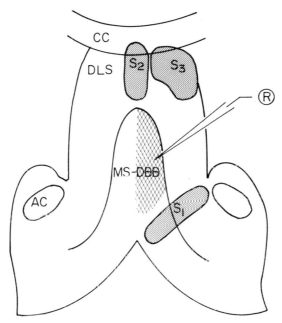

Fig. 12.2. Coronal section of the septal slice preparation indicating the location of recording electrode Ⓡ in the medial septal–diagonal band region (MS–DBB) and stimulating zones S_1, in the area at the median forebrain bundle; S_2, in the dorsomedial and S_3, in the dorsolateral septum (DLS). Abbreviations: AC, anterior commissure; CC, corpus callosum.

Increasing the stimulus intensity (1.5–2.5 T) elicited a characteristic sharp negative or negative–positive field potential which was recorded with varying amplitudes throughout the depth of a particular electrode tract (Fig. 12.3A). The correlation between the latency of this field response and the spike activation, together with the fact that the number of cells evoked and the amplitude of the initial negative-going waveform covaried with the stimulus intensity, suggests that this potential reflects the synchronous discharge of a population of striatal neurons. Reduction of the Ca^{2+} concentration in the perfusion medium from the normal level of 2.0 mM to 0.2 mM, decreased or eliminated this population response 5–8 minutes after the medium change (Fig. 12.3B). During the initial period of perfusion with low Ca^{2+} there was a significant increase in both the number of cells evoked and in their spontaneous activity. These responses were also eliminated 6–10 minutes after the perfusion onset. The population response and spike activation recovered 3–5

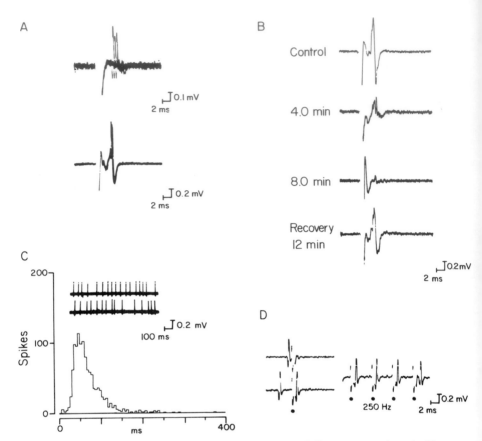

Fig. 12.3. Neuronal activity recorded in the striatal slice preparation. A, Upper trace indicates three oscilloscope sweeps of single spikes evoked by local stimuli at low stimulus intensity. Lower trace indicates characteristic population response evoked by higher intensity with spikes superimposed on the negative field potential (negative is up in this and all subsequent records). The traces in B, show the diminution of the evoked population response by perfusion with Ca^{2+}-free medium for 8 min. C, Interval histogram indicating characteristic irregular discharge pattern of spontaneously firing striatal neurone. (Bin width 10 ms, 1000 counts.) D, Cell activated antidromically by local stimulation on side towards the confluence of the internal capsule (see position (2) in Fig. 1B). Traces on the left indicate collision-extinction with spontaneously occurring spike, on the right, frequency following to 250 Hz. Dots indicate stimulus artifacts.

minutes after re-exposure to the standard medium. The fact that the locally evoked unitary activity and population response are eliminated in the Ca^{2+}-free medium together with the findings that they have minimal latency variability and follow relatively high frequencies of stimulation suggests that these responses are elicited by a monosynaptic excitatory input.

A short latency field (1.3–3.4 ms) or spike discharge (1.4–3.9 ms) was evoked when stimulating electrodes were positioned on the side of the recording placement which was towards the confluence of the internal capsule. Stimulation on the cortical side failed to evoke this response unless the stimulating and recording electrodes were in close proximity to one another (< 1.0 mm). Both the evoked field potential and single-cell discharge displayed characteristics of antidromicity; a consistent latency at threshold intensities, the ability to follow high frequency (> 200 Hz), and when cells were spontaneously firing, a collision–extinction at critical intervals (Fig. 12.3D). In addition, these responses were unaffected by perfusion with a Ca^{2+}-free medium. Since this activity could only be recorded from placements in the cellular regions and from neurons that could also be depolarized by the iontophoretic application of glutamate, it is unlikely that they represent the discharge of fibers coursing through the striatum. Few of these responses (3/47) were orthodromically activated from another stimulating site in the tissue, indicating that the synaptically evoked population response and these antidromic cells reflect activity in two different groups of striatal neurons.

Following the initial activation, local stimulation resulted in a period of inhibition ranging from 20 to 260 ms on glutamate-induced or spontaneously active neurons (Fig. 12.4). The duration of this inhibitory period was directly related to stimulus intensity; low levels (T–1.5T) producing minimal inhibition whereas high levels (> 2.0T) elicited a maximal effect. A pure inhibitory response was seldom observed in the absence of any prior activation. Since there were relatively few spontaneously active neurons in the striatal tissue with which to assess these inhibitory processes, the effects of a preceding conditioning pulse on the population spike evoked by local stimulation was examined. At intensities up to 1.5 × threshold, this response was potentiated when preceded by a conditioning pulse at intervals of 10 to 100 ms (Fig. 12.4B). A similar potentiation of single-unit activity was observed with subthreshold stimulation. At higher intensities (> 2.0T) a potent depression of the population response was produced when preceded by a conditioning pulse at intervals of 10–200 ms (Fig. 12.4C). The maximal depression occurred at interpulse intervals of 10–50 ms.

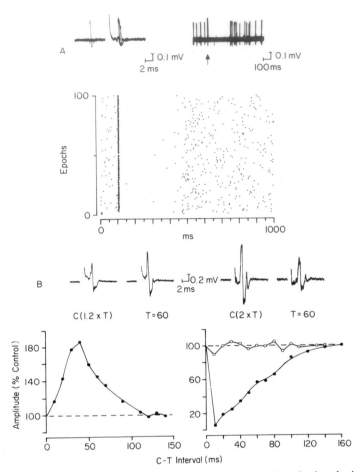

Fig. 12.4. A Activation–inhibition sequence evoked by local stimulation on a spontaneously firing striatal neuron. Oscilloscope traces at top left indicate variability in latency of spike activation and on the right the inhibitory component at a slower sweep speed. A rastered display of the poststimulus response is shown at the bottom. Stimulus artifacts and spike activation are indicated by arrows and as the vertical line in the raster. B, Effects of paired-pulse stimulation on the locally evoked population response. Oscilloscope records of conditioning (C) and test (T) responses evoked by low (top left) and high (top right) stimulus intensities indicate the respective potentiation and depression of the test response produced at a C–T interval of 60 ms. At the bottom are graphs illustrating the duration of potentiation and depression of the test response at low and high intensities respectively. Note also elimination of inhibition in Cl⁻-free (o) compared with standard (●) medium.

When the tissue slice was perfused with a Cl⁻-free medium, the population spike gradually increased in amplitude (20–50%) as did the number of cells evoked by a particular stimulus. Within 4–5 minutes of the onset of perfusion in a Cl⁻-free medium, the depression of the population spike resulting from a preceding conditioning pulse was reduced or eliminated (Fig. 12.4C). On replacement of the standard medium, the inhibition of the test response recovered to normal levels. Perfusion with a Cl⁻-free medium produced a similar effect on the inhibitory component of two spontaneously firing cells which displayed the characteristic activation–inhibition sequence following local stimulation.

2. *Local evoked responses in the chronically isolated striatum*

The electrical activity of striatal slices obtained from chronic preparations in which this structure was isolated from surrounding tissues was examined in order to determine whether the evoked excitation and inhibition produced by local stimulation is dependent on intrinsic synaptic networks or afferent projections to this region. Unilateral transections at the level of the anterior thalamic nuclei, which eliminates the major afferents from the thalamic and brainstem regions, and semi-circular cuts separating the striatum from the overlying cortex were made in six animals. Following a recovery period of 2–3 weeks, the levels of spontaneous activity recorded in the striatal slices ipsilateral to the transection were not changed, nor were the characteristic excitatory and inhibitory responses induced by local stimulation.

3. *Extrinsic evoked responses*

Since the striatum lacks a morphological organization that would allow for visual placement of stimulating electrodes into identifiable fiber systems innervating neurons in this structure, attempts were made to preserve, within an individual section, anatomical regions that contain either cells of origin for afferents to the striatum (cortex) or target neurons of efferents from the striatum (globus pallidus).

Stimulating electrodes positioned in the overlying cortex along either the dorsal or lateral edges of the tissue slice evoked single or multiple spikes in the cellular regions of the striatum at latencies of 3.1–4.8 ms (Fig. 12.5A, B). The excitability characteristics of these cortical-evoked striatal neurons were similar to those evoked by local stimuli. A subthreshold stimulus elicited a spike activation when preceded by a relatively low-intensity conditioning pulse at intervals of 10–50 ms.

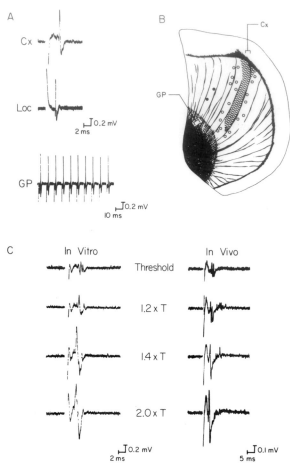

Fig. 12.5. Typical responses of striatal neurones to stimulation of the cortex (Cx) and globus pallidus (GP). A, Orthodromic activation of cell following single-pulse stimulation of overlying Cx and local (LOC) stimulation. High frequency following of a different cell to 100 Hz stimulation of the GP. B, Location of stimulation sites in the GP and Cx. The sites of two different cells antidromically evoked by GP stimulation are shown. Stimulation of the Cx at the indicated site evoked cell responses or field potentials only in the cross-hatched zone. Cells identified by local stimulation in adjacent regions (o) were unaffected by this Cx site. C, Comparison of cortical-evoked population response recorded *in vitro* and *in vivo* showing increasing amplitude of characteristic waveform with increased stimulus intensities.

Perfusion with low Ca^{2+} or Ca^{2+}-free medium eliminates the activation, which is reversed on returning to the normal solution. These data, together with the fact that there is latency variability in the evoked spikes at threshold intensities and frequency following up to 50–75 Hz, all suggest that the cortical-evoked activity is mediated by a direct monosynaptic input. Of 26 cells tested by both cortical and local stimulation, 19 showed a convergence indicating that the same population of neurons was being activated from both sources (Fig. 12.5A). Unitary activity elicited from a particular cortical stimulation site was restricted to a narrow band or strip of cellular tissue arranged longitudinally between the distinctive fibers coursing through the area (Fig. 12.5B). When a section was cut so that a clearly defined strip extended from the cortical to the medial aspect of the slice in the region of the confluence of the internal capsule, the stimulation was effective in activating neurons up to 5–6 mm distant. Few, if any, responses were obtained from recording sites in parallel strips, except when recording from regions in close proximity to the stimulation site. In addition to this cellular activation, cortical stimuli of sufficient intensity evoked a negative or negative–positive-going field response similar to that elicited by local stimulation and to that recorded in the intact preparation. A comparison of the *in vitro* and *in vivo* cortical-evoked population response is shown in Fig. 12.5C. Stimulation of the prefrontal cortex evokes single or multiple spikes at latencies of 4–8 ms in the *in vivo* preparation. As the intensity is increased, a field response having essentially the same configuration as that obtained in the *in vitro* slice is recorded. Previous analysis of the excitability properties of this cortical-evoked population response in the intact preparation has shown an intensity-dependent potentiation or depression comparable to that elicited in the *in vitro* slice following paired-pulse stimulation (Miller and Rutherford, 1978). The only significant difference between these responses is that the latency obtained in the intact preparation is slightly longer, which would be expected because of the greater conduction distance.

Sections of tissue removed from the middle one-third of the striatum included an area that could be histologically identified as the globus pallidus (see Fig. 12.1B and 12.5B). Stimulating electrodes positioned in this region evoked spikes in the striatum at latencies of 1.5–2.9 ms. The constant latency of these responses to threshold stimuli and their ability to follow high-frequency volleys (Fig. 12.5A) suggests that they are antidromically evoked. Very few neurons were orthodromically activated by stimuli applied in this region, unlike sites that were located either within the striatal tissue proper or in sections that did not contain portions of the pallidum. Of 18 cells identified as being antidromically activated, local or cortical stimulating sites failed to evoke any of them, suggesting that the target neurons of the cortico-striatal input are distinct from the striopallidal output cells.

4. *Actions of acetylcholine and γ-aminobutyric acid*

Biochemical and histochemical studies have indicated that acetylcholine (ACh) and γ-aminobutyric acid (GABA) are present in relatively

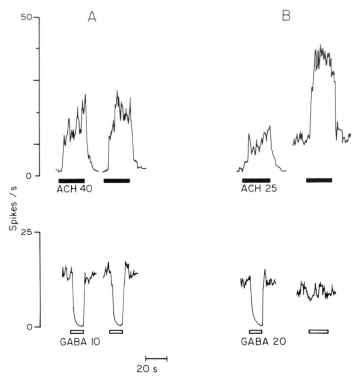

Fig. 12.6. Ratemeter records of changes in firing rate of two locally evoked striatal neurons in response to iontophoretic application of acetylcholine (ACH) and GABA. A, Excitatory effect of ACH applied at 40 nA and inhibition by GABA (10 nA). The latter effect was observed on a cell made to discharge by constant glutamate application at 5 nA. B, Effects of perfusion with Cl^--free medium. The first traces indicate typical responses to ACH and GABA at 25 and 20 nA respectively. After 11 minutes of perfusion, the effect of ACH is enhanced and GABA has no effect. Note that the increased 'spontaneous' activity in Cl^--free medium made it unnecessary to use glutamate-induced firing to demonstrate the change in the normal inhibitory action of GABA.

high concentrations in the striatum and further that these compounds are presumably localized within the soma and processes of intrinsic interneurons in this structure (Lynch *et al.*, 1972; McGeer and McGeer, 1975). While the iontophoretic application of GABA consistently produces an inhibition of cells in the striatum (Bernardi *et al.*, 1975), ACh elicits both inhibition and excitation (Bloom *et al.*, 1965; Herz and Zieglgansberger, 1968). Since these effects have, in most cases, been examined on unidentified striatal cells, it is difficult to determine whether these actions are mediated directly or indirectly through interneurons. In the present study, the actions of ACh and GABA were tested on 46 striatal cells identified by local stimulation. The iontophoretic application of ACh at currents of 5–80nA produced an excitatory effect on 36 of these but the remaining 10 were unaffected (Fig. 12.6A). The depolarizing action of ACh was relatively slow in onset and was sometimes preceded by a transient depression of the spontaneous or glutamate-induced firing. The application of GABA at currents of 3–20 nA produced a rapid onset of depression on all cells tested (Fig. 12.6A). The effects of ACh and GABA were also examined on neurons recorded from slices perfused with a Cl^--free medium (Cl^- replaced with propionate). On five cells tested, the GABA-induced inhibition was either markedly diminished or completely eliminated following a 10–15 minute perfusion with Cl^--free medium (Fig. 12.6B). On the other hand, the excitation produced by ACh was enhanced as was the baseline discharge rate. Following re-exposure to the standard medium, the responses to both ACh and GABA returned to control values.

B. Septum

1. *Spontaneous activity*

A total of 350 neurons were recorded from coronal sections of the septal area. In contrast to the 'silent' or infrequent activity of striatal neurons, the majority of these cells (61%) exhibited spontaneous activity and the remaining 39% were evoked by local stimulation. Although neurons have been recorded throughout the medial and lateral divisions of the septum, the present results are based on 215 cells restricted to the medial septal–diagonal band region (see Fig. 12.2). Spike-train analysis of the discharge pattern and rate of these neurons indicates that at least three distinct populations may be identified.

About 35% of the neurons fired in an irregular or random pattern at rates between 2 and 35 Hz ($\bar{X} = 11.5$). The histograms indicate a wide variance in the interspike interval and a brief inhibitory period (20–40

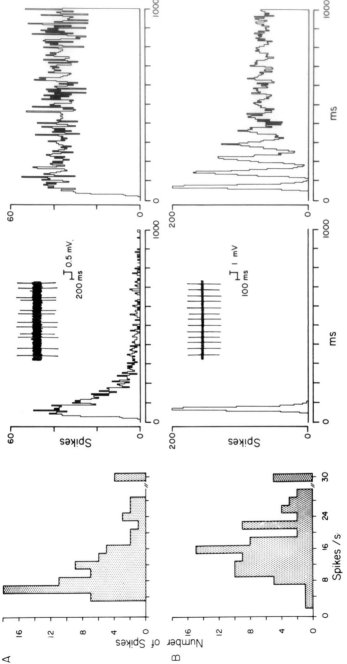

Fig. 12.7. Characteristics of medial septal–diagonal band neurons discharging in an irregular (A) or regular (B) pattern. The frequency distribution of cells in each population are shown in the left panels. Interval histograms and autocorrelograms are shown in the middle and right panels respectively. Insert records indicate single oscilloscope sweeps of discharging cells. Each histogram consists of 1000 spikes and the bin width is 10 ms.

ms) after the initial spike discharge, which is then followed by recovery to baseline (Fig. 12.7A). The second population, which accounted for 42% of the neurons, displayed a remarkable degree of regularity, indicated by the low variance in their discharge pattern (Fig. 12.7B). The rate of firing ranged from 4 to 48 Hz with a mean frequency of 16.6 Hz. Following an initial inhibitory period, the duration of which was inversely related to the firing rate, the discharge pattern revealed by the autocorrelogram was typically sinusoidal, consisting of four to nine waves of decreasing amplitude, the intervals of which corresponded to the early mode. The third group of neurons, which accounts for 23% of the total, displayed a rhythmic pattern of bursts consisting of two to five spikes and discharging between 3 and 12 Hz. The typical bimodal distribution of the interval histogram and rhythmical firing pattern of these bursting neurons are shown in Fig. 12.8. Although it was not usual for a neuron discharging with one pattern to change to one of the other patterns during the recording period, there were several instances in

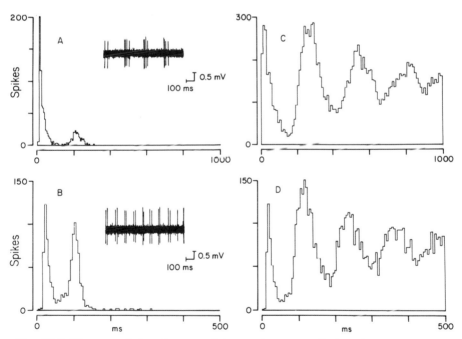

Fig. 12.8. Characteristics of two medial septal neurons discharging with a burst pattern. Interval histograms (AB) indicate the typical bimodal distribution of firing the autocorrelograms (CD) the rhythmical pattern. Insert records show single oscilloscope sweeps of the two cells. Note the differences in burst discharge frequency.

which bursting neurons altered their firing to an irregular pattern or vice versa. These two patterns may therefore reflect different states of activity in the same population of neurons. Similar characteristic patterns have been described for medial septal–diagonal band neurons recorded *in vivo* and are functionally associated with the generation of synchronized and desynchronized activity in the hippocampal region (Assaf and Miller, 1978).

2. *Local and extrinsic evoked activity*

Single-pulse stimulation of each of the three anatomically distinct regions of the septum (see Fig. 12.2) evoked single or multiple spike activity in silent or spontaneously discharging neurons recorded from the medial septal–diagonal band region.

Stimulation of the S_1 zone, which includes fibers of the ascending median forebrain bundle, activated a total of 37 neurons with latencies which varied between 2.1 to 5.0 ms (Fig. 12.9A). These responses followed stimulation frequencies of 50–80 Hz and were evoked up to a distance of 3 mm from the stimulation site suggesting that they are monosynaptically activated. Neurons that characteristically fired in an irregular pattern were inhibited for periods of up to 400 ms following the initial excitatory response. Relatively few of the regular firing cells were influenced by stimulation of the S_1 zone. However, those that were ($n = 11$) exhibited an initial brief inhibitory period (20–50 ms) followed by a synchronization of their discharge pattern.

Stimulation of the S_2 zone evoked synaptic and antidromic responses in the medial septal–diagonal band region. Irregular-firing neurons displayed an activation or activation–inhibition sequence, the latency and duration of which was essentially the same as those obtained after stimulation of the S_1 zone. On the other hand, single-pulse stimulation resulted in a pure inhibitory response ranging from 80 to 310 ms on cells exhibiting the regular discharge pattern (Fig. 12.9B). Both 'silent' and irregular firing cells were antidromically evoked by stimulation of the S_2 region and this response was followed by a period of inhibition ranging from 40–190 ms when the cells discharged spontaneously (Fig. 12.9C). At intensities that were below threshold for evoking the antidromic spike, this period of inhibition was significantly reduced or absent suggesting that the latter component of the response is mediated by a recurrent collateral inhibitory system. In several cases ($n = 8$) a high-frequency cellular barrage was recorded either simultaneously with an antidromic spike, as is shown in Fig. 12.9D, or by itself. This response consisted of 4–12 spikes of relatively small amplitude which sometimes

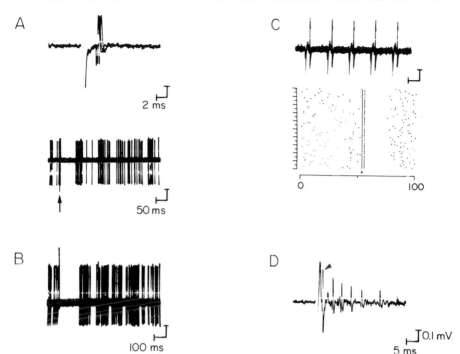

Fig. 12.9. Responses of medial septal neurons to stimulation of various zones within the slice preparation. A, Orthodromic activation -inhibition sequence of an irregular firing cell following single pulse stimulation of the S_1 zone (see Fig. 12.2). Top trace indicates variability in latency of activation and lower trace at slower sweep speed indicates the inhibitory period. B, Inhibition and synchronization of discharge of a regular firing neuron following stimulation of the S_2 zone (25 superimposed sweeps). C, An irregular firing cell antidromically evoked by S_2 stimulation. Top trace indicates high frequency following to 100 Hz. Rastered display shows collision–extinction of antidromic evoked spike with spontaneously firing cell within critical period and the inhibitory period that follows. D, Single trace indicating an antidromic spike (arrow) evoked by S_2 stimulation and the high-frequency discharge of a small amplitude cell in the period immediately following this response.

decremented during the burst. The correlation in latency between these discharges and the onset of inhibition of the irregular firing neurons suggests that these may reflect activity of inhibitory interneurons.

Stimulation of the S_3 region elicited an activation or activation–inhibition sequence on 'silent' and spontaneously firing neurons with an irregular pattern at latencies similar to those evoked from the S_2 zone.

The regular firing cell group was unaffected from this site of stimulation and relatively few neurons were evoked antidromically.

3. *Actions of substance P*

Recent studies have indicated a wide distribution of various peptides throughout the central nervous system and have suggested that these compounds may play a significant role in the process of synaptic transmission (Emson, 1979). Several investigations have shown that the septal area contains a significant level of one of these peptides, substance P, and that it is largely contained within intrinsic neurons of this structure (Leung *et al.*, 1981). In the present study, the effects of perfusion of known concentrations of this peptide were tested on 28 identified cells recorded from the medial septal–diagonal band region (Fig. 12.10). Substance P applied in concentrations of 100 nM resulted in a potent excitation of irregular firing neurons ($n = 12/15$). This

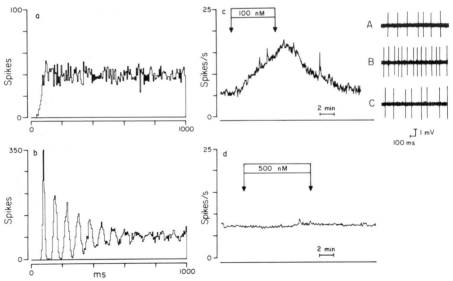

Fig. 12.10. Effect of substance P (sP) perfusion on medial septal–diagonal band neurons. Autocorrelograms (a, b) indicate two different cells displaying an irregular (a) and regular (b) discharge pattern. Rate recordings in c and d indicate excitatory effect of 100 nM of sP on irregular cell (c) but no effect on regular (d) firing cell. Insert oscilloscope records indicate cellular discharge of irregular neuron before (A), 8 minutes 'onset' (B) and 25 minutes after termination (C) of sP perfusion.

response is relatively slow in onset and persists for long periods (10–12 minutes) following termination of the drug perfusion. At higher concentrations (500, 1000 nM), the magnitude and duration of this effect was increased in a dose-related manner. On three neurons, repeated applications had little or no effect on the discharge rate indicating a response tachyphylaxis. Concentrations of substance P up to ten times greater than that required to produce excitation of the irregular firing cells had no effect on the discharge rate of neurons exhibiting the regular pattern (Fig. 12.10). These data indicate that not only are neurons in the medial septal–diagonal band region extremely sensitive to the application of this peptide but that its excitatory action is relatively specific on a particular population of neurons.

IV. Discussion

On the basis of several electrophysiological criteria, including spontaneous activity, synaptically evoked responses from local and extrinsic inputs and sensitivity of these responses to ionic or pharmacological manipulations, the results of the present study have shown the viability of both striatal and septal tissue slices maintained in vitro. The results also indicate many close similarities in electrical activity to that recorded from intact preparations.

A. Similarities to intact tissue

In the striatum, local and cortical stimulation evoked action potentials in both silent and spontaneously firing neurons. At sufficiently high stimulus intensities, a negative field response was elicited at a latency that coincided with the spike activation. Since this field potential appeared to reflect the synchronous discharge of a number of cells in the immediate vicinity of the recording electrode, it is referred to as a population response. The fact that the response to local stimulation persists after chronic deafferentation of the tissue, indicates that a major intrastriatal excitatory network is functional in the slice. Furthermore, the cortico-striatal excitatory input converges on the same neural elements. A similar field response and cellular activation has been recorded in vivo following stimuli applied both locally and to cortical afferents innervating the striatum (Marco et al., 1973; Miller and Rutherford, 1978). With the exception of the raphe-striatal system (Miller et al., 1975), relatively few inputs result in an initial inhibition of neuronal activity. In the present study, the inhibition of spontaneously firing cells and test population response, following either local or cortical stimulation, was always preceded by an activation. Since direct inhibitory effects were not observed, the data suggest that the neuronal processes underlying this response are dependent on collateral interneurons associated with the primary cells excited by local or extrinsic inputs. The present data, however, do not exclude the possibility that in the *in vitro* striatal slice a significant proportion of the trajectories of intrinsic inhibitory neurons are destroyed by cutting and preparing the tissue. This, however, appears unlikely since elimination of the synaptically evoked inhibition in the slice following perfusion with a Cl^--free medium resulted in a dramatic increase in both spontaneous activity and amplitude of evoked population responses, suggesting that neural elements in the tissue are normally subjected to a tonic inhibitory influence. This may also account for the relatively low incidence of spontaneously firing neurons recorded in both *in vitro* or *in vivo* striatal preparations.

Neuronal activity recorded from the *in vitro* septal slice indicates that both the rate of firing and the discharge patterns of cells in the medial septal region are directly comparable to those reported for intact preparations (Assaf and Miller, 1978). Of particular interest are the two groups of cells that fire with either a rhythmical burst or irregular discharge. Not only do the rhythmically firing cells discharge in bursts at a rate that is within the normal range for that recorded from the intact preparation (4–12 Hz), but many of these were also shown to fire in an irregular pattern, suggesting that they may be the same population of neurons reflecting different states of ongoing activity. Previous studies have demonstrated that the bursting and irregular firing cells in the medial septal region serve as 'pacemakers' for the generation of synchronized and desynchronized hippocampal electrical activity (McLennan and Miller, 1974; Assaf and Miller, 1978). The fact that these patterns of septal neuronal activity are maintained in the *in vitro* slice demonstrates that the substrates within the septum, which are required to initiate the contrasting states of hippocampal activity, are present and functional in this tissue. Although the intraseptal mechanisms that produce these activities are not yet clear, a previous study (McLennan and Miller, 1976) postulated a frequency-related mechanism within the lateral septum that could evoke either an irregular or bursting mode of discharge in the medial septal. The present findings, which indicate that local stimulation of the lateral septum elicits both orthodromic and antidromic responses in the medial septum, provide the necessary evidence to suggest that a feedback loop between these two regions of the septum exists. Depending on the level of afferent information impinging on this intrinsic loop, either pattern of neuronal discharge would be generated.

The group of spontaneously firing septal neurons that displayed an extremely regular discharge pattern are not typical of those recorded from the intact preparation. Although a regular firing pattern is sometimes indicative of cellular damage, this appears to be unlikely since the rate of discharge of these cells was not abnormally high, in addition to which, they could be recorded for long periods. As was suggested previously to account for the lack of inhibition in striatal slices, tissue sectioning may have damaged some of the inhibitory processes within the septum. This also appears unlikely because the sinusoidal discharge exhibited in the autocorrelelogram is presumably due to an 'active' inhibitory process. Furthermore, local stimulation resulted in an inhibition of the regular firing indicating that inhibitory mechanisms are still present in the slice. Previous studies have demonstrated that certain neurons in the medial septum are influenced by inhibitory inputs arising in various brainstem and hypothalamic regions (Assaf and Miller, 1978; Wilson et al., 1976). As the results of the present study have shown, stimulation in the region of the median forebrain bundle, through which many of these inputs are mediated, resulted in a pure inhibition of these regular firing cells. It is therefore possible that this discharge pattern results from the removal of a tonic inhibitory input.

B. Intrinsic circuitry

The visual placement of stimulating and recording electrodes in a tissue slice allows for unprecedented flexibility in the examination of the intrinsic synaptic organization of both the septal and striatal regions. Anatomically, the septal area is divided into several distinct nuclei and although it is not as well organized as the hippocampus, it is nevertheless possible to determine the interaction between these various parts as, for example, was indicated between the lateral and medial septal regions. The striatum, on the other hand, has been described as being markedly homogeneous in its organization making it extremely difficult to examine the interactions of the intrinsic circuitry of this structure. The present findings have, however, provided preliminary electrophysiological evidence to suggest that the striatum is organized into distinct functional units. Local or cortical stimulation elicits an initial excitation of striatal neurons which is mediated by intrinsic circuitry. When the radially oriented cellular arrays are continuous through the tissue section, it is possible to evoke this response through the full extent of the slice as long as the stimulating and recording electrodes are aligned in the same cellular array. Stimulation of juxtaposed strips separated by axon fascicles results in little, if any, activity on these neurons. These data suggest that there are extensive interconnections between particular populations of neurons which are arranged in columns traversing the striatal slice in a plane parallel to the cortico-spinal fascicles. Although the present findings do not demonstrate how the neural elements that comprise such columns interact, it is evident that both intrinsic excitatory and inhibitory processes are present. Several recent anatomical investigations have described various cyto-architectural arrangements in the striatum which may form the structural basis for this type of micro-organization within this region. Ring-like clusters of cells and dendrites are detected in striatal sections cut perpendicular to the axon fascicles, whereas sections parallel to the fibers show columns of cells with extensive dendritic intertwining (Mensah, 1977; Chronister *et al.*, 1976). Autoradiographic studies have also indicated that the terminal distribution of cortical and thalamic afferent fibers in the striatum are organized into patch-like regions which, if sectioned in a particular orientation, also give the impression of columns (Kalil, 1978; Goldman and Nauta, 1977). Although further experiments are required to identify the neural elements associated with these columns and their mechanisms of interaction, it is clear that the striatum is organized into functional units and is not, as was previously suggested, a homogeneous structure.

C. Pharmacological properties

The preliminary pharmacological data presented in this report indicate several distinct advantages to using the *in vitro* preparation for investigating the actions of various chemical compounds on neuronal activity in the striatum and septal regions. Although an extensive literature exists on the effects of a wide variety of putative transmitters on cells recorded in these areas, there is relatively little known about either the site or the mode of action of many of these compounds. This is largely due to the fact that few reports have adequately identified the cell populations being tested, which is an essential criterion in structures like the striatum or septum which have complex local circuits. As is shown in the present study, intrinsic neurons identified by electrophysiological criteria have been examined for their responsiveness to several pharmacological agents. In the striatum, the major population of identified neurons were shown to be excited by the microiontophoretic application of ACh. These results are in contrast to most other *in vivo* studies, which have demonstrated a mixture of both excitatory and inhibitory effects on striatal cells. The major factor which may account for this discrepancy is that cells tested in the present experiments all belong to the same synaptically identified population, whereas previous data report effects on a heterogeneous population.

In contrast to the striatum, a large proportion of the neurons recorded from the septal area exhibit a number of characteristic discharge patterns which allows for the identification of different populations of cells. Although thus far only the effects of substance P have been examined on identified septal cells, the data illustrate the advantages of perfusing known concentrations of a compound on neuronal activity. This is particularly relevant when dealing with substances for which it is difficult to determine the amount being extruded from micropipettes.

Another advantage of the *in vitro* preparation for the study of the pharmacological action of various transmitters is that the ionic mechanisms underlying these processes may be determined. The results have shown that in the striatum both the synaptic and GABA-induced inhibition, which are presumably mediated by recurrent collateral interneurons, are decreased or eliminated when the tissue is perfused with a Cl^--free medium. These data indicate that not only are the observed inhibitory processes Cl^- dependent but also indirectly suggest that the inhibition recorded in the *in vitro* striatal slice is GABA mediated. The possibility that other compounds alter Cl^- permeability to produce inhibition cannot be excluded.

It is apparent from these data that further electrophysiological analysis of local circuit neurons in the *in vitro* striatal and septal preparations will facilitate investigations of the pharmacological properties of neurons in these regions.

V. Conclusions

1. Extracellular recordings were made from slices of striatal and septal tissue maintained *in vitro*.

2. Different cell populations were identified by their spontaneous discharge patterns or by their response to orthodromic and antidromic stimulation of intrinsic and extrinsic pathways.

3. The similarities to that described for the intact preparation indicate a preservation of the functional integrity of the intrinsic cellular networks in these tissues.

4. The iontophoretic application of ACh (5–80 nA) produced excitations of most striatal cells tested.

5. The application of GABA (3–20 nA) produced a rapid onset of depression on all cells tested. On five of these striatal cells perfused with Cl^--free medium, the GABA-induced inhibition was markedly decreased or completely eliminated.

6. Single-pulse stimulation of each of the three anatomically distinct regions of the septum evoked single or multiple spike activity in 'silent' or spontaneously discharging neurons recorded from the medial septal–diagonal band region.

7. The data indicated that not only are neurons in the medial septal–diagonal band region extremely sensitive to the application of substance P, but that its excitatory action is relatively specific on a particular population of neurons.

8. Although the electrophysiological and pharmacological investigations of these networks have not yet achieved the same level of sophistication as that for some other tissues examined *in vitro*, the data reported here demonstrate that both the striatum and septum are suitable models for analyzing some of the complexities of these regions, which have been difficult to approach using the more conventional intact preparation.

9. Future experiments employing a combination of multidisciplinary techniques to these *in vitro* preparations will become increasingly productive in the elucidation of cellular mechanisms underlying normal and abnormal functions of the mammalian central nervous system.

Acknowledgements

The author is indebted to D. P. Rutherford and P. Y. Leung for their contributions to these studies and to H. Brandejs and K. Henze for their skilful assistance. This work was supported by the Medical Research Council of Canada.

References

Alger, B. E. and Teyler, T. J. (1976). Long-term and short-term plasticity in the CA1, CA3, and dentate regions of the rat hippocampal slice. *Brain Res.* **110,** 463–480.

Assaf, S. Y. and Miller, J. J. (1978). The role of a raphe-serotonin system in the control of septal unit activity and hippocampal desynchronization. *Neuroscience* **3**, 539–550.

Bernardi, G., Marciani, M. G., Morocutti, C. and Giacomini, P. (1975). The action of GABA on rat caudate neurones recorded intracellularly. *Brain Res.* **92**, 511–515.

Bliss, T. V. P. and Richards, C. D. (1971). Some experiments with *in vitro* hippocampal slices. *J. Physiol. (Lond.)* **232**, 331–356.

Bloom, F. E., Costa, E. and Salmoiraghi, G. C. (1965). Anaesthesia and the responsiveness of individual neurones of the caudate nucleus of the cat to acetylcholine, norepinephrine and dopamine administered by microelectrophoresis. *J. Pharmacol. exp. Ther.* **150**, 244–252.

Carpenter, M. B. (1976) Anatomical organization of the corpus striatum and related nuclei. *In* "The Basal Ganglia" (M. D. Yahr, ed.), pp. 1–36. Raven Press, New York.

Chronister, R. B., Farnell, K. E., Marco, L. A. and White, L. E. (1976). The rodent neostriatum: a golgi analysis. *Brain Res.* **108**, 37–46.

Emson, P. C. (1979). Peptides as neurotransmitter candidates in the mammalian CNS. *Prog. Neurobiol.* **13**, 61–116.

Goldman, P. S. and Nauta, W. J. H. (1977). An intricately patterned prefronto-caudate projection in the rhesus monkey. *J. Comp. Neurol.* **171**, 369–386.

Hassler, R., Chung, J. W., Wagner, A. and Rinne, U. (1977). Experimental demonstration of intrinsic synapses in cat's caudate nucleus. *Neurosci. Letters*, **5**, 117–121.

Herz, A. and Zeiglgänsberger, W. (1968). The influence of microelectrophoretically applied biogenic amines, cholinomimetics and procaine on synaptic excitation in the corpus striatum. *Int. J. Neuropharmacol.* **7**, 221–230.

Kalil, K. (1978). Patch-like termination of thalamic fibers in the putamen of the rhesus monkey: an autoradiographic study. *Brain Res.* **140**, 333–339.

Kemp. J. M. and Powell, T. P. (1971). The synaptic organization of the caudate nucleus. *Phil. Trans. B.* **262**, 403–412.

Kocsis, J. D., Sugimori, M. and Kitai, S. T. (1977). Convergence of excitatory synaptic inputs to caudate spiny neurons. *Brain Res.* **124**, 403–413.

Leung, P. Y., McIntosh, C. and Miller, J. J. (1981). Localization and action of substance P on neurones in the septal area of the rat. *Neuropeptides*, in press.

Lynch, G. S., Lucas, P. A. and Deadwyler, S. A. (1972). The demonstration of acetylcholinesterase containing neurons in the caudate nucleus of the rat. *Brain Res.* **45**, 617–621.

McIntosh, C. and Miller, J. J. (1981). Localization and action of substance P on neurones in the septal area of the rat. *Neuropeptides*, in press.

Marco, L. A., Copack, P., Edelson, A. M. and Gilman, S. (1973). Intrinsic connections of caudate neurons. I. Locally evoked field potentials and extracellular unitary activity. *Brain Res.* **53**, 291–305.

McGeer, P. L. and McGeer, E. G. (1975). Evidence for glutamic acid decarboxylase containing interneurons in the neostriatum. *Brain Res.* **91**, 331–335.

McLennan, H. and Miller, J. J. (1974). The hippocampal control of neuronal discharges in the septum of the rat. *J. Physiol. (Lond.)* **237**, 607–624.

McLennan, H. and Miller, J. J. (1976). Frequency related inhibitory mechanisms controlling rhythmical activity in the septal area. *J. Physiol. (Lond.)* **254**, 827–841.

Mensah, P. (1977). The internal organization of the mouse caudate nucleus: evidence for cell clustering and regional variation. *Brain Res.* **137**, 53–66.

Miller, J. J. and Rutherford, D. P. (1978). Electrophysiological analysis of the synaptic organization of afferent projections to the corpus striatum. *Proc. Can. Fed. Biol. Soc.* **21**, 55.

Miller, J. J., Richardson, T. L., Fibiger, H. C. and McLennan, H. (1975). Anatomical and electrophysiological identification of a projection from the mesencephalic raphe to the caudate-putamen in the rat. *Brain Res.* **97**, 133–138.

Pasik, P., Pasik, T. and Difiglia, M. (1976). Quantitative aspects of neuronal organization in the neostriatum of the macaque monkey. *In* "The Basal Ganglia" (M. D. Yahr, ed.), pp. 57–90. Raven Press, New York.

Purpura, D. P. (1976). Physiological organization of the basal ganglia. *In* "The Basal Ganglia" (M. D. Yahr, ed.), pp. 91–114. Raven Press, New York.

Richardson, T. L., Miller, J. J. and McLennan, H. (1977). Mechanisms of excitation and inhibition in the nigrostriatal system. *Brain Res.* **127**, 219–234.

Schwartzkroin, P. A. (1975). Characteristics of CA1 neurons recorded intracellularly in the hippocampal *in vitro* slice preparation. *Brain Res.* **85**, 423–436.

Schwartzkroin, P. A. (1977). Further characteristics of hippocampal CA1 cells *in vitro*. *Brain Res.* **128**, 53–68.

Schwartzkroin, P. A. and Wester, K. (1975). Long-lasting facilitation of a synaptic potential following tetanization in the *in vitro* hippocampal slice. *Brain Res.* **89**, 107–119.

Skrede, K. K., and Westgaard, R. H. (1971). The transverse hippocampal slice: A well defined cortical structure maintained *in vitro*. *Brain Res.* **35**, 589–593.

Swanson, L. W. (1978). The anatomical organization of septo-hippocampal projections. *In* "Functions of the Septo-Hippocampal System" (K. Elliott and J. Whelan, eds.), pp. 25–43. Elsevier Press, London.

Swanson, L. W. and Cowan, W. M. (1976). Autoradiographic studies of the development and connections of the septal area in the rat. *In* "The Septal Nuclei" (J. F. DeFrance, ed.), pp. 37–64. Plenum Press, New York.

Wilson, C. L., Motter, B. C. and Lindsley, D. B. (1976). Influences of hypothalamic stimulation upon septal and hippocampal electrical activity in the cat. *Brain Res.* **107**, 55–68.

Yamamoto, C. (1972). Activation of hippocampal neurons by mossy fiber stimulation in thin brain sections *in vitro*. *Exp. Brain Res.* **14**, 423–435.

13

An *In Vitro* Preparation of the Spinal Cord of the Mouse

J. BAGUST AND G. A. KERKUT

Department of Neurophysiology, School of Biochemical and Physiological Sciences, Southampton University, Southampton SO9 3TU, U.K.

I. Introduction

Studies of the neuropharmacology of the mammalian spinal cord have been largely conducted on *in vivo* preparations in which compounds have been applied in the close vicinity of cells by iontophoresis or pressure ejection from micropipettes inserted into the cords of anaesthetized animals (Curtis and Eccles, 1958; Curtis *et al.*, 1961; Curtis, 1969; Biscoe *et al.*, 1976; McLennan and Wheal, 1976; Headly *et al.*, 1978; Martin *et al.*, 1978; McLennan and Lodge, 1979). In addition to considerable practical difficulties in maintaining the stability of such preparations to allow good recordings to be obtained, the possible effects of the anaesthetic agents employed have to be considered. For example, the barbiturates are known to depress spinal activity (Weakly, 1969) and have recently been shown to enhance the activity of GABA (Barker and Ransom, 1978; Evans, 1979). α-Chloralose, another commonly used anaesthetic agent, produces a state of heightened reflex activity and can cause convulsions in the intact animal. Muscle relaxants, which interfere with synaptic transmission at the neuromuscular junction, are also often employed to suppress spontaneous respiratory movements, thereby allowing the animal to be ventilated in a controlled way.

The application of test substances to *in vivo* preparations is largely confined to localized ejection from micropipettes placed close to a particular cell. This technique, while providing much information about the receptive properties of individual nerve cells, suffers from several drawbacks (see Bloom, 1974). In addition to technical problems, such methods necessarily select a very localized area to release the drugs, whereas the effect on the whole system is often of more interest. The administration of substances by injection into the cardiovascular system allows single-dose experiments to be performed, but unless the drugs are rapidly metabolized they cannot be washed out and there is also the problem of the extent to which they penetrate the blood–brain barrier.

The development of *in vitro* techniques for thin sections of mammalian nervous system has provided a way of overcoming some of the problems of *in vivo* preparations. These systems can be made very stable, and individual cells can be held for long periods of time while the composition of the surrounding medium is changed, allowing known concentrations of test substances to be applied, buffered to physiological pH. Drugs can be washed out of the bath easily and the effects of agonists and antagonists on responses examined. These techniques can be combined with iontophoretic application of substances to individual cells, providing more information than could be obtained from preparations *in situ*.

Cutting thin (200–500 μm) transverse sections of spinal cords has proved technically difficult. The presence of the tough pial membrane results in considerable damage due to crushing of the softer inner tissue, although Takahashi (1978) has reduced such damage by embedding the cord in agar prior to cutting. The cord, however, is essentially a longitudinally arranged structure with activity passing up and down its length along the bundles of nerves in the white matter, and the value of transverse sections must therefore be limited.

Much work has been done on the isolated amphibian spinal cord, hemisected sagitally to allow adequate penetration of nutrients and test chemicals (for a review, see Kudo, 1978). In mammals, the use of such preparations has been confined to the examination of isolated cords from very immature animals, usually rats or cats (Otsuka and Konishi, 1974; Evans, 1978; Preston and Wallis, 1979; Shapovalov et al., 1979). In more mature animals, the cord is too large to allow adequate penetration of oxygen to keep the tissue alive. However, the nervous systems of neonatal animals are not fully developed. They are blind and only capable of crude reflex movements. Morphological studies have shown the myelination of the spinal white matter is incomplete and some projections to the spinal cord from the dorsal roots and the brain are not complete until the third week of life (Skoglund, 1969; Donatelle, 1977; Gilbert and Stelzner, 1979).

We describe here an isolated preparation of the spinal cord of juvenile mice. At the ages used in this study, 4–6 weeks, the animals eyes are open, and they are fully active. Myelination is more advanced than in the neonatal animal, and yet the cord is small enough to allow oxygen to penetrate sufficiently to keep it alive.

II. Methods

A. Dissection

Albino mice of the Porton strain were obtained from the Southampton University colony. Animals of all ages from newborn to fully mature (30–40 g) have been used, but the majority of results presented here were obtained from 4–6 week-old animals (12–20 g) (Fig. 13.1). They were decapitated, without anaesthetic, and the vertebral column was removed by cutting up either side of the back from the anus to the neck. This tissue was rapidly cooled by plunging into ice-cold Ringer solution, and the rest of the dissection was carried out in a bath perfused with Ringer solution at 4°C.

The spinal cord was exposed by removing laminae from the ventral

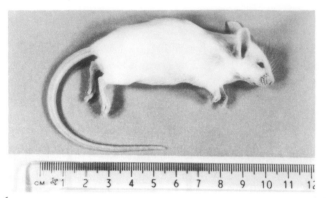

Fig. 13.1. A mouse of the Porton strain weighing 20 g, typical of the size of animal used in this study.

surface using a pair of fine scissors, working from rostral to caudal. After cutting the spinal nerves peripheral to the dorsal root ganglia, the cord was floated free of the vertebrae. Any adhering dura was carefully dissected away, and the ventral and dorsal roots in the lumbar region were separated. The cord was then split along the midline using a knife made from a fragment of a razor blade, and one-half mounted in the experimental bath, the other half being kept in oxygenated Ringer solution at room temperature for later use if needed.

In some experiments the animals were anaesthetized with sodium pentobarbitone (1 mg i.p.) or halothane (2%) and cooled by immersion in ice for 15 minutes. A mixture of 95% O_2/5% CO_2 was administered until respiration ceased, when the animal was completely immersed. The rest of the dissection was performed as with the decapitated animal, but the precooling allowed peripheral nerves or structures in the brainstem to be removed still attached to the spinal cord.

B. Experimental bath

The experimental bath consisted of a groove 1 cm wide, milled in a block of perspex. The base of this channel was filled with a silicone rubber compound (Dow Corning Encapsulant 3112 TRV) providing a soft floor into which pins could be driven. Perfusion fluid flowed into the chamber at a low level at one end (Fig. 13.2) and was sucked out of a reservoir at the other end, the level of the fluid in the bath being determined by the level of the J-shaped outlet suction probe.

Inflow into the bath was by gravity feed from a 1-litre aspirator providing a pressure head of about 20 cm of water. This lower reservoir was fed continuously from an upper airtight reservoir, the outlet of

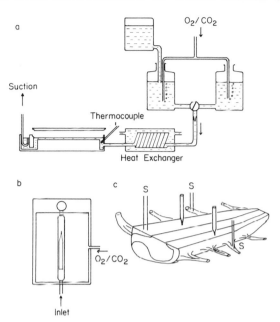

Fig. 13.2. a, Diagram of the arrangement for continuously perfusing the experimental bath with oxygenated Ringer solution. b, Plan of the recording chamber, showing a hemisected cord pinned into the central slot. c, Diagram of a hemisected cord with micropipette recording electrodes penetrating the dorsal and ventral horns, and bipolar wire stimulating electrodes (S) placed on the dorsal and ventral roots and the dorsal tracts.

which dipped into the fluid in the lower chamber. Solution could only pass from the upper to the lower vessel when the tip of the upper outlet pipe was exposed to the air, thereby maintaining a constant level in the lower aspirator. The Ringer solution in the lower reservoir was bubbled with a 95% O_2/5% CO_2 gas mixture.

The flow rate into the bath was determined by a constriction in the pipe (a 25-gauge hypodermic needle) giving a flow of about 10 ml min^{-1}. A heat exchanger supplied with warm water controlled the temperature of the solution entering the bath, which was measured by a thermocouple. The volume of the horizontal section of the bath was 1 ml which, with a flow rate of 10 ml min^{-1}, provided 10 changes of bath contents per minute.

A system of three-way stopcock taps enabled test solutions to be run through the bath from a second reservoir with minimal changes in flow rate.

C. Stimulation and recording

The hemisected spinal cord was pinned to the bottom of the perfusion chamber using fine stainless-steel or tungsten pins, and the lumbar roots were separated for stimulation. Square-wave stimuli from two Grass S44 stimulators were applied to the roots or spinal tracts through pairs of electrodes made from teflon-coated stainless-steel wire, having a tip separation of about 250 μm. The usual arrangement of stimulating electrodes is shown in Fig. 13.2, one pair of electrodes being placed on the dorsal roots L 2–4, one pair on the equivalent ventral roots and one pair on the exposed surface of the dorsal tracts in the thoracic region.

Recordings were made using glass micropipettes broken back to a diameter of 5–10 μm. Leakage of electrolyte was prevented by plugging the tips of the electrodes with 2.5 M NaCl in 1% agar before filling the shanks of the electrodes with 2.5 M NaCl. For extracellular recordings from individual cells, electrodes having finer tips filled with 2.5 M NaCl (resistance 2–6 MΩ) were used. Signals were amplified using a Neurolog NL 100 pre-amplifier and a NL 105 AC amplifier and displayed on an oscilloscope or averaged using a NL 750 averager (Neurolog, Digitimer Ltd.). The contents of the averager could be output onto a Servoscribe flat-bed recorder and a Hewlett-Packard pressurized ink chart recorder for later analysis. Single cell spikes were counted using a Digitimer spike processor D130.

D. Composition of Ringer solution

Normal Ringer: NaCl, 118 mM; KCl, 2.0 mM; NaHCO$_3$, 24 mM; glucose, 12 mM; CaCl$_2$, 2.5 mM; pH 7.4. In some experiments on the ventral horn, KCl was raised to 5.0 mM. To block synaptic responses the CaCl$_2$ was replaced by 2 mM MnCl$_2$.

Low chloride Ringer: sodium acetate, 118 mM; potassium acetate, 2.0 mM; sodium bicarbonate, 24 mM; glucose, 12 mM; calcium acetate, 2.5 mM. All solutions were gassed with 95% O$_2$/5% CO$_2$.

III. Results

Most of the results reported here have been obtained when recording in the region of the lumbar expansion of the cord. Although similar activity can be detected at higher levels, the shortness of the spinal roots in the thoracic and cervical regions makes stimulus spread direct to the cord a problem.

A. Field potentials

Stimulation of the dorsal roots evoked field potentials of up to 2 mV in the dorsal horn near the point of entry of the root, and smaller potentials in the ventral part of the cord (Fig. 13.3a). The dorsal-horn potentials consisted of a short latency (approx. 1 ms) biphasic positive–negative wave followed by a slower more complex positive–negative waveform having a latency of 2–4 ms. This later wave reached a maximum in the region of the dorsal horn close to the dorsal white matter, and became less clear towards the lateral edge of the cord, where smaller, longer latency potentials were observed. The short latency potential reached a maximum at deeper levels in the dorsal horn (approx. 400 μm) in the medial aspect of the dorsal horn.

In the intermediate and central parts of the cord, dorsal root stimulation produced only small potentials, which, with increased gain, could be seen to have two components with latencies similar to the main potentials observed in the dorsal horn (Fig. 13.3c). The possibility that these were the result of passive current effects produced by the relatively intense potentials in the dorsal horn cannot be eliminated. On occasions, individual cells have been found in the ventral horn which responded to single stimuli to the dorsal roots by firing an action potential, indicating that a part, at least, of these potentials was of synaptic origin.

In addition to these short latency potentials, a slow negative potential was recorded along the dorsal edge of the dorsal horn (Fig. 13.3b). This potential was small on the cut surface of the dorsal funiculus, but in the dorsal grey matter it could reach a size of 1 mV in amplitude, the peak having a latency of approximately 20 ms. The total duration of the potential was 300–400 ms. In the deeper layers of the dorsal horn a corresponding slow positive potential was recorded, indicating that this region was acting as the current source for the sink in the more dorsal regions. No slow potentials were detected in the ventral horn.

The spread along the cord of the effects of dorsal-root stimulation was investigated by moving the recording electrode 1 mm and 2 mm rostrally and recording the responses in the dorsal layers of the dorsal horn (Fig. 13.3d and 13.3e). Short latency potentials were greatly reduced in amplitude at 1 mm, and absent at 2 mm, except in the most medial recording in the dorsal white matter where small biphasic potentials were detected, which probably resulted from activity in the ascending branches of stimulated afferent fibres. The longer-latency negative potential spread further up the cord, and was still detectable 2 mm rostral to the point of entry of the stimulated dorsal root.

Stimulation of the dorsal tracts 3–4 segments rostral to the recording

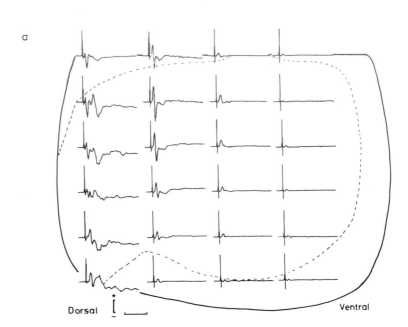

Dorsal   Ventral

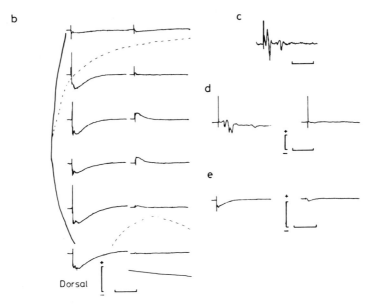

Dorsal

site evoked a similar pattern of responses to that produced by dorsal-root stimulation, but with a more prolonged time course. Potentials evoked in the ventral horn by stimulation of the dorsal tracts were usually larger than those produced by stimulation of the dorsal roots. In addition to antidromically exciting ascending branches of dorsal root afferent fibres, stimulation of the dorsal tracts would also have activated the descending corticospinal tract fibres found in the dorsal funiculus in rodents (Schoen, 1964; Brown, 1971).

These dorsal horn potentials recorded from the isolated mouse spinal cord are similar in time course and polarity to those recorded *in situ* from cat spinal cords on stimulation of afferent nerves and dorsal roots (Coombs *et al.*, 1956; Fernandez de Molina and Grey, 1957; Eccles *et al.*, 1962b; Wall, 1964) or the motor cortex (Vasilenko and Kostyuk, 1966).

B. Effects of temperature and perfusion rate

The effects of changing the temperature on the preparation were measured by averaging dorsal horn responses for the 25 ms following stimulation and recording the averaged signal on paper. At the end of the experiment the bath was perfused with 10 mM procaine, to block all activity and an average obtained of the stimulus artefact. By cutting out the area between the dorsal horn signal and the stimulus artefact and weighing the paper an estimate was made of the area of the evoked response.

Early experiments were conducted with a perfusion rate of 2 ml min^{-1} through the bath. Under these conditions, the area under the curve rose steadily to a maximum value between 25 and 30°C before falling sharply at 35°C to less than 60% of the maximum achieved (Fig. 13.4). When the flow rate was increased to 10 ml min^{-1}, there was little change in the area under the curve between 20 and 35°C and above this temperature a decline in the area was again observed.

Fig. 13.3 Short latency (a) and long latency (b) potentials recorded across a hemisected mouse spinal cord at the point of entry of dorsal root L4 to stimulation of that root. Recordings were made at steps of 200 μm in depth from the cut medial surface, and at distances of 0.1 mm, 0.4 mm, 0.7 mm and 1.0 mm, from the dorsal edge. Each trace is the average of 16 successive responses. c Shows the short latency response recorded in the ventral horn at a depth of 800 μm from the medial surface with the amplification increased by a factor of 32, average of 128 responses. In d and e, responses recorded 1 mm and 2 mm rostral to the recording sites shown in a and b. Horizontal callibration for a, c and d − 5 ms, for b and e − 50 ms. Vertical callibration − 1 mV.

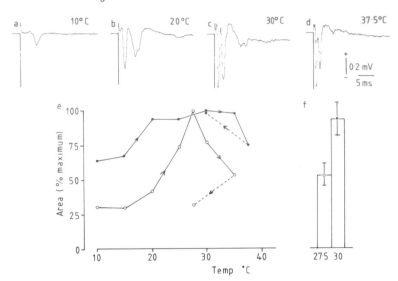

Fig. 13.4. a–d, Responses recorded in the dorsal horn to stimulation of dorsal root L4 at different temperatures. Each recording is the average of 16 successive responses. e, The effect of temperature on the area under the curve of responses such as those in a–d with flow rates through the bath of 10 ml min^{-1} (●) and 2 ml min^{-1} (○), the arrows indicate the sequence of the temperature changes. f, Shows the effects of the flow rate on the recovery of the signal (integrated area) following heating above its optimum temperature, and return to 27.5°C (2 ml min^{-1}) or 30°C (10 ml min^{-1}). Values are means ± s.D. of five experiments.

This technique did not allow a detailed examination to be made of the effects of temperature since as the temperature was increased the potentials became briefer and the areas under them decreased. The maintainance of the area of the signal obtained with the higher flow rate was largely due to an increased contribution from long-latency potentials (Fig. 13.4c), but it indicated that the tissue was more viable at higher temperatures with a flow rate of 10 ml min^{-1}. This was reinforced by the recovery of the preparation from heating beyond its optimum temperature (Fig. 13.4f). On returning the temperature of four low-flow-rate preparations from 35°C to 27.5°C, the area under the curve was only 54 (s.E. ± 8)% of its previous value, indicating that some tissue damage had occurred, whereas cooling five high-flow-rate preparations from 37.5°C to 30°C resulted in almost complete restoration of the previous level of activity (mean 95 ± 12%).

All subsequent experiments were therefore carried out at temperatures of 33–35°C and a bath perfusion rate of 10 ml min^{-1}. Under these conditions, good field responses could be obtained from preparations for up to 12 hours after dissection. When longer survival times were required, it was necessary to reduce the temperature, sacrificing part of the synaptic response. Cords have been kept alive for as long as 50 hours at room temperature (18–20°C).

C. Responses of single cells in the dorsal horn

Using NaCl-filled electrodes with resistances in the range 2–5 MΩ, extracellular recordings could be made from cells in the dorsal horn of the cord. Spontaneous firing was seldom encountered, most cells had to be driven by stimulating either the dorsal roots or dorsal tracts supplying that or adjacent segments rostrally.

Although no systematic investigation of the properties and distribution of dorsal horn unit responses has been undertaken, the cells encountered fell into three groups. The most common type were cells that responded to a weak stimulus to either the dorsal roots, or dorsal tracts, or sometimes to both with a single action potential having a variable, and often long, latency (10–20 ms). With higher stimulus intensities, the latency was shortened and additional action potentials were evoked, producing a burst of 2–7 action potentials lasting up to 50 ms (Fig. 13.5a and b). The amplitude of the later action potentials was often smaller than that of the initial spike. When trains of stimuli were applied, these cells were unable to follow frequencies greater than 50 Hz (Fig. 13.5c), and were considered to be synaptically driven by afferent fibres excited by stimulation of the dorsal root or dorsal tracts.

Stimulation of the fibres of the dorsal tracts 2–3 segments rostral to the recording site caused some cells to be invaded by action potentials that appeared to be antidromically evoked (Fig. 13.5d). These had short (0.5–2 ms) and constant latencies, could follow trains of stimuli at frequencies of more than 100 Hz (often up to 250 Hz), and were not blocked by the replacement of the calcium in the Ringer solution by 2 mM MnCl$_2$ (see next section). Such cells were most frequently found on the dorsal edge of the cord, where the action potentials recorded extracellularly were sometimes in excess of 10 mV. In addition to the single antidromic spike, longer latency groups of action potentials were also observed, but the stimulation threshold for these was different from that for the short latency response, and they were considered to be the result of synaptic activity of the cell caused by simultaneous activation of the cell axon and excitatory afferent fibres synapsing with the cell (Fig. 13.5e).

The third category, of which only three examples have so far been

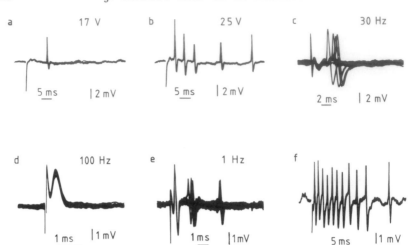

Fig. 13.5. Responses of single cells recorded in the lumbar dorsal horn to stimulation of dorsal root L4 (a, b) or the dorsal tracts in the lower-thoracic region (c–f). a and b, Responses of the same cell at different stimulus intensities (stimulus duration 0.05 ms); c, 10 superimposed responses to a train of stimuli applied at 30 Hz, illustrating synaptic failure; d, a cell presumed to be antidromically activated faithfully following 10 stimuli at 100 Hz; e, a cell showing both short-latency antidromic action potentials and synaptically generated action potentials (four superimposed responses); f, a cell responding to a single dorsal tract stimulus with a high-frequency burst of action potentials.

encountered, are cells that respond to stimulation of the dorsal tracts with a high-frequency burst of action potentials lasting for up to 50 ms. Frequencies as high as 500 Hz have been observed. This is very similar to the pattern of firing seen in the Renshaw cells of the ventral horn which are responsible for recurrent inhibition of the motoneurons, and perhaps corresponds to the C or D cells of Eccles *et al.* (1960, 1962b) which it is suggested are involved in presynaptic inhibition in the dorsal horn (Eccles, 1964).

D. Synaptic blockade by cations

For many purposes, it is desirable to be able to separate activity due to synaptic transmission from that due to direct conduction to the recording site along the stimulated nerve fibres. The release of transmitter substances from presynaptic nerve endings is calcium dependent (see Hubbard, 1970) and the replacement of calcium by cations such as magnesium, manganese, nickel, cobalt or lanthinum, all of which interfere with calcium transport across membranes, has been used to bring about the failure of synaptic transmission (Hutter and Kostill, 1954; Jenkinson, 1957; Miledi, 1966; Richards and Sercombe, 1970; Meiri and Rahamimoff, 1972; Baker *et al.*, 1973; Berry and Pentreath, 1976; Shapovalov *et al.*, 1979).

To determine the most effective synaptic blocking agent, a series of experiments was conducted in which stimuli were applied to the dorsal white matter in the mid-thoracic region, and two electrodes were used to record simultaneously the response in the dorsal tracts, and dorsal horn in the lumbar region. The dorsal horn electrode recorded a complex wave consisting of an early directly conducted wave, and a later series of responses presumed to be synaptically generated (Fig. 13.6b). The electrode on the dorsal tracts recorded only a short-latency biphasic

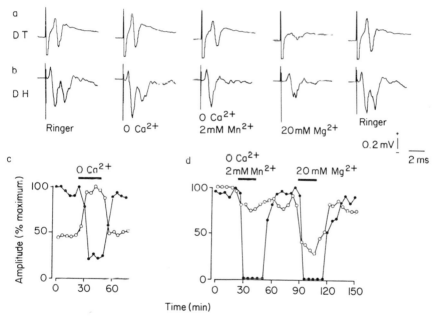

Fig 13.6. a and b, Responses recorded from the dorsal tracts (DT) and dorsal horn (DH) in the mid-lumbar region of a hemisected mouse spinal cord to stimulation of the dorsal tracts in the mid-thoracic region. (Temperature 25°C.) Each trace is the average of 16 successive responses recorded after 20 minutes exposure to Ringer solution containing altered concentrations of calcium, manganese and magnesium. c and d, Experiments illustrating the effects of lowered calcium and raised manganese and magnesium concentrations on the amplitude of the responses evoked in the dorsal tracts (o) and the longer latency (about 4 ms) responses in the dorsal horn (●) by stimulation of the dorsal tracts.

wave, due to the propagated action potential in the axons excited by the stimulating electrodes. By measuring the amplitude of the dorsal tract signal, and one of the longer latency waves in the dorsal horn signal, it was possible to monitor independently the effects of applied substances on both propagated activity and synaptic potentials. (These experiments were all conducted at 27°C with a bath flow rate of about 2 ml min^{-1}, but they have been confirmed on preparations at 35°C and 10 ml min^{-1} flow rate.)

Omission of calcium from the perfusion medium resulted in a large reduction in the amplitude of the longer latency waves generated in the dorsal horn, although seldom complete suppression (Fig. 13.6c), whereas the directly conducted activity in the dorsal tracts was often increased, probably as a result of a reduction in the threshold of fibres in the dorsal tracts that had not previously been caused to fire. Replacement of the calcium by 2 mM manganese chloride caused total suppression of synaptic activity, with little effect on the dorsal tract response. Nickel, cobalt and 20 mM magnesium (sulphate) all effectively blocked synaptic activity, but also reduced the size of the directly conducted

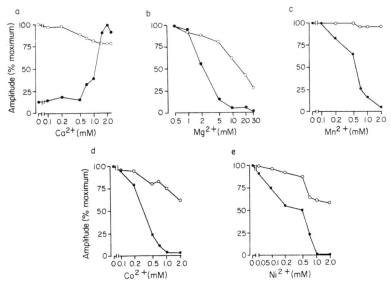

Fig. 13.7. Dose-related effects of $CaCl_2$, $MgSO_4$, $MnCl_2$, $CoCl_2$ and $NiCl_2$ on the responses evoked in the dorsal tracts (o) and the dorsal horn (●) in the lumbar region of the hemisected mouse spinal cord by stimulation of dorsal tracts in the mid-thoracic region. Each point represents the average of 16 responses evoked after 20 minutes exposure to the test solution.

wave to the dorsal tracts. Recovery was seldom complete after exposure to magnesium for more than 20 minutes. The addition of 40 mM sucrose to the Ringer solution had little effect on either potential, showing that the effects of magnesium were not due to changes in osmotic pressure.

These observations were confirmed by plotting dose–response curves for the various ions, 2 mM $MnCl_2$ produced a complete block of the long-latency dorsal horn response with little effect on the dorsal white matter response (Fig. 13.7). Cobalt, nickel and magnesium all produced some reduction in the dorsal tract signal at the concentrations necessary for suppression of the dorsal horn response. Fig. 13.8 shows the effects of manganese and magnesium on the responses of a dorsal horn cell that produced both antidromic and synaptically generated action potentials on stimulation of the dorsal tracts. Replacement of the calcium in the

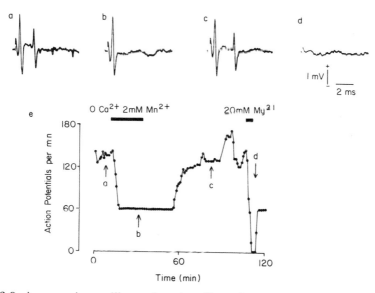

Fig. 13.8. An experiment illustrating the effect of manganese and magnesium on synaptic and antidromically evoked responses. a–d, Show responses evoked from a dorsal horn cell by stimulation of the dorsal tracts. The short-latency action potential had a constant latency and followed stimulation frequencies in excess of 200 Hz. The longer-latency action potentials (occasionally a pair) had a variable latency and failed when the stimulation frequency was raised above 30 Hz. Replacement of the calcium in the Ringer solution by manganese caused the failure of the long-latency spike and 20 mM Mg^{2+} caused both short- and long-latency responses to fail. e, Shows the number of action potentials evoked per minute by stimulation at 1 Hz during the course of this experiment, the arrows indicate the times at which the traces a–d were recorded.

Ringer solution by 2 mM $MnCl_2$ blocked the synaptically generated response but had no effect on the antidromic action potential, even after 30 minutes. In Ringer solution containing 20 mM magnesium, all action potentials were blocked after 4 minutes exposure. In all subsequent work, replacement of the calcium in the Ringer solution by 2 mM $MnCl_2$ has been used when it was necessary to block synaptic transmission. Lanthinum proved to be too insoluble in the Ringer solution to be used.

E. Inhibition in the dorsal horn

The slow negative wave evoked in the dorsal horn to stimulation of the dorsal roots or dorsal tracts is similar to the P-wave that can be recorded from the dorsal surface of the cord and the slow negative wave of the dorsal root potential (Barron and Matthews, 1938). These are believed to be an expression of the depolarization of primary efferent fibres by the incoming volley, resulting in presynaptic inhibition (Eccles *et al.*, 1962a).

In the isolated cord, when a conditioning stimulus was applied to the dorsal tracts before a test stimulus to the dorsal roots, the response to the dorsal root stimulus was reduced compared to that produced by the test stimulus alone. By varying the interval between the conditioning and test stimulus, the time course of the inhibition was plotted (Fig. 13.9). Maximum inhibition was reached approximately 20 ms after the conditioning stimulus. This was followed by an initial rapid recovery of the test response during the first 100 ms, and a slower recovery phase lasting up to 1.5 seconds after the conditioning stimulus. The time course of the inhibition was similar to that found by Eccles *et al.* (1961) for the reduction in size of EPSPs (excitatory postsynaptic potentials) in motoneurons due to presynaptic inhibition, and reached a peak at about the same time as the maximum of the slow negative potential in this preparation.

Inhibition of dorsal horn cell firing patterns has also been observed. It was most easily demonstrated in cells that were firing spontaneously like those in Fig. 13.10. A single stimulus to the dorsal tracts evoked one or two action potentials, followed by total suppression of activity for up to 300 ms, followed by an irregular burst of action potentials on recovery. Such inhibitory responses could be evoked on stimulation of either the dorsal roots or the dorsal tracts in the lumbar region.

F. Ventral horn responses

Although strong responses to stimulation were obtained in the dorsal

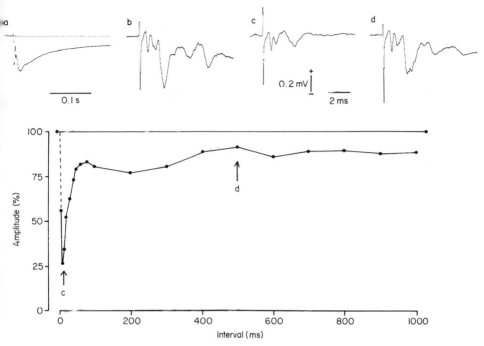

Fig. 13.9. Inhibition of the response evoked in the lumbar dorsal horn by dorsal root stimulation (b) caused by a conditioning stimulus applied to the dorsal tracts. a, Shows the slow negative wave produced in the dorsal horn by the conditioning stimulus, and c and d the response evoked by stimulating the dorsal roots 20 ms and 500 ms after the conditioning stimulus. The graph shows the time course of the inhibition and the arrows indicate the times at which the traces c and d were recorded.

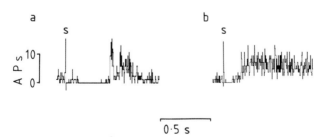

Fig. 13.10. Poststimulus time histograms showing inhibition of two spontaneously active cells in the dorsal horn on stimulation (S) of the dorsal tracts.

horn, ventral horn responses to either dorsal root or dorsal tract stimulation were small, the field potentials seldom exceeding 0.1 mV in amplitude. It also proved difficult to evoke antidromic responses in the ventral horn to stimulation of the ventral roots. Small short-latency (0.5–1 ms) negative or positive–negative potentials were observed which were partially obscured by the stimulus artefact (Fig. 13.11a). These were considered to be antidromic volleys entering the cord along the motor axons, but few spikes due to the firing of cell somas were found. However, preparations have frequently been observed in which ventral horn cells could be activated by stimulation of either the ventral or dorsal tracts, indicating that some ventral horn cells were alive, but antidromic activation of the motoneurons was not successful.

The ventral horn is surrounded by a thick layer of white matter, and this might limit oxygen uptake, resulting in the failure of the more vulnerable cells. Several procedures have been used in an attempt to improve the penetration of oxygen into the ventral horn. In some preparations, much of the ventral white matter was dissected away, or oblique sections were cut across the cord in the lumbar region to expose a larger area of gray matter. Others have used immature mice (5–10 g) in which the white matter is poorly developed. None of these procedures produced any increase in the ventral horn activity.

G. Spinal Shock

Transection of the spinal cord in vertebrates produces a state of "spinal shock" in which all reflexes below the level of the section are absent for periods of time ranging from a few minutes in the frog to several weeks in man. The mechanisms underlying this state are not known, but it is caused by the lack of descending activity from the brain, and Barnes *et al.* (1962) have shown that during the cold block of the thoracic spinal cord in the cat the motoneurons in the lumbar region hyperpolarize by 2–6 mV. In rat motoneurons, the soma-dendritic component of the action potential has sometimes been observed to fail despite depolarization of the initial segment by an antidromic volley. Transection of the cord and the removal of all afferent input, as occurs when the cord is removed from the body, might therefore result in a large proportion of the motoneurons being sufficiently hyperpolarized to cause the failure of the soma-dendritic spike.

If this was the case in these preparations, then increasing the potassium concentration in the Ringer solution should have caused the cells to depolarize and fire more readily. Alternatively, if the motoneurons were dead or already depolarized due to annoxia or other trauma, changing the potassium concentration would have had no effect on the antidromic activity recorded in the ventral horn. Fig. 13.11 shows that the amplitude of the ventral horn response to ventral root stimulation was increased by increasing the potassium concentration from 2 mM to 5 mM. Dose–response experiments (Fig. 13.11f)showed that the optimum potassium levels were between 5 mM and 7 mM KCl. At concentrations above 10 mM KCl, all activity in both dorsal and ventral horns was suppressed. These effects were fully reversible. Despite these findings, in 5 mM and 7 mM KCl no action potentials were detected which could be ascribed to motoneuron somas firing.

In view of the demonstration of inhibition in the dorsal horn, the possibility that the motoneurons might be hyperpolarized due to a continuous inhibitory input was considered. The ions most commonly involved in inhibiting processes are K^+ and Cl^-, and so the effects of reducing the Cl^- ion concentration were investigated. Total replacement of Cl^- ions in the Ringer by acetate ions caused an increase in the amplitude of the ventral horn response to stimulation of the ventral roots, by as much as 250% (Fig. 13.11d). In addition, unitary responses from individual cells could be detected. With fine electrodes (5 MΩ), extracellular recordings have been made from single motoneurons (Fig.

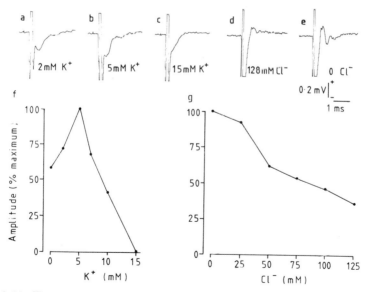

Fig. 13.11. Two experiments showing the effects of various concentrations of K^+ (a, b, c and f) and Cl^- (d, e and g) ions on the response evoked in the lumbar ventral horn by stimulation of the lumbar ventral roots.

13.12a–c), although using NaCl-filled electrodes the soma-spike usually disappears after about 5 minutes.

The effect of changing the concentration of Cl⁻ ions in the Ringer solution on an antidromically evoked ventral horn cells is shown in Fig. 13.12d–f. This cell was located in low-chloride Ringer solution, and on changing to normal Ringer solution no activity could be detected, but a return to the low-chloride solution caused the cell to fire again. Dose–response plots for the effects Cl⁻ ion on the antidromic ventral horn response show an almost linear relationship (Fig. 13.11g). Similar increases in signal amplitude have been observed in the dorsal horn in low-chloride Ringer solution.

Attempts to reproduce these effects of Cl⁻ ions by using the inhibitory blocking agents strychnine (2 μm) and picrotoxin (100 mg l⁻¹) have failed to show any change in the antidromic ventral horn response, and cause doubts about the effects of low-chloride concentrations being solely due to suppression of inhibition.

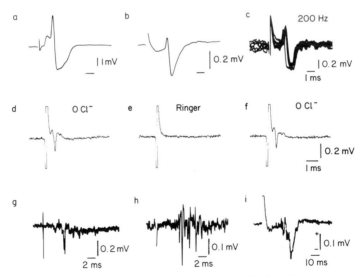

Fig. 13.12. a–f, Single-cell responses recorded in the lumbar ventral horn on stimulation of the ventral roots. c, Shows 10 superimposed responses of a cell to stimulation at 200 Hz. d–f, The effect of changing from low-chloride Ringer solution to normal Ringer solution on a small unit in the ventral horn responding to ventral root stimulation, average of eight responses. g, Response recorded in dorsal horn in the lumbar region to stimulation of the dorsal tracts in the cervical region. h, Response recorded in the nucleus gracilis to a single stimulus applied to dorsal root L4. i, Response recorded in the dorsal horn of the thoracic cord to stimulation of the eleventh intercostal nerve.

The loss of Cl^- ions from cells in the low-chloride solution would also be expected to cause depolarization of the cell membranes and this may have been the more important factor in causing the motoneurons to fire. But even with the use of a Ringer solution containing 5 mM K^+ and no Cl^-, active motoneurons were not found in every preparation and some doubt persists about the viability of the ventral horn cells.

H. Conduction along the cord

Stimulation of the longitudinal tracts produces compound action potentials detectable for up to 2 cm from the point of stimulation, but above this distance they are too small and asynchronous to detect. It is possible, however, to detect the effects of such stimulation on cellular responses, and Fig. 13.12 shows the excitatory effects of stimulation of the dorsal tracts in the mid-cervical region on cells recorded in the dorsal horn in the lower lumbar region, a distance of more than 5 cm. In a similar way, the firing of cells in the nucleus gracilis in the medulla to stimulation of the ipsilateral dorsal roots in the lumbar region has been demonstrated (Fig. 13.12h), showing that longitudinal conduction both up and down the cord is possible.

Attempts have also been made to dissect out peripheral nerves attached to the cord and stimulate these. Stimulation of the sciatic nerve seldom proved successful in evoking any responses from the cord, most probably due to damage to the roots where they emerge between the vertebrae during the dissection. At this point they are firmly attached by connective tissue to the bone but at an acute angle, making them prone to kinking during the dissection. In addition, the large diameter of the sciatic nerve, composed of three major spinal roots, and its thick connective tissue sheath probably made the penetration of oxygen difficult.

In contrast to the sciatic nerve, the spinal nerves in the thoracic region are of moderate diameter, easy to dissect and emerge from between the vertebrae perpendicular to the cord. Responses have been obtained in the thoracic cord to stimulation of the lower intercostal nerves (Fig. 13.12i).

I. Anaesthetic effects

Experiments involving lengthy peripheral dissection or removal of the medulla as well as the spinal cord have been carried out on mice that were cooled in ice and anaesthetized with halothane or pentobarbitone, allowing more time for the removal of the tissue from the body. It was assumed that once the cord was in the bath the anaesthetic would rapidly be washed out of the tissue, and there would be no prolonged effect. To test this, cords obtained from decapitated non-anaesthetized mice were subjected to anaesthetic added to the bath.

The longer latency, synaptically generated waves in the dorsal horn were found to be most susceptable to anaesthetic agents, and the effects on these of increasing concentrations of urethane, sodium pentobarbitone, halothane and the local anaesthetic agent procaine are shown in Fig. 13.13. Each dose level was applied for 20 minutes, and the different shape of the curves probably reflects differing rates of penetration of the drugs into the tissue. The effects of urethane were observed within 5 minutes, whereas sodium pentobarbitone took over 15 minutes to produce its maximal effect. Halothane and procaine were intermediate between these two.

The concentrations of anaesthetic needed to produce a reduction of the synaptic response in the cord were similar to those used to bring about anaesthesia in the intact animal. Assuming that the anaesthetic agent is distributed throughout the extracellular fluid (approximately 20% of the body mass) the anaesthetic dose for sodium pentobarbitone (50 mg kg body weight^{-1}) represents a concentration of 250 mg l^{-1}, and that for urethane (1.5 g kg^{-1}) is equivalent to 7.5 g l^{-1}. At these concentrations, both compounds produced a suppression of the evoked dorsal horn response of greater than 50%. Halothane was added to the gas bubbling through the perfusion fluid from a Fluotec 3 diluter, but

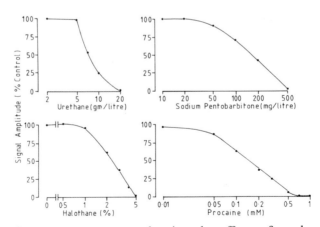

Fig. 13.13. Dose–response curves showing the effects of urethane, sodium pentobarbitone, halothane and procaine on the amplitude of the synaptically generated dorsal horn wave resulting from stimulation of the dorsal tracts. Each point represents the average of 16 successive responses after 20 minutes exposure to Ringer solution containing the test compound.

being volatile its final concentration in the bath was likely to be below the concentration set on the diluter. In spite of this, the concentration normally used to maintain anaesthesia (2% halothane) caused a 40% reduction in the dorsal horn response.

Two experiments illustrating the time course of the onset and recovery of dorsal horn responses to halothane, urethane and sodium pentobarbitone are shown in Figs. 13.14 and 13.15. After 40 minutes exposure to halothane, the amplitude of the synaptically generated dorsal horn potential was reduced by 75% and on return to normal Ringer solution, 90% recovery was obtained after 15 minutes. Increasing the halothane concentration to 5% resulted in total suppression of the synaptically generated activity, and a reduction in the amplitude of the short latency afferent potential (Fig. 13.14d). Recovery from this level of depression produced by high concentrations of halothane was always incomplete.

Urethane (10 g l^{-1}) caused a rapid reduction in the amplitude of the

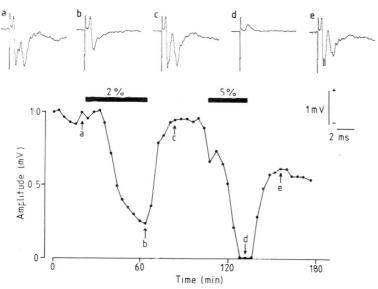

Fig. 13.14. An experiment showing the effects of bubbling 2% and 5% halothane through the Ringer solution perfusing the bath, on the response evoked in the dorsal horn of the lumbar cord by stimulation of the dorsal tracts in the thoracic region. The graph plots the amplitude of the negative wave having a latency of 2 ms, and the arrows indicate the times of recording of the corresponding traces a–e.

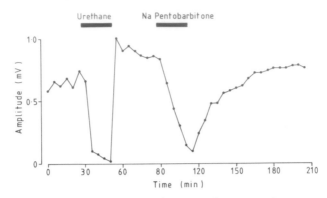

Fig. 13.15. An experiment showing the rate of onset and recovery from the effects of urethane (10 g l^{-1}) and sodium pentobarbitone (250 mg l^{-1}) on the amplitude of the synaptically generated response evoked in the dorsal horn of the lumbar cord by stimulation of the dorsal tracts in the thoracic region. Each point represents the average of 16 successive responses.

dorsal horn response and within 10 minutes it was down to 10% of the control size. Recovery was equally fast, being complete within 5 minutes of exposure to normal Ringer solution. After treatment with urethane, the amplitude of the dorsal horn synaptic response was often found to be larger than the preceding control values, as shown in the experiment illustrated in Fig. 13.15.

The time course of onset and recovery from the effects of 250 mg l^{-1} sodium pentobaributone was found to be slower than either halothane or urethane and in two of the four cases examined was incomplete 1 hour after return to normal Ringer solution.

Successful spinal cord preparations have been obtained using halothane, sodium pentobarbitone or urethane as the anaesthetic agent, but these results indicate that urethane is the most suitable. A single intraperitoneal injection produces long-lasting anaesthesia, without the depression of respiration associated with the barbiturates, and once the tissue is placed in the experimental bath the anaesthetic will be rapidly washed out.

IV. Discussion

The object of this paper was to demonstrate an *in vitro* preparation of mammalian spinal cords from more mature animals than are usually employed for this purpose. Until now most of the work in this field has employed very immature animals, either neonatal rats (Otsuka and Konishi, 1974; Evans, 1978; Preston and Wallis, 1979) or 19–21-day-old kittens (Shapovalov *et al.*, 1979). For most of the work reported here, juvenile mice (4–6 weeks old) were used. At this age, although not sexually mature, the animals have their eyes open and are fully active. In some experiments we have used adult animals with weights ranging up to 40 g and obtained essentially similar results to those from juvenile mice.

The similarity of the responses recorded in the dorsal horn to those seen in *in vivo* preparations, and the presence of both excitatory and inhibitory effects on cell-firing patterns indicate the viability of this part of the preparation. Responses from the ventral horn of the cords were weaker than expected, but the ability of raised concentrations of K^+ ions, and especially lowered Cl^- ion concentrations, to cause motoneurons to fire action potentials antidromically on stimulation of the ventral roots, perhaps indicates that the ventral horn cells were alive but not producing somadendritic spikes. Intracellular recording from motoneurons would resolve this problem.

Using dorsal horn responses, we have attempted to define some of the criteria necessary for obtaining optimum responses. During the dissection of the cord it was essential to keep the tissue cool to reduce oxygen needs. Our dissecting bath was perfused with Ringer solution at 4°C. Where long dissections were necessary, such as in dissecting out peripheral structures or the medulla attached to the cord, it was felt desirable to pre-cool the anaesthetized animal in ice.

Once the hemisected cord was in the experimental bath the survival time of the preparation was principally determined by the temperature and flow rate of the superfusing fluid. At 20°C, the preparation could be kept alive for up to 2 days with a flow rate of 2 ml min^{-1} through the bath, but to obtain maximum synaptic activity in the cord, a temperature of 35°C was necessary, requiring a flow rate of approximately 10 ml min^{-1}. Even with the higher flow rates, responses usually began to deteriorate after 8–10 hours at 35°C.

High flow rates through the experimental bath had the advantage of reducing the "dead time" on changing the composition of the Ringer solution, and speeding up the rate of washout of agents from the bath. For example, with a flow rate of 10 ml min^{-1} synaptic blocking agents such as manganese were usually observed to be having an effect after 2 minutes, whereas, with the lower flow rate the "dead time" was in excess of 10 minutes.

The investigation of the use of various inorganic cations to replace calcium and block synapses showed the 2 mM manganese was the agent of choice. The other metals tested, cobalt, nickel and magnesium, all had some effect on nerve fibre conduction as well as synaptic transmission (see also Bagust and Kerkut, 1979 and 1980). Where it was necessary to block all activity, both conducted and synaptic, 15 mM KCl or 1 mM procaine were very effective and fully reversible providing the tissue was not exposed to them for longer than necessary.

We feel that this preparation should prove useful in many aspects of cord physiology and pharmacology. The ease with which the medium around the tissue can be altered makes it a suitable preparation for the investigation of transmitter pharmacology. Agonists and antagonists can be added to the bath fluid as well as directly applied on to cells by iontophoresis or pressure ejection. It should also prove useful for the screening of novel pharmacological agents for possible effects on nervous tissues. Preservation of the longitudinal spinal tracts, and the possibility of the inclusion of peripheral or brainstem structures using cooled, anaesthetized animals will make the long uncrossed pathways more accessible to examination than is the case with transverse slice preparations of the cord.

V. Conclusions

1. The apparatus and procedure for setting up *in vitro* a hemisected spinal cord preparation from juvenile and adult mice (weight range 12–40 g), are described. Using extracellular recording in the lumbar cord, field potentials of up to 2 mV in amplitude were detected in the dorsal horn on stimulation of the dorsal roots or dorsal tracts. Extracellular action potentials of up to 10 mV in amplitude have also been recorded from single cells.

2. Optimum responses were obtained with a flow rate through the bath of 10 ml min^{-1} at a temperature of 30–35°C.

3. Synaptically generated responses were blocked by replacing the calcium in the perfusion medium with magnesium, cobalt, nickel or manganese, but only 2 mM manganese did not also reduce conducted activity.

4. Inhibition of the dorsal horn responses was produced by conditioning stimuli applied to the dorsal roots or dorsal tracts. The duration of this inhibition was in excess of 1 second, and it was associated with a slow negative wave in the dorsal horn, similar to that described for dorsal horn presynaptic inhibition *in vivo*.

5. Responses in the ventral horn evoked by dorsal root stimulation were smaller than those detected in the dorsal horn. Antidromic responses to stimulation of the ventral roots were also small and it proved difficult to find individual motoneurons. Increasing the K^+ and decreasing the Cl^- concentrations in the Ringer solution resulted in more motoneurons being detected, indicating that they might be in a hyperpolarized state, perhaps as a result of "spinal shock", but some doubt remains about the viability of the ventral horn cells in these preparations.

6. A technique is described for using cooled, anaesthetized animals to allow the dissection of peripheral nerves or the brainstem attached to the cord, and evidence is presented of conduction from the cord to the medulla.

7. The rate at which some anaesthetic agents were washed out of the preparation have been investigated, and urethane was found to be preferable to sodium pentobarbitone for this type of preparation.

References

Bagust, J. and Kerkut, G. A. (1979). Some effects of magnesium ions upon conduction and synaptic activity in the isolated spinal cord of the mouse. *Brain Res.* **177,** 410–413.

Bagust, J. and Kerkut, G. A. (1980). The use of the transition elements manganese, cobalt and nickel as synaptic blocking agents on isolated, hemisected, mouse spinal cord. *Brain Res.* **182,** 474–477.

Baker, P. F., Meves, H. and Ridgeway, E. B. (1973). Effects of manganese and other agents upon the calcium uptake that follows depolarization of squid axons. *J. Physiol. (Lond.)* **231,** 511–526.

Barker, J. L. and Ransom, B. R. (1978). Pentobarbitone pharmacology of mammalian central neurones grown in tissue culture. *J. Physiol. (Lond.)* **280,** 355–372.

Barnes, C. D., Joynt, R. J. and Schottelius, B. A. (1962). Motoneurone resting potentials in spinal shock. *Am. J. Physiol.* **203,** 1113–1116.

Barron, D. H. and Matthews, B. H. C. (1938). The interpretation of potential changes in the spinal cord. *J. Physiol. (Lond.)* **92,** 273–321.

Berry, M. S. and Pentreath, V. W. (1976). Criteria for distinguishing between monosynaptic and polysynaptic transmission. *Brain Res.* **105,** 1–20.

Biscoe, T. J., Evans, R. H., Headley, P. M., Martin, M. R. and Watkins, J. C. (1976). Structure–activity relations of excitatory amino acids on frog and rat spinal neurones. *Br. J. Pharmacol.* **58,** 373–382.

Bloom, F. (1974). To spritz or not to spritz; the doubtful value of aimless iontophoresis. *Life Sci.* **14,** 1819–1834.

Brown, L. T. (1971). Projections and terminations from the corticospinal tract in rodents. *Exp. Brain Res.* **13**, 432–450.

Coombs, J. S., Curtis, D. R. and Landeren, S. (1956). Spinal cord potentials generated by impulses in muscle and cutaneous afferent fibres. *J. Neurophysiol.* **19**, 452–467.

Curtis, D. R. (1969). The pharmacology of spinal postsynaptic inhibition. *Prog. Brain Res.* **31**, 171–189.

Curtis, D. R. and Eccles, R. M. (1958). The excitation of renshaw cells by pharmacological agents applied electrophoretically. *J. Physiol. (Lond.)* **141**, 435–445.

Curtis, D. R., Phillis, J. W. and Watkins, J. C. (1961). Cholinergic and non-cholinergic transmission in the mammalian spinal cord. *J. Physiol. (Lond.)* **158**, 296–323.

Donatelle, J. M. (1977). Growth of the corticospinal tract and the development of placing reactions in the postnatal rat. *J. Comp. Neurol.* **175**, 207–232.

Eccles, J. C. (1964). Presynaptic inhibition in the spinal cord. *Prog. Brain Res.* **12**, 65–91.

Eccles, J. C., Eccles, R. M. and Lundberg, A. (1960). Types of neurone in and around the intermediate nucleus of the lumbosacral cord. *J. Physiol. (Lond.)* **154**, 89–114.

Eccles, J. C., Eccles, R. M. and Magni, F. (1961). Central inhibitory action attributable to presynaptic depolarization produced by muscle afferent valleys. *J. Physiol. (Lond.)* **159**, 147–166.

Eccles, J. C., Magni, F. and Willis, W. D. (1962a). Depolarization of central terminals of group I afferent fibres from muscles. *J. Physiol. (Lond.)* **160**, 62–93.

Eccles, J. C., Kostyuk, P. E. and Schmidt, R. F. (1962b). Central pathways responsible for depolarization of primary afferent fibres. *J. Physiol. (Lond.)* **161**, 237–257.

Evans, R. H. (1978). The effects of amino-acids and antagonists on the isolated hemisected spinal cord of the immature rat. *Br. J. Pharmacol.* **62**, 171–176.

Evans, R. H. (1979). Potentiation of the effects of GABA by pentobarbitone. *Brain Res.* **171**, 113–120.

Fernandez De Molina, A. and Gray, J. A. B. (1957). Activity in the dorsal spinal grey matter after stimulation of cutaneous nerves. *J. Physiol. (Lond.)* **137**, 126–140.

Gilbert, M. and Stelzner, D. J. (1979). The development of descending and dorsal root connections in the lumbosacral spinal cord of the postnatal rat. *J. Comp. Neurol.* **184**, 821–838.

Headley, P. M., Duggan, A. W. and Griersmith, B. T. (1978). Selective reduction by noradrenaline and 5-hydroxytryptamine of nocioceptive responses of cat dorsal horn neurones. *Brain Res.* **145**, 185–189.

Hubbard, J. I. (1970). Mechanisms of transmitter release. *Prog. Biophys. Mol. Biol.* **21**, 35–124.

Hutter, O. F. and Kostill, K. (1954). Effect of magnesium and calcium ions on the release of acetylcholine. *J. Physiol. (Lond.)* **124**, 234–241.

Jenkinson, D. H. (1957). The nature of the antagonism between calcium and magnesium ions at the neuromuscular junction. *J. Physiol. (Lond.)* **138**, 434–444.

Kudo, Y. (1978). The pharmacology of the amphibian spinal cord. *Progr. Neurobiol.* **11**, 1–76.

Martin, M. R., McHanwell, S. and Biscoe, T. J. (1978). A micro-electrophoretic and electrophysiological study on normal and jimpy mutant mouse spinal neurones and reflexes. *Brain Res.* **151**, 225–233.

McLennan, H. and Lodge, D. (1979). The antagonism of amino acid induced excitation of spinal neurones in the cat. *Brain Res.* **169**, 83–90.

McLennan, H. and Wheal, H. V. (1976). The specificity of action of three possible antagonists of amino acid induced neuronal excitations. *Neuropharmacology* **15**, 709–712.

Meiri, U. and Rahamimoff, R. (1972). Neuromuscular transmission: inhibition by manganese. *Science* **176**, 308–309.

Miledi, R. (1966). Strontium as a substitute for calcium in the process of transmitter release at the neuromuscular junction. *Nature (Lond.)* **212**, 1233–1234.

Otsuka, M. and Konishi, S. (1974). Electrophysiology of mammalian spinal cord *in vitro*. *Nature (Lond.)* **252**, 733–734.

Preston, P. R. and Wallis, D. I. (1979). Dorsal root potentials recorded in the isolated spinal cord of the neonate rat. *J. Physiol (Lond.)* **289**, 84P.

Richards, C. D. and Sercombe, R. (1970). Calcium, magnesium and the electrical activity of guinea-pig olfactory cortex *in vitro*. *J. Physiol. (Lond.)* **211**, 571–584.

Schoen, J. H. R. (1974). Comparative aspects of the descending fibre systems in the spinal cord. *Prog. Brain Res.* **11**, 203–222.

Shapovalov, A. I., Shiriaev, B. I. and Tamarova, Z. A. (1979). Synaptic activity in motoneurones of the immature cat spinal cord *in vitro*—effects of manganese and tetrodotoxin. *Brain Res.* **160**, 524–528.

Skoglund, S. (1969). Growth and differentiation with special emphasis on the central nervous system. *Ann. Rev. Physiol.* **31**, 19–42.

Takahashi, T. (1978). Intracellular recording from visually identified motoneurones in rat spinal cord slices. *Proc. Roy. Soc. Ser B.* **202**, 417–421.

Vasilenko, D. A. and Kostyuk, P. G. (1966). Activation of various groups of spinal neurones on stimulation of cat sensorimotor cortex. *Fed. Proc. Trans. Suppl.* **25**, T569–T573.

Wall, P. D. (1964). Presynaptic control of impulses at the first central synapse in the cutaneous pathway. *Prog. Brain Res.* **12**, 92–118.

Weakly, J. N. (1969). Effect of barbiturates on quantal synaptic transmission in spinal motoneurones. *J. Physiol. (Lond.)* **204**, 63–77.

14

A Study of Neuronal Activity of Mammalian Superfused or Intra-arterially Perfused CNS Preparations

A. I. SHAPOVALOV, B. I. SHIRIAEV AND
Z. A. TAMAROVA

Laboratory of Physiology of the Nerve Cell, Sechenov Institute of the Academy of Sciences of USSR, Thorez Pr 44, Leningrad 194223, U.S.S.R.

I. Introduction

The isolation and perfusion of the central nervous system provides many new possibilities for electrophysiological, biochemical and developmental neurobiology. It offers an important means for the efficient control of the extracellular ionic medium of neurons and for ease and efficacy of drug application, since selected agents can be brought to act on the cells without mediation by other tissues. On the other hand, the need to study the unanaesthetized nervous system can be easily met by the perfused preparations. The ability to study drastic changes in the ionic environment on synaptic transmission and on the synaptic and nonsynaptic membranes of the nerve cells is especially important for identifying the mode of junctional transmission at particular central synapses and for analysis of the ionic channels responsible for a variety of electrical responses.

In the last decade there has been a growing crescendo of research on the isolated perfused nervous system preparation. Up to very recently, most of this work had been done on the preparations of the isolated superfused spinal cord of the cold-blooded animals which survive especially well in isolation (Araki et al., 1953; Babsky and Kirillova, 1938; Barker and Nicoll, 1973; Barrett and Barrett, 1976; Batueva and Shapovalov, 1977; Brookhart et al., 1959; Eccles, 1946; Grinnell, 1966; Curtis et al., 1961; Davidoff, 1972; Shapovalov et al., 1978; Sonnhof et al., 1976; Tebecis and Phillis, 1967).

Most published work on the isolated mammalian CNS preparations deals with the thin slices prepared from the spinal cord or different brain regions (Brown et al., 1979; Okada and Saito, 1979; Schwartzcroin, 1975; Takahashi, 1978; Yamamoto, 1972) or with tissue cultures (Crain, 1966; Nelson and Peacock, 1973). An obvious limitation of slice preparations is in difficulties with selective activation of identified synaptic inputs and the possibility of local neuronal damage. In the case of cultured cells, there is some question as to whether their properties and behaviour reflect what happens in the living animals.

The use of the isolated spinal cord of newborn rats as an experimental preparation (Otsuka and Konishi, 1974; Evans, 1978; Bagust et al., 1979) has opened up a new possibility of investigation of the mammalian CNS. These investigations were concerned, however, primarily with extracellular recording. The need to investigate the activity of single neurons under controlled environmental conditions and with preserved synaptic inputs has prompted us to modify the technique of the superfusion of the rat spinal cord and to develop a new preparation, the isolated spinal cord of the kitten. A technique of perfusion of mammalian CNS through arteries was also developed. The latter approach appeared to be especially promising for investigation of the brain, since it offers the possibility to study synaptic and cellular mechanisms at conditions that had previously been possible only with lower vertebrates. In the present paper, we describe the work on three different mammalian CNS preparations: the isolated superfused hemicords, the spinal cord and the hind limb perfused through arteries and the intra-arterially perfused medulla.

II. Superfusion of the Isolated Spinal Cord *In Vitro*

Previous work on the superfused spinal cord of the neonatal rats is based on the dissection of few spinal segments with their subsequent incubation in a small chamber superfused with oxygenated physiological saline (Otsuka and Konishi, 1974; Evans *et al.*, 1977). This approach is very similar to that usually employed for isolation and perfusion of amphibian cord. Some modifications of this technique prolonged the survival of the superfused mammalian preparations and provided the possibility of stable recording from motoneurons of rats 9–17 days of age, in contrast to previous studies in which only younger animals were used. The spinal cord of rats 16–20 days after birth is in many respects similar to that of adult animals (Bursian, 1977).

The same approach appeared to be useful for the isolation and superfusion of the spinal cord of kittens. The perfusion of feline spinal cord is of special interest since from a viewpoint of cellular neurobiology, cat motoneurons and their synaptic inputs are the best studied class of mammalian nerve elements and they are usually used as a model for analysis of synaptic events. This report, therefore, accentuates the results of a study of a new preparation—the isolated superfused spinal cord of the kitten.

A. Methods

1. *Preparation of hemicords*

The technique of preparing the isolated spinal cord of young rats and kittens was similar. It was somewhat different from that employed by Evans *et al.* (1977) who removed a block of tissue containing the spine after decapitation and dissected the spinal cord in a dish. The spinal cord was exposed from the dorsal surface. The dorsal approach made it possible to preserve spinal lumbar arteries. All surgical procedures, including the saggital hemisection of lumbar cord and transverse sections at lower lumbar and upper lumbar levels, as well as the dissection of the final block of tissue from the body, were conducted not in the chamber but *in situ*. The natural blood supply, therefore, was maintained up to the final stage of complete isolation. The main surgery was performed as follows. After anaesthesia has been induced (urethane 1.5–2.3 g kg^{-1} i.p.), the laminectomy was carefully performed. After laminectomy and dissection of dura, the sixth and seventh lumbar ventral and dorsal spinal roots were cut on the left side. Then the cord was hemisected sagitally, a little bit to the right from the dorsal vein and by two consecutive transverse sections (firstly, caudal to the seventh lumbar segment and then rostral to the fifth or sixth lumbar segments). The final block, together with corresponding spinal roots, was completely isolated. During all these procedures the lumbar region was continuously superfused by standard solution equilibrated with 95% O_2/5% CO_2 cooled to 15–17°C.

The solution reached the cord surface through the glass cannula. It prevented the development of blood clotting and at the same time supplemented the oxygen supply as the natural blood supply was reduced step by step.

2. Superfusion

Immediately after complete isolation from the body, the excised segments together with spinal roots were transferred into the experimental chamber: the inclined groove in a plexiglass block. In the chamber, the cord was placed on two or three layers of filter paper so that its lateral surface was upwards (Fig. 14.1). Then the cord was covered by one layer of filter paper, on which the perfusing fluid was continuously delivered through the glass cannula. The solution flowed round the cord and through the filter paper and passed into a special compartment from which it was sucked out by the water pump.

The perfusion rate was adjusted to give a flow of approximately 2–4 ml min^{-1}. The standard superfusion medium contained (mM): NaCl, 118.0; KCl, 2.0; NaHCO$_3$, 24.0–30.0; CaCl$_2$, 2.0; glucose, 12.0; pH 7.4–7.6. The modified solutions containing zero Ca^{2+}, 1.5–2.0 mM Mn^{2+} or 2–5 mM Mg^{2+}, or drugs (tetrodotoxin, 4-aminopyridine) could be applied to the spinal cord by turning a stopcock which replaced the flow of one solution with the other. The diffusion of drugs applied in the bathing medium, as well as nutrients and oxygen, into the depths of

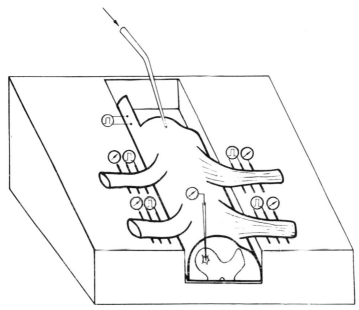

Fig. 14.1. An arrangement of the superfusion of the hemisected segments of the spinal cord in an experimental chamber.

the spinal cord was facilitated as the diffusion barriers had been partly removed by the hemisection.

3. Stimulation and recording

Ventral and dorsal spinal roots were placed on gold wires on both sides of the cord and covered by a thin layer of vaseline.

These electrodes were used for stimulation and recording (Fig. 14.1). The reference electrode (Ag/AgCl) was placed on the filter paper under the cord. In some experiments the ventral column in the region of the fourth to fifth lumbar segments were dissected while the stimulating electrode was placed in the ventral white matter. This stimulating electrode was bipolar (platinum wires insulated but for the tips). The duration of the stimulating pulses was 0.05–0.1 ms.

Intracellular recording from motoneurons was accomplished with the aid of glass micropipettes filled with 3 M KCl or 2 M potassium citrate. Tip resistances of microelectrodes were in the range 20–50 MΩ. A bridge circuit was used to pass current through the recording electrode. Part of the pia covering the lumbar segments was removed to facilitate insertion of a microelectrode into the spinal cord. The recorded potentials were photographed on the oscilloscope either directly or after averaging with a DIDAC-4000 computer (Intertechnique).

4. *Functional state of the preparation*

The rats used were 9–17 days old, whereas kittens were 3–22 days old. In the properly treated preparations, the uniform levels of the neuronal activity could be observed for the duration of experiments (10–12 hours). During this period the hemicords continued to show good evoked and spontaneous activity. The ventral root responses produced by stimulation of dorsal roots or ventral column showed no signs of depression, as well as potentials recorded from individual motoneurons.

The optimal temperature for this preparation was around 20–22°C. Warming usually decreased the amplitude of evoked responses, probably because of insufficient oxygenation. Fig. 14.2 illustrates the potentials recorded from the seventh lumbar ventral root in response to single dorsal root volley in a 16-day-old rat at different temperatures. It may be seen that the amplitude of response rises up at 22°C, but with further increase in temperature, it declines, especially at 28°C. Therefore we usually kept the temperature of the superfused preparations in the range 18–26°C.

The maximal duration of intracellular recording from a single motoneuron was 40 minutes in the young rats, but it reached 4–5 hours in the kittens. Therefore, many successful attempts could be made to change the ionic composition of the bathing solution during the course

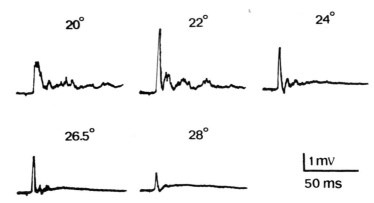

Fig. 14.2. Effects of changes in temperature (°C) on the ventral root response derived from the dorsal root of a 16-day-old rat. The temperature is indicated above each record.

of a single penetration. In kittens, the resting membrane potential of impaled motoneurons varied from -50 to -75 mV and the amplitude of the antidromic action potential ranged from 50 to 113 mV. These values suggest that the cells used for recording were in a reasonably good condition.

B. Results

1. Mechanisms of junctional transmission

The junctional mechanisms even at the best studied synapses of vertebrates—the Ia synapses on mammalian motoneurons—are still incompletely understood. Recently, there has been some doubt placed on the chemical model for Ia E.P.S.P. (excitatory postsynaptic potentials) production in the cat (Edwards et al., 1976; Werman and Carlen, 1976). In experiments carried out on the cat's spinal cord in situ, Mn^{2+} and Co^{2+} released from micropipettes in the vicinity of motoneurons depressed the monosynaptic EPSPs evoked by stimulation of afferent fibres from hind-limb muscles (Krnjević et al., 1978, 1979). However, the block so induced was not total, probably because of the considerable distance between the iontophoretic electrode and some Ia synapses.

The perfusion of the mammalian cord with Ca^{2+}-deficient, Mn^{2+} or Mg^{2+}-containing medium may overcome these difficulties (Shapovalov et al., 1978, 1979; Tamarova et al., 1978).

In experiments carried out both on young rats and kittens, it was found that ventral root potentials produced by dorsal root volley were completely and reversibly abolished when the standard solution in the bath was replaced by the test solution without Ca^{2+}, containing 1.5–2.0 mM Mn^{2+} or 2.0–5.0 mM Mg^{2+}. It generally takes between 10 and 15 minutes to produce a significant depression and the full block was observed after 20–25 minutes of perfusion with Mn^{2+}-containing solution. The time course of changes in synaptically induced responses was slower in the case of Mg^{2+} (usually 40–50 minutes). In contrast to postsynaptic response, the presynaptic component of the ventral root potential was not affected (Fig. 14.3B).

Administration of 4-aminopyridine (4-AP), which had been shown to potentiate the electrically mediated synaptic transmission (Shapovalov and Shiriaev, 1978), did not restore dorsal to ventral root transmission suppressed by Ca^{2+}-deficient, Mn^{2+}-supplemented perfusion media. On returning to standard solution, the full recovery was observed even after prolonged (60–90 minutes) perfusion with blocking saline. It generally took 3–5 minutes to produce a partial restoration. The effects

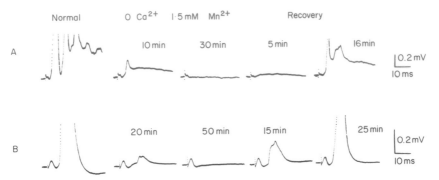

Fig. 14.3. Potentials evoked in the ventral root in response to dorsal root stimuli before and at different times (in min) after Ca^{2+}-free solution containing Mn^{2+} (1.5 mM) was admitted to the bath. A, 13-day-old rat; B, 9-day-old kitten. Each record represents an average built up by the computer over 20 sweeps.

of Ca^{2+}-deficient, Mn^{2+}-supplemented medium on the ventral root responses in 13-day-old rats and 9-day-old kittens are shown in Fig. 14.3.

Similar results were obtained by application of the technique of intracellular recording from individual motoneurons. Superfusion with Ca^{2+}-deficient, Mn^{2+}-supplemented solutions depressed EPSPs produced by stimulation of dorsal roots and ventral columns in both rat and kitten spinal cord. The marked depression was already visible in the first 3–5 minutes of exposure to this solution and the full block was observed 15–20 minutes later. The amplitude of the antidromic spike was not affected. Fig. 14.4A shows the abolition of dorsal root EPSP in a motoneuron of a 16-day-old rat by the test solution containing zero Ca^{2+}, 2.5 mM Mg^{2+} and 2.0 mM Mn^{2+}. The subtraction of the field potential recorded after withdrawal of the microelectrode from motoneuron, with the aid of a digital computer, demonstrated the complete absence of transmembrane response in a motoneuron. Similar effects were observed in the kitten. Fig. 14.4B illustrates the depression of the dorsal root EPSP in the same motoneuron of a 4-day-old-kitten in Mn^{2+}-supplemented medium and the restoration of this response when the standard solution was readmitted. Similar results were observed in all motoneurons under investigation (26 rat motoneurons and 31 kitten motoneurons).

Fig. 14.5 shows EPSPs produced by descending volley and the directly evoked spikes in a motoneuron of 16-day-old kittens. This motoneuron was kept in a stable condition for 5 hours and, as may be

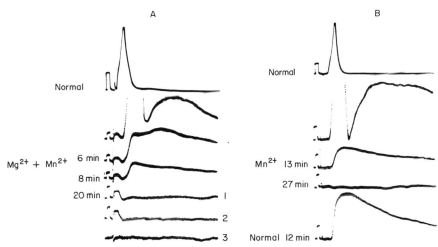

Fig. 14.4. Potentials evoked in motoneurons of 16-day-old rat (A) and 4-day-old kitten (B) in response to ventral root stimuli and effects of Ca²⁺-deficient, Mg²⁺- and Mn²⁺-supplemented medium on the EPSPs produced by dorsal root volley. In A, upper trace is an antidromic action potential (calibration pulse 20 mV, 1 ms). All other traces are averaged records of EPSPs produced by dorsal root stimulation in standard saline and at different times (in min) after perfusion with test solution (0.2 mM Ca²⁺, 2.5 mM Mg²⁺ and 2.0 mM Mn²⁺). Trace 3 shows the result of subtraction of the extracellular response (2) from the intracellular one (1). Calibration pulse 0.5 mV, 1 ms (subtracted by computer in trace 3). In B, upper trace is an antidromic action potential (calibration pulse 20 mV, 1 ms). All other traces are averaged records of EPSPs produced by dorsal root stimulation in standard saline and at different times after perfusion with test solution (0.2 mM Ca²⁺ and 2.0 mM Mn²⁺) and after returning to standard saline. Calibration pulse 1 mV, 1 ms.

seen in Fig. 14.5, the EPSP blocked by substitution of Mn²⁺ for Ca²⁺ were later completely recovered after superfusion with normal solution. Subsequent administration of tetrodotoxin abolished not only the EPSPs but also the motoneuronal action potentials. Washing out the tetrodotoxin effect was very slow. However, 88 minutes after returning to standard saline the partial restoration of both EPSP and the directly evoked spike is quite obvious. Thus, the possibility of electrical transmission in mammalian spinal cord receives no support in the present work. Moreover, the information already at hand indicates that synaptic actions produced in mammalian motoneurons from dorsal root afferents and descending pathways are chemically mediated.

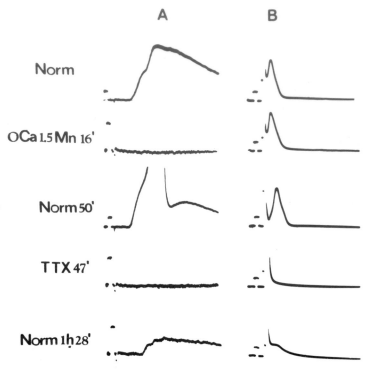

Fig. 14.5. Potentials evoked in the same motoneuron of a 16-day-old kitten in response to VC stimuli (A) and to intracellularly injected currents (B) in standard solution (Norm) and after perfusion with Ca^{2+}-free-Mn^{2+}-supplemented solution, or standard solution containing tetrodotoxin (TTX; 0.15 μg ml^{-1}). Calibration pulse 1 ms, 2 mV (A) and 20 mV (B).

2. *Spontaneous synaptic activity*

Another unresolved problem of the synaptic mechanisms in the mammalian spinal cord is to what extent the spontaneous synaptic activity recorded from spinal neurons represents the spontaneous, nonimpulse related release of neurotransmitter. The most concentrated efforts to investigate this problem were made by Hubbard *et al.* (1967) and Blankenship and Kuno (1968) who locally applied the tetrodotoxin in experiments on cat spinal cord *in situ*. But their results were inconclusive because of the limitations concerned with diffusion of the drug within the spinal gray matter. Electrophysiological analysis of the nature of spontaneous PSPs (postsynaptic potentials) has evolved around the technique of selective blockage of evoked release. Substances known to block nervous conduction or to inhibit the membrane fluxes of Ca^{2+} ions should not have any effect on the "true" miniature synaptic potentials reflecting the spontaneous release of quanta of transmitter.

In standard solution, practically all motoneurons recorded from either rat or kitten spinal cord exhibited a marked spontaneous synaptic activity. The amplitude of the individual spontaneous potentials varied a great deal and reached 2–4 mV, and their frequency was usually 20–30 s^{-1}. The substitution of Mn^{2+} for Ca^{2+} usually markedly reduced the frequency of spontaneous synaptic activity (Fig. 14.6). However, when all evoked synaptic responses derived either from dorsal roots or ventral columns were completely lost, relatively large spontaneous potentials still could be encountered. On returning to standard perfusing solution, the spontaneous synaptic activity returned to its initial values although the rate of recovery was relatively slow. These data suggest that only a relatively small part of spontaneous synaptic activity reflects the non-impulse-related release of quanta of neurotransmitter. This conclusion is further supported by experiments with administration of tetrodotoxin. An example of the effects of tetrodotoxin on the spontaneous synaptic potentials and on the directly evoked action potentials of the same motoneuron is shown in Fig. 14.7. It is clear from this illustration that at the time when the mechanism of the spike generation is completely abolished, the spontaneous synaptic potentials still may be observed. Late in an exposure to standard solution the spike starts to recover as well as the frequency of the spontaneous synaptic potentials.

The histograms of frequency distribution constructed for all populations of motoneurons recorded from are shown in Fig. 14.8C. They further indicate that relatively large spontaneous synaptic actions can still be observed after the suppression of evoked synaptic responses by Mn^{2+} or tetrodotoxin. It cannot be excluded, however, that at least part of these potentials were in fact inversed IPSPs (inhibitory postsynaptic potentials), since the recording micropipettes were frequently filled with 3M KCl. Amplitude and half-width distributions of spontaneous synaptic potentials of the same motoneurons in standard solution and after adding tetrodotoxin are plotted in Figs. 14.8A and 14.8B. These histograms indicate that the time course and the maximum amplitude of these potentials fall in the same range of values.

Thus, as in the isolated amphibian cord (Colomo and Erulkar, 1968; Katz and Miledi, 1963) spontaneous synaptic potentials resistant to Ca^{2+} lack and to addition of Mn^{2+} or tetrodotoxin represent effects evoked by single quanta of transmitter. The large size of some miniature potentials recorded from kitten's motoneurons may, at least in part, be due to the high input resistance of these cells. According to our measurements, the input resistance of motoneurons of the superfused isolated

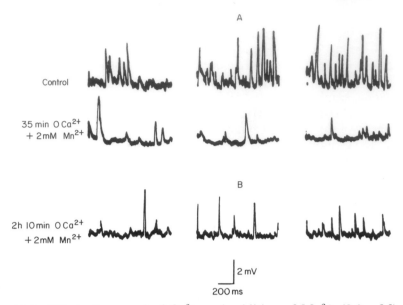

Fig. 14.6. Effect of removal of Ca^{2+} and addition of Mn^{2+} (2.0 mM) on spontaneous synaptic activity of two different motoneurons (A and B) of 11-day-old kitten (Shapovalov et al., 1979).

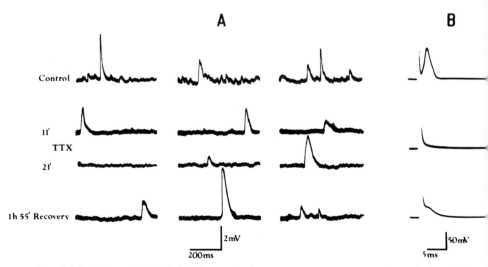

Fig. 14.7. Effect of TTX (0.15 μg ml^{-1}) on spontaneous synaptic activity (A) and directly evoked spike (B) recorded from the same motoneuron of a 9-day-old kitten (Shapovalov et al., 1979).

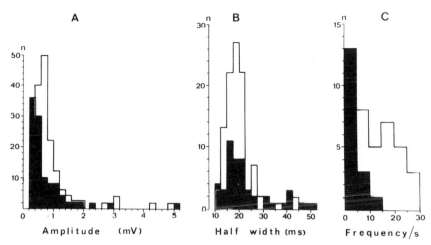

Fig. 14.8. Histograms of amplitude (A), half-width (B) and frequency (C) distribution of spontaneous potentials recorded from motoneurons of isolated superfused spinal cord of kittens. A and B, spontaneous EPSPs of the same motoneuron of 9-day-old kitten in standard solution (white columns) and after addition of TTX (0.15 μg ml^{-1}) (black columns). Ordinate, number of spontaneous EPSPs. C, Maximal frequency of spontaneous EPSPs recorded from kitten motoneurons in standard solution (white columns) and after Ca^{2+}-free solution containing 2.0 mM Mn^{2+}, or standard solution containing TTX (0.15 μg ml^{-1}) (black columns). Ordinate, number of motoneurons.

spinal cord of kittens ranged from 3.5 to 10.6 MΩ (mean 5.93 MΩ). In contrast, the input resistance of motoneurons of adult cats was shown to be about 0.8 MΩ (Eccles, 1957).

III. Intra-arterial Perfusion of Kitten Spinal Cord

Although isolated superfused spinal cord preparations have provided very useful experimental conditions, they have some important limitations. One disadvantage of such preparations is the possibility of tissue damage due to the procedure of hemisection. As the dendrites of the motoneurons and some internuncial cells may extend into the contralateral half of the spinal cord, the saggital hemisection may inflict some damage on motoneurons and the synaptic systems around them. The other disadvantage is concerned with the absence of peripheral nerves. Many particular synaptic inputs, therefore, become inaccessible for electrophysiological investigation.

The perfusion of the spinal cord of cats and rabbits *in situ* through the central canal (Fagg *et al.*, 1978; Morton *et al.*, 1969, 1977; Sverdlov *et al.*, 1978) cannot provide conditions for the efficient manipulation of the humoral environment of nerve cells.

We developed, therefore, a method of perfusion of mammalian spinal cord through arteries. Although in this case, in contrast to the super-fused hemicords, the constituents of the perfusing medium cannot all reach the nerve cells and synapses within the blood–brain barrier, the experimental results suggest cells can be manipulated at will. Moreover, the parallel perfusion of the hind limb provides the availability of the peripheral nerves for stimulation (Radicheva *et al.*, 1980).

A. Methods

1. Technical procedures

A schematic diagram of the intra-arterial perfusion of the kitten spinal cord together with the left hind limb is shown in Fig. 14.9. The perfusion medium flows through the aorta to the spinal cord via the spinal arteries. To achieve an appropriate intra-arterial pressure, all large arteries to the abdomen and the contralateral limb were ligated.

The preliminary surgery was performed as follows. After anaesthesia (urethane, 1.5–2.0 g kg^{-1}, i.p.) had been induced, the abdomen was opened and the coeliac, cranial mesenteric, suprarenal, renal, sper-matic, caudal mesenteric, iliolumbar, internal and external right iliac arteries were dissected and prepared for ligation (Fig. 14.9). Next the thorax was exposed and a polyethylene cannula was inserted into the thoracic aorta allowing the flow of oxygenated standard solution into the circulation. The composition of the standard perfusion solution was similar to that used for superfusion of the isolated spinal cord. A thermis-tor probe was attached near the cannula to monitor the fluid tempera-ture. The latter was maintained in the range 15–22°C.

Immediately after the aorta was cannulated, all the arteries listed above were ligated, and the perfusion solution could reach the spinal cord through seven pairs of vertebral arteries and the left hind limb through the left internal and external iliac arteries. The outflow of perfusate was accomplished through the polyethylene cannula inserted into the rostral and caudal ends of the transected caudal vena cava. The rate of perfusion was adjusted to 2–5 ml min^{-1}.

Lumbosacral cord was exposed by a conventional laminectomy (Th$_{8-10}$–S$_{1,2}$). The left hind limb was prepared by dissecting out the branches of the posterior biceps-semi-tendinosus, the medial and lateral

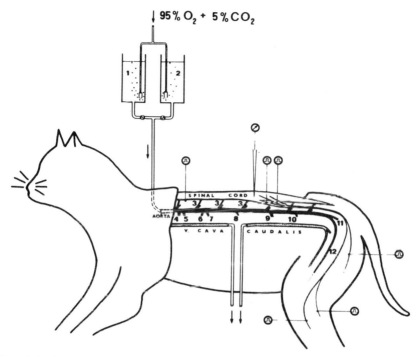

Fig. 14.9. An arrangement of intra-arterial perfusion of kitten spinal cord and the left hind limb. 1, reservoir filled with standard solution; 2, reservoir filled with test solution; 3, spinal arteries; 4, coeliac artery; 5, cranial mesenteric artery; 6, supraspinal arteries; 7, renal arteries; 8, spermatic arteries; 9, caudal mesenteric artery; 10, iliolumbar arteries; 11, right internal and external iliac arteries; 12, left internal and external iliac arteries.

gastrocnemius-soleus and the common peroneal, tibial and plantar nerves.

The kitten was then mounted in a rigid steel frame. The dura was opened and the fine dissection of the spinal cord was then performed. The dissection involved separating the dorsal and ventral roots L_7 and S_1, and placing stimulating electrodes on the ventral columns and exposed roots. The nerves were positioned under paraffin on the bipolar stimulating electrodes.

After continuous perfusion for several hours, the preparation sometimes became oedematous, but retained its usual activity.

2. Stimulation and recording

The nerves of the left hind limb of the kitten were stimulated with single and double current shocks. The duration of the stimulating pulses was 0.05–0.1 ms.

Intracellular recordings were made from motoneurons using glass microelectrodes filled with 3 M KCl or 2 M potassium citrate. The electric currents could be passed through the recording electrode by using a conventional DC current balanced-bridge circuit for offsetting the voltages developed across an electrode resistance, while the current was applied via the electrode. An extracellular average was also recorded of the voltage response of motoneuron. This average was later subtracted from the intracellular average to obtain the membrane potential transient. A pen writer provided a continuous record of the resting membrane potential. Recordings from the dorsal root were used in deciding the monosynaptic nature of intracellular responses.

3. *Functional state of the preparation*

The spinal cord perfused in this way retained a high level of activity for 10–12 hours after the onset of perfusion. When the standard solution was employed, stimulation of the peripheral nerves evoked muscle jerks. Spontaneous movements were also frequently observed. D-Tubocurarine $(3–5 \text{ mg l}^{-1})$ was therefore added to the perfusing medium to obtain proper immobilization.

In experiments carried out with intra-arterial perfusion, it was not always possible to keep the impaled motoneurons for long periods. Only in 28 motoneurons out of a total of 179 cells did the duration of recording exceed 20 minutes. In seven cells, it was 40 minutes, and only two motoneurons could be kept longer than 1 hour. Apparently, the less stable recording, compared with the isolated superfused hemicords of the kittens of the same age, may be caused by inadequate fixation of the preparation. It was rarely possible, therefore, to follow the same cell during changes in the composition of perfusing fluid and as a rule motoneurons were sampled at different times throughout the perfusion.

B. Results

1. *Action potentials and input resistance*

The amplitude of the antidromic action potentials varied in different cells between 54 and 103 mV and the overshoot was always visible, unless the cell was severely damaged. The maximal overshoot reached 30 mV. The depolarizing and hyperpolarizing afterpotentials usually followed the spike. By superimposing an antidromic action potential on the current step, the resistance was calculated from the change in the amplitude of the action potential (Frank and Fuortes, 1956). The input resistance of 22 motoneurons so calculated was found to vary between 1.3 and 15.1 MΩ, mean 5.89 MΩ. These values fit well with those obtained on motoneurons of superfused isolated spinal cord of the same animals.

2. *Synaptic activity*

In the spinal cord perfused through arteries, it was possible, in addition to effects produced by stimulation of the dorsal roots or ventral columns, also to record responses derived from individual nerves of hind limb. Both EPSPs and IPSPs could be easily encountered during perfusion with standard solution (Fig. 14.10). It may be noted that clear IPSPs were observed even in motoneurons of the youngest kittens under study. The earliest component of the EPSP appeared 1.5–1.8 ms after arrival of the presynaptic volley. Such large values of central delay are explain-

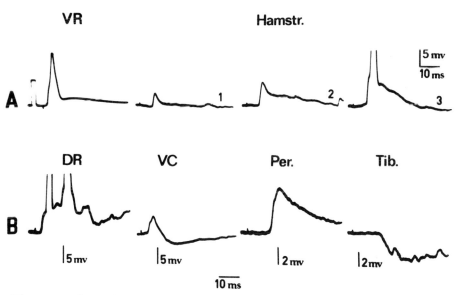

Fig. 14.10. Potentials evoked in motoneurons of the intra-arterially perfused spinal cord of the kittens by stimulation of ventral (VR) and dorsal (DR) roots, ventral column (VC), common peroneal (Per) and common tibial (Tib) nerves. A, Records from the same motoneuron of 15-day-old kitten. Calibration pulse on the first record, 1 ms, 20 mV. B, Records from four different motoneurons of 18 , 17-, 5- and 7-day-old kittens, respectively.

able taking into account the low temperature of the perfusion medium.

If cords were perfused with Ca^{2+}-free solution containing 1.5–2.0 mM Mn^{2+}, all evoked synaptic activity was completely but reversibly blocked. These results agree with those obtained on the superfused isolated spinal cord of young rats and kittens (Tamarova et al., 1978; Shapovalov et al., 1978, 1979) and indicate the chemical mode of junctional transmission at synapses between afferent and descending fibres and feline motoneurons. It should be stressed, however, that the time of onset of block in the isolated hemicords and in the spinal cord perfused through arteries differed considerably. In the latter case, it took usually not less than 20–40 minutes, and sometimes even 60–90 minutes to achieve a complete block. On the other hand, in a few cells of the intra-arterially perfused preparation, the complete block could be obtained already within 8–14 minutes of perfusion with Ca^{2+}-free, Mn^{2+}-supplemented media. It may be speculated that such wide variations in time of onset may be caused by different distances between the particular intraspinal capillaries and the motoneurons recorded from. This suggestion agrees with the observation about the marked difference between the susceptability to block different motoneurons of the same kitten. No such difference was found in the isolated hemicords.

Fig. 14.11A illustrates the averaged EPSP recorded from the same motoneuron. The first records were obtained 64 minutes after the test solution was admitted. Under these conditions, only antidromic action potential, but not the EPSP, could be produced in a motoneuron. When the standard solution was readmitted, the recovery began within 5–8 minutes after the start of perfusion with normal solution. The second application of Ca^{2+}-free, Mn^{2+}-supplemented medium rapidly depressed the synaptic responses.

The polysynaptic responses were depressed, as a rule, earlier than monosynaptic reaction. In motoneurons with slow onset of depression, it was possible to observe monosynaptic EPSPs without any contamination by polysynaptic components (Fig. 14.11B). Such uncontaminated monosynaptic EPSPs revealed frequency potentiation when the stimuli were repeated. This facilitation agrees with the chemical nature of transmission at 1a synapses.

As in the isolated superfused hemicords, Ca^{2+} depletion and addition of Mn^{2+} markedly reduced the frequency of spontaneous postsynaptic potentials while some of them were preserved throughout the perfusion with the test solution (Fig. 14.11). Comparing Fig. 14.11 with Fig. 14.6, which shows results from the isolated hemicord, it can be seen that Mn^{2+} affects spontaneous synaptic activity in a similar way in both preparations.

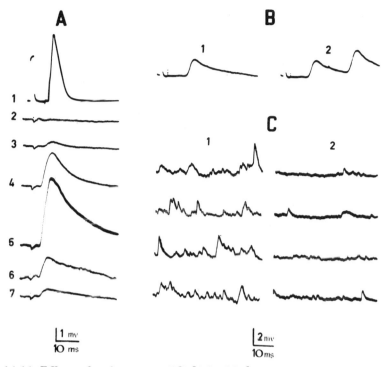

Fig. 14.11. Effects of replacement of Ca^{2+} by Mn^{2+} on evoked and spontaneous synaptic activity in three different motoneurons (A, B, C) of the intra-arterially perfused spinal cord of the kittens. A, Responses in the same motoneuron of 12-day-old kitten on ventral root (1) and dorsal root (2–7) stimulation. 1, 2, 64 minutes after perfusion with zero Ca^{2+} and 1.5 mM Mn^{2+}-containing solution; 3, 4, 5, recovery of responses after 5, 8 and 15 minutes perfusion with standard solution, respectively; 6, 7, responses obtained 4 and 7 minutes after the second replacement by Ca^{2+}-free, Mn^{2+}-containing medium. Calibration pulse on the first record 50 mV, 1 ms. B, Monosynaptic EPSPs evoked in a motoneuron of 17-day-old kitten by supramaximal dorsal root stimulation. 1, single shock; 2, double shock stimulation. The spinal cord was perfused with Ca^{2+}-free solution containing 1.5 mM Mn^{2+} for 63 minutes. Calibration pulse 2mV, 1 ms. C, Spontaneous activity in a motoneuron of 12-day-old kitten during perfusion with standard solution (1) and 30 minutes after perfusion with Ca^{2+}-free solution containing 1.5 mM Mn^{2+} (2) was started.

3. After-potentials in motoneurons

With intra-arterial perfusion it is also possible to study ionic dependence of postspike hyperpolarization. Action potentials recorded from kitten motoneurons, as in adult cats, revealed both depolarizing and hyperpolarizing after-potentials. When in normal solution, the amplitude of depolarizing after-potentials varied between 1.5 and 19.0 mV (mean 8.9 mV) and that of the hyperpolarizing after-potential ranged from 1.2 to 18.0 mV (mean 5.8 mV). When Ca^{2+}-free, Mn^{2+}-containing solution was perfused through the cord, the depolarizing after-potentials were not affected. However, the hyperpolarizing after-potentials were reduced markedly and they were almost completely abolished after treatment with the test solutions for 40–60 minutes. Fig. 14.12 shows the consecutive records taken from four different motoneurons of the 12-day-old kitten, firstly in standard solution, and then after different periods of perfusion by Ca^{2+}-free, 1.5 mM Mn^{2+}-containing medium. The antidromic spike potentials and synaptic responses derived from dorsal root can be compared in the same cells. It may be seen that at a time when hyperpolarizing after-potentials are still maintained, although greatly attenuated, the postsynaptic responses were markedly suppressed (Fig. 14.12B). At 1 hour 30 minutes after this period, both the hyperpolarizing after-potentials and postsynaptic responses were completely lost (Fig. 14.12C). The same situation was found in all other impaled motoneurons of this preparation. At 15 minutes after the readmittance of standard solution, both the hyperpolarizing after-potential and synaptically induced activity had recovered (Fig. 14.12D).

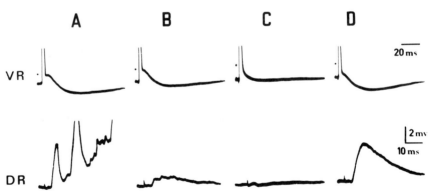

Fig. 14.12. Effect of replacement of Ca^{2+} by Mn^{2+} on hyperpolarizing after-potentials and EPSPs evoked in the same motoneurone (A, B, C, D) by ventral (VR) and dorsal root (DR) stimulation of a 12-day-old kitten. A, records taken in standard solution; B, 63 minutes after starting perfusion with solution containing no Ca^{2+} and 2.0 mM Mn^{2+}; C, after 150 minutes in the test solution; D, 15 minutes after returning to standard solution. Calibration pulse for VR responses 5 mV.

It is well established that ionic mechanisms underlying prolonged hyperpolarization after the motoneuronal spike involve increased potassium conductance. The abolition of this potential by Ca^{2+} lack and by blockers of Ca^{2+} conductance agrees with the proposed role of Ca^{2+} entry into the cell for the generation of postspike hyperpolarization (Barrett and Barrett, 1976; Krnjević et al., 1978) and coincides with results obtained with extracellular iontophoretic application of Mn^{2+} to motoneurons of adult cats (Krnjević et al., 1978, 1979).

Thus, the perfusion of the kitten spinal cord through arteries makes it possible to maintain the preparation in good condition for many hours and to study the effects of different perfusing solutions both on particular synaptic inputs and intrinsic properties of the nerve cells.

IV. Intra-arterial Perfusion of Kitten Brain

The successful intra-arterial perfusion of the spinal cord prompted us to develop a method of perfusing the mammalian brain in contrast to previous authors who studied neuronal activity of brain slices.

A. Methods

1. Technical procedures

The kittens used were 9–21 days old. After anaesthesia had been induced (urethane 1.5–2.5 g kg^{-1}, i.p.) the thorax was opened and the polyethylene cannula was inserted into the aorta as shown in Fig. 14.13. The composition of the standard perfusion solution was similar to that used for perfusion of the spinal cord and it was gassed with a mixture of CO_2 (2% or 5%) and O_2 (98% or 95%). The temperature of the perfusion medium was maintained between 17 and 24°C, pH 7.4–7.6.

The solution reached the brain through the carotid and vertebral arteries. The right and left subclavian arteries (distal to the vertebral arteries) and the intercostal and aortic arteries (in the close vicinity to the heart near the right and left coronary artery) were ligated. This ligation of the arteries mentioned above excluded them from the perfusion and increased the flow through the brain. The exit for the perfusate was through a polyethylene cannula inserted into the cranial vena cava.

The rate of perfusion was adjusted to 4–6 ml min^{-1}. In some experiments, both carotid arteries were ligated and the perfusion was made

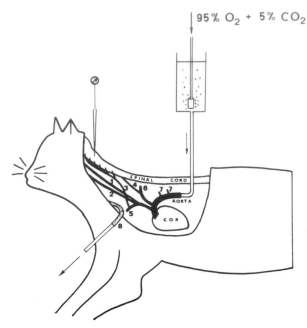

Fig. 14.13. An arrangement for intra-arterial perfusion of kitten brain. 1, right common carotid artery; 2, left common carotid artery; 3, right vertebral artery; 4, left vertebral artery; 5, right subclavian artery; 6, left subclavian artery; 7, intercostal arteries; 8, cranial vena cava.

exclusively through the vertebral arteries. This latter procedure improved the functional state of the medulla. The schematic diagram of perfusion is shown in Fig. 14.13.

When the dissection was complete, the animal was placed in a special frame allowing us to clamp the head and cervical vertebrae. The animal was paralyzed by the addition of D-tubocurarine to the perfusion saline. Without curare the movements of the head and shoulders prevented stable recording from medullary neurons. After proper fixation, immobilization and craniotomy, the cerebellum was sectioned and the medulla was exposed.

2. *Stimulation and recording*

Monopolar or bipolar electrodes, insulated but for the tips, were used for stimulation of MLF and vestibular nerve.

Spontaneous and evoked activity of neurons was recorded using a conventional microelectrode technique. The micropipettes were filled with 3M KCl or 2M potassium citrate. Their resistance ranged between 80 and 100 MΩ.

B. Results

The preparations of perfused brain survived for a period of 3–9 hours from the beginning of perfusion. In five experiments, 26 neurons with a resting membrane potential of 27–65 mV were impaled. The maximum duration of recording, however, did not exceed 12 minutes. We did not try, therefore, to change the perfusion medium in the course of a single penetration.

Fig. 14.14 illustrates the EPSP evoked in different neurons by stimulations of vestibular nerve (A), MLF (B) and the antidromic spike in response to MLF volley (C). The amplitude of the action potential suggests that the functional condition of the neuron is reasonably good.

Many cells also showed considerable spontaneous synaptic activity (Fig. 14.15). The amplitude of the spontaneous EPSPs reached 4–5 mV, as in experiments on the hippocampal slices (Brown *et al.*, 1979).

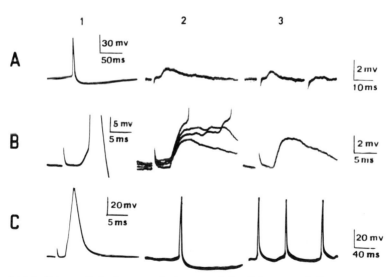

Fig. 14.14. Intracellularly recorded potentials of three medullary neurons (A, B, C) of intra-arterially perfused brain of kittens. A, Spontaneous action potential (1) and EPSPs produced by stimulation of vestibular nerve (2, 3) of a 9-day-old kitten. B, EPSPs evoked by stimulation of MLF (1–3) of a 17-day-old kitten. Neuron of the nucleus RGC. C, Antidromic action potential (1) produced by stimulation of MLF (1) and spontaneous action potentials (2, 3) of the neuron of the nucleus RGC in an 11-day-old kitten.

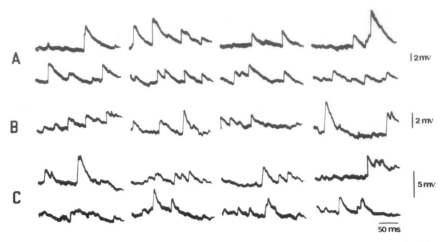

Fig. 14.15. Spontaneous synaptic activity recorded from three different medullary neurons (A, B, C) of intra-arterially perfused brain of 11-day-old kitten.

The data presented above demonstrate the availability of the mammalian brain perfused through arteries for electrophysiological analysis. Further work in this direction may be promising.

V. Conclusions

Three different preparations are described; (a) the isolated perfused spinal cord; (b) the *in situ* arterially perfused spinal cord and hind limb; (c) the *in situ* arterially perfused medulla.

A. *In vitro* hemi-spinal cord

1. Intracellular recordings could be maintained from motoneurons for up to 40 minutes in young rats and for 4–5 hours in kittens.

2. The resting potential was from -50 to -75 mV and antidromic action potentials ranged from 50 mV to 113 mV. The input resistance ranged from 3.5–10.6 MΩ (mean 5.9 MΩ).

3. Dorsal-root stimulation produced a ventral root potential that was blocked in Ca^{2+}-free Ringer containing 2 mM Mn^{2+} or Mg^{2+}. The presynaptic responses were not blocked.

4. The EPSPs were reversibly blocked in Ca^{2+}-free, Mn^{2+}-supplemented Ringer. The antidromic spike was not affected.

5. Tetrodotoxin blocked EPSP and antidromic action potentials.

6. There is no evidence for electrical transmission at the synapses in the isolated mammalian spinal cord.

7. In standard Ringer, all the motoneurons showed spontaneous activity of 2–4 mV amplitude at a frequency of 20–30 s^{-1}. Mn^{2+} reduced the activity. These potentials could still be seen when tetrodotoxin blocked the action potentials and so could be due to spontaneous presynaptic release of transmitter.

B. *In situ* perfused kitten spinal cord

1. Recordings could be made from impaled motoneurons for about 20 minutes to a maximum of 1 hour. The preparation was mechanically less stable than the isolated hemi-spinal cord.

2. The antidromic actions potentials ranged from 54 to 103 mV with a maximal overshoot of 30 mV. The input resistance was between 1.3 and 15.1 MΩ (mean 5.89 MΩ).

3. EPSP and IPSP occurred and were blocked by perfusion with Ca^{2+}-free, Mn^{2+}-supplemented Ringer. It took longer for the block to occur in the *in situ* preparation than in the *in vitro* hemi-spinal cord.

4. Mn^{2+} depressed the polysynaptic responses earlier than the mono-synaptic responses.

5. In normal Ringer, the depolarizing after-potentials of the motoneuron varied between 1.5 and 19.0 mV (mean 8.9 mV) and the hyperpolarizing after-potential ranged between 1.2 and 18.0 mV (mean 5.8 mV).

6. Ca^{2+}-free, Mn^{2+}-supplemented Ringer did not affect the depolarizing after-potential, but the hyperpolarizing after-potential was reduced and almost abolished after 40–60 minutes perfusion. They reappeared after 15 minutes perfusion with standard Ringer.

C. Perfused kitten brain

1. The preparation survived for 3–9 hours after the perfusion had been started.

2. 26 neurons have been impaled and the resting potential varied between 27 65 mV.

3. Stimulation of the vestibular nerve or the MLF led to EPSPs in the impaled neurons.

4. Spontaneous synaptic activity was noted and it could be as large as 4–5 mV.

D. General conclusions

The studies in this chapter indicate that it is possible to develop a series of preparations that allow greater control of the ionic and chemical environment of the CNS. From studies of the normal intact *in situ* CNS, we have developed preparations that allow the *in situ* perfusion of the brain and cranial nerves, the spinal cord and hind limbs, and the isolated hemi-spinal cord and spinal roots. Each preparation has its own advantages and allows it to act as a control for the other preparations. The combined use of these systems should provide greater understanding of the manner in which the neurons in the CNS perform and interact.

References

Araki, T., Otani, T. and Furukawa, T. (1953). The electrical activities of single motoneurones in toad's spinal cord, recorded with intracellular electrodes. *Jap. J. Physiol.* **3,** 254–267.
Babsky, E. B. and Kirillova, A. A. (1938). The action of acetylcholine on the excitability of the central nervous system. *Bull. Exp. Biol. Med.* **6,** 174–176. (In Russian.)
Bagust, J. and Kerkut, G. A. (1979). Some effects of magnesium ions upon conduction and synaptic activity in the isolated spinal cord of the mouse. *Brain Res.* **177,** 410–413.
Barker, J. L. and Nicoll, R. A. (1973). The pharmacology and ionic dependency of amino acid responses in the frog spinal cord. *J. Physiol. (Lond.)* **228,** 259–277.
Barrett, E. F. and Barrett, J. N. (1976). Separation of two voltage-sensitive potassium currents, and demonstration of a tetrodotoxin-resistant calcium current in frog motoneurones. *J. Physiol. (Lond.)* **255,** 737–774.
Batueva, I. V. and Shapovalov, A. I. (1977). Electrical and chemical EPSPs evoked in lamprey motoneurons by stimulation of descending tract and dorsal root afferents. *Neurophysiologia (USSR)* **9,** 512–517.
Blankenship, J. E. and Kuno, M. (1968). Analysis of spontaneous subthreshold activity in spinal motoneurones of the cat. *J. Neurophysiol.* **31,** 195–209.
Brookhart, J. M., Machne, X. and Fadiga, E. (1959). Patterns of motor neurone discharge in the frog. *Arch. Ital. Biol.* **97,** 53–67.
Brown, T. H., Wong, R. K. S. and Prince, D. A. (1979). Spontaneous miniature synaptic potentials in hippocampal neurons. *Brain Res.* **177,** 194–199.

Bursian, A. V. (1977). Evolution of segmental motor apparatus function in early ontogenesis of warm-blooded animals. Ph.D. Thesis: Leningrad. (In Russian.)

Colomo, F. and Erulkar, S. D. (1968). Miniature synaptic potentials at frog spinal neurons in the presence of tetrodotoxin. *J. Physiol. (Lond.)* **199**, 205–221.

Crain, S. M. (1966). Development of organotypic bioelectrical activity in central nervous tissues during maturation in culture. *Int. Rev. Neurobiol.* **9**, 1–43.

Curtis, D. R., Phillis, J. W. and Watkins, J. C. (1961). Actions of amino-acids on the isolated hemisected spinal cord of the toad. *Br. J. Pharmacol.* **16**, 262–283.

Davidoff, R. A. (1972). GABA antagonism and presynaptic inhibition in the frog spinal cord. *Science* **175**, 331–333.

Eccles, J. C. (1946). Synaptic potentials of motoneurones. *J. Neurophysiol.* **9**, 87–120.

Eccles, J. C. (1957). "The Physiology of Nerve Cells." The Johns Hopkins Press, Baltimore.

Edwards, F. R., Redman, S. J. and Walmsley, B. (1976). The effect of polarizing currents on unitary Ia excitatory postsynaptic potentials evoked in spinal motoneurons. *J. Physiol. (Lond.)* **259**, 705–724.

Evans, R. H. (1978). The effects of amino acids and antagonists on the isolated hemisected spinal cord of the immature rat. *Br. J. Pharmacol.* **62**, 171–176.

Evans, R. H., Francis, A. A. and Watkins, J. C. (1977). Effects of monovalent cations on the responses of motoneurones to different groups of amino acid excitant in frog and rat spinal cord. *Experientia* **33**, 246–248.

Fagg, G. E., Jordan, C. C. and Webster, R. A. (1978). Descending fibre mediated release of endogenous glutamate and glycine from the perfused cat spinal cord *in vivo*. *Brain Res.* **158**, 159–170.

Frank, K. and Fuortes, M. J. F. (1956). Stimulation of spinal motoneurones with intracellular electrodes. *J. Physiol. (Lond.)* **134**, 451–470.

Grinnell, A. D. (1966). A study of the interaction between motoneurones in the frog spinal cord. *J. Physiol. (Lond.)* **182**, 612–648.

Hubbard, J. I., Stenhouse, D. and Eccles, R. M. (1967). Origin of synaptic noise. *Science* **157**, 330–331.

Katz, B. and Miledi, R. (1963). A study of spontaneous miniature potentials in spinal motoneurones. *J. Physiol. (Lond.)* **168**, 389–422.

Krnjević, K., Lamour, I., MacDonald, J. F. and Nistri, A. (1978). Motoneuronal after-potentials and extracellular divalent cations. *Can. J. Physiol. Pharmacol.* **56**, 516–520.

Krnjević, K., Lamour, I., MacDonald, J. F. and Nistri, A. (1979). Effects of some divalent cations on motoneurones in cats. *Can. J. Physiol. Pharmacol.* **57**, 944–956.

Morton, J. K. M., Stagg, C. J. and Webster, R. A. (1969). Perfusion of the sub-arachnoid space and central canal of the mammalian spinal cord. *J. Physiol. (Lond.)* **202**, 72P.

Morton, J. K. M., Stagg, C. J. and Webster, R. A. (1977). Perfusion of the central canal and subarachnoid space of the cat and rabbit spinal cord *in vivo*. *Neuropharmacol.* **16,** 1–6.

Nelson, P. G. and Peacock, J. H. (1973). Electrical activity in dissociated cell cultures from fetal mouse cerebellum. *Brain Res.* **61,** 163–174.

Okada, I. and Saito, M. (1979). Stimulation-dependant release of possible transmitter substances from hippocampal slices studied with localized perfusion. *Brain Res.* **160,** 368–371.

Otsuka, M. and Konishi, S. (1974). Electrophysiology of mammalian spinal cord *in vitro*. *Nature (Lond.)* **252,** 733–734.

Radicheva, N. I., Tamarova, Z. A., Shapovalov, A. I. and Shiriaev, B. I. (1980). Electrical activity of the motoneurons of the kitten spinal cord perfused through the vessels. *Sechenov J. Physiol. (USSR)* **66, N4,** 489–496.

Schwartzkroin, P. A. (1975). Characteristics of CA 1 neurons recorded intracellularly in the hippocampal *in vitro* slice preparation. *Brain Res.* **85,** 423–436.

Shapovalov, A. I. and Shiriaev, B. I. (1978). Modulation of transmission in different electrotonic junction by aminopyridine. *Experientia* **34,** 67–68.

Shapovalov, A. I., Shiriaev, B. I. and Tamarova, Z. A. (1978). Electrical and chemical synaptic transmission in the vertebrate spinal cord. *Neurosci. Letters Suppl.* **1,** 305.

Shapovalov, A. I., Shiriaev, B. I. and Tamarova, Z. A. (1979). Synaptic activity in motoneurons of the immature cat spinal cord *in vitro*. Effects of manganese and tetrodotoxin. *Brain Res.* **160,** 524–528.

Sonnhof, U., Richter, D. W. and Taugner, R. (1976). Electrotonic coupling between frog spinal motoneurons. An electrophysiological and morphological study. *Brain Res.* **138,** 197–215.

Sverdlov, Yu. S., Ruchinskaya, T. Yu. and Erzina, G. A. (1979). New evidence of depression of depolarization of primary afferents with ammonium ions. *Bull. Exp. Biol. Med.* **88,** 387–389. (In Russian.)

Takahashi, T. (1978). Intracellular recording from visually identified motoneurones in rat spinal cord slices. *Proc. Roy. Soc. Lond. Ser. B.* **202,** 417–421.

Tamarova, Z. A., Shapovalov, A. I. and Shiriaev, B. I. (1978). Action of magnesium and manganese ions and calcium deficiency on the synaptic transmission in the isolated perfused rat spinal cord. *Neurophysiologia (USSR)* **10,** 530–533.

Tebecis, A. K. and Phillis, J. W. (1976). The effects of topically applied biogenic monoamines on the isolated toad spinal cord. *Comp. Biochem. Physiol.* **23,** 553–563.

Werman, R. and Carlen, P. L. (1976). Unusual behavior of the 1a EPSP in cat spinal motoneurons. *Brain Res.* **112,** 395–401.

Yamamoto, C. (1972). Activation of hippocampal neurons by mossy fibre stimulation in thin brain sections *in vitro*. *Exp. Brain Res.* **14,** 423–435.

Index

395